T0135302

Advances in Intelligent Systems and Computing

Volume 1386

The series "Advances in Intelligent Systems and Computing" contains publications on theory, applications, and design methods of Intelligent Systems and Intelligent Computing. Virtually all disciplines such as engineering, natural sciences, computer and information science, ICT, economics, business, e-commerce, environment, healthcare, life science are covered. The list of topics spans all the areas of modern intelligent systems and computing such as: computational intelligence, soft computing including neural networks, fuzzy systems, evolutionary computing and the fusion of these paradigms, social intelligence, ambient intelligence, computational neuroscience, artificial life, virtual worlds and society, cognitive science and systems, Perception and Vision, DNA and immune based systems, self-organizing and adaptive systems, e-Learning and teaching, human-centered and human-centric computing, recommender systems, intelligent control, robotics and mechatronics including human-machine teaming, knowledge-based paradigms, learning paradigms, machine ethics, intelligent data analysis, knowledge management, intelligent agents, intelligent decision making and support, intelligent network security, trust management, interactive entertainment, Web intelligence and multimedia.

The publications within "Advances in Intelligent Systems and Computing" are primarily proceedings of important conferences, symposia and congresses. They cover significant recent developments in the field, both of a foundational and applicable character. An important characteristic feature of the series is the short publication time and world-wide distribution. This permits a rapid and broad dissemination of research results.

Indexed by DBLP, INSPEC, WTI Frankfurt eG, zbMATH, Japanese Science and Technology Agency (JST).

All books published in the series are submitted for consideration in Web of Science.

More information about this series at http://www.springer.com/series/11156

Pushparaj Shetty D. · Surendra Shetty
Editors

Recent Advances in Artificial Intelligence and Data Engineering

Select Proceedings of AIDE 2020

Editors
Pushparaj Shetty D.
Department of Mathematical
and Computational Sciences
National Institute of Technology Karnataka
(NITK)
Mangalore, India

Surendra Shetty
Department of MCA
NMAM Institute of Technology
Karkala, India

ISSN 2194-5357   ISSN 2194-5365 (electronic)
Advances in Intelligent Systems and Computing
ISBN 978-981-16-3344-7   ISBN 978-981-16-3342-3 (eBook)
https://doi.org/10.1007/978-981-16-3342-3

Preface

This volume of *Artificial Engineering in Data Engineering* contains the proceedings of the 2nd International Conference on Emerging Trends in Engineering—ICETE 2020 being held in NMAM Institute of Technology, Nitte, Karkala, Udupi, India, during December 22 and 23, 2020. Some of the best researchers in the field delivered keynote address on the theme areas of the conference. This gave an opportunity to the delegates to interact with these experts and to address some of the challenging interdisciplinary problems in the areas of artificial intelligence and data engineering.

AIDE 2020 attracted over 120 submissions. Through rigorous peer review, 41 high-quality papers were recommended by the international program committee to be presented at the conference and included in this volume. The AIDE 2020 conference is the second series of the international conference in artificial intelligence and data engineering that focuses on the recent technologies and trends in artificial intelligence and data engineering field in general.

AIDE 2020 was devoted to novel methods in the fields of machine learning, IoT, artificial intelligence, deep learning, computer networks, cloud computing, and data mining. The conference features several keynote addresses in the area of advanced computer science and information technology processing. These areas have been recognized to be the key technologies poised to shape modern society in the upcoming decade.

On behalf of the organizing committee, we would like to acknowledge the support from sponsors who helped in one way or the other to achieve our goals for the conference. We wish to express our appreciation to Springer for publishing the proceedings of AIDE 2020. We also wish to acknowledge the dedication and commitment of the AIDE Editorial Staff. We would like to thank the authors for submitting their work, as well as the technical program committee members and reviewers for their enthusiasm, time, and valuable suggestions. The invaluable help of the members from the organizing committee and volunteers in setting up and maintaining the online submission systems, assigning papers to the reviewers, and preparing the camera-ready version of the proceedings is highly appreciated. We would like to profusely thank them for making AIDE 2020 a success.

Mangalore, India	Dr. Surendra Shetty
Karkala, India	Dr. Pushparaj Shetty D.
January 2021	

Contents

Contents

About the Editors

Dr. Pushparaj Shetty D. is working as a faculty at National Institute of Technology Karnataka (NITK) Surathkal since 2000. He is presently holding the position of Associate Professor in department of Mathematical and computational sciences at NITK Surathkal. He obtained his B.E. in Computer Science and Engineering in 1999 from Mangalore University and M.E. in Computer Science and Engineering in 2005 from Indian Institute of Engineering Science and Technology, Shibpur, Kolkata (formerly Bengal Engineering and Science University). He obtained his Ph.D. from the Computer Science and Applications group, Department of Mathematics at the Indian Institute of Technology Delhi (July 2014). His research interests are in the area of wireless sensor networks, graph algorithms and their applications in engineering, cloud computing and high performance computing. He has co-authored more than 25 research papers in peer-reviewed journals and conferences including few book chapters. He is a senior member of IEEE, senior member of ACM, and life member of Computer Society of India (CSI) and Institution of Engineers India (IEI). Dr. Pushparaj is volunteering as the Chair of IEEE Mangalore Subsection for the year 2021.

Dr. Surendra Shetty is currently a Professor and Head, Department of Master of Computer Applications, NMAM Institute of Technology, Nitte. He had completed his B.Sc. in 2001 and he obtained his Master of Computer Applications in 2004 from Visvesvaraya Technological University, Belagavi. Dr. Surendra Shetty had been awarded his doctoral degree for his research work "Audio Data Mining Using Machine Learning Techniques" in 2013 from University of Mangalore. He has published more than 30 research papers in different international journals and conferences. He is currently guiding six research scholars. Dr. Surendra Shetty authored two book chapters in different publications entitled *Machine Learning Approach for Carnatic Music Analysis* and *Applications of Unsupervised Techniques for Clustering of Audio Data*. He has received research grant of 20 lakhs from VGST (GoK) for carrying out research on "Automatic Natural Language Processing and Speech

Disorder Problems in Kannada Language". He has 16 years of teaching experience. His research areas of interest are cryptography, data mining, pattern recognition, speech recognition, MIS, software engineering and testing.

Smart Environment and Network Issues

Machine Learning-Based Ensemble Network Security System

Prashanth P. Wagle, Shobha Rani, Suhas B. Kowligi, B. H. Suman,
B. Pramodh, Pranaw Kumar, Srinivasa Raghavan, K. Aditya Shastry⑩,
H. A. Sanjay⑩, Manoj Kumar⑩, K. Nagaraj, and C. Subhash

Abstract A vital issue related to security for systems that are interconnected is the undesirable intrusions by agents. Various types of attacks are possible by the intruders. The attacks are associated with network security since the entry to the system can be accomplished through a network. Hence, in a typical organization, attacks can be external or internal attacks. In this work, the possibility of using Machine Learning (ML) techniques to detect network intrusions in combination with other techniques is explored. The outcome of this research will be the automatic application of firewall rules according to the present state of the network to prevent intrusions. Various classifiers are explored and then the Decision Tree (DT) classifier is implemented that functions as a signature-based system. For classification, the network traffic and logs of system in a UNIX based machine are captured. The frequent connections are identified, so that any anomaly in resource usage can be detected. This along with the classifier functions as an ensemble system to prevent network intrusions.

P. P. Wagle · S. Rani · S. B. Kowligi · B. H. Suman · K. A. Shastry (✉) · H. A. Sanjay · M. Kumar
Nitte Meenakshi Institute of Technology, Banglore 560064, India
e-mail: adityashastry.k@nmit.ac.in

H. A. Sanjay
e-mail: sanjay.ha@nmit.ac.in

M. Kumar
e-mail: manoj.kumar@nmit.ac.in

B. Pramodh · P. Kumar · S. Raghavan · K. Nagaraj · C. Subhash
UNISYS, Banglore 560035, India
e-mail: Pramodh.Bettadapura@in.unisys.com

P. Kumar
e-mail: pranaw.kumar@in.unisys.com

S. Raghavan
e-mail: srinivasa.Raghavan@in.unisys.com

K. Nagaraj
e-mail: Nagaraja.k@in.unisys.com

C. Subhash
e-mail: subhash.chanda@in.unisys.com

Keywords Network intrusion detection · Signature based · Decision tree · Firewalls · Parameters

1 Introduction

An intrusion detection System (IDS) represents a software application/device used for supervising a network/system to detect anomalies/violations of policy. Decision Tree algorithm denotes a technique for predictive modeling. Classification trees are the tree models in which the target variable is having discrete values. Historic information is taken by the predictive models for forecasting future values in an inexpensive manner. These models assist humans in taking critical decisions, making them more effective or in some instances the whole process of taking decisions can be automated [1].

The motivation behind this work is to explore predictive modeling techniques to accurately classify the malicious packets from the non-malicious ones and to choose one model out of Decision Tree (DT), Naive Bayes (NB), K-Nearest Neighbor (k-NN), and Logistic Regression (LR) with appropriate reasoning. After trying out the various models, the DT models were found to perform better with respect to accuracy and precision.

In this paper, the NSL-KDD dataset (NKD) [2] is used that denotes the improvised version of the KDD CUP 1999 dataset which is a database of connections. This database comprises of standard dataset that needs to be examined for several simulated intrusion types. A connection represents TCP packet sequence that begins and ends at certain definite times, amid which flow of data occurs to and from source IP address to a target IP address through a standard protocol. The connection is categorized as belonging to normal class or attack class. No connection is labeled with multiple attack types. Every connection instance comprises of approximately 100 bytes. Out of all the connections, exploratory data analysis (EDA) is done to choose only 10 attributes out of the 41 attributes of each connection.

The paper contents are as follows: Related work is discussed in Sect. 2. The data source with the associated parameters is described in Sect. 3. Section 4 examines the EDA with the implementation and the reasoning for choosing the DT algorithm over the others. Section 5 demonstrates the experimental setup and results. Section 6 concludes the research work done followed by references.

2 Related Work

The previous works for classifying packets as malicious or not using different ML models are mentioned in this section.

Kumar et al. [3] proposed a ML based IDS, mNIDS (Mobile-Network IDS) using tree and rule-based algorithms such as J48, Random Forest, RIDOR, JRIP, and PART.

Although mNIDS produced high accuracy, it was found that it can be further enhanced when used in conjunction with the traditional IDS such as knowledge-based anomaly detection systems and hybrid systems based on networks.

Yang Xin et al. [4] proposed a review of related work in ML and deep learning (DL) techniques for applications related to cybersecurity. Several commonly used datasets like DARPA, KDD Cup 99, NSL-KDD, and ADFA datasets were recommended. The authors also summarized outcomes of applying various ML and DL algorithms on these datasets.

Rishabh Das et al. [5] evaluated four ML models based on MODBUS data gathered from a gas pipeline. Numerous attacks have been categorized utilizing the ML algorithms, and lastly, the algorithm's performance has been measured. It was concluded that the RF algorithm might be more suitable as a core IDS algorithm for its optimal real-time performance.

Nutan Farah Haq et al. [6] conducted a survey on the designs of solitary, hybrid, and ensemble systems. The survey comprised of comparison of classification techniques in a statistical fashion. Along with the analysis of datasets, feature selection (FS) was also performed. It was found that utilizing diverse FS techniques and operating with the superlative technique was helpful for the classification and enhanced the deliberation of FS steps in identifying the intrusions. Additional examination demonstrated that solitary / baseline classifiers in measurement of performance can be substituted by ensemble classifiers.

Almseidin et al. [7] performed and evaluated several experiments to assess various ML classification algorithms on KDD intrusion dataset. They mainly focused on false negative (F_N) and false positive (F_P) to enhance the detection rate of the IDS. Experiments by authors revealed that the RF algorithm attained the highest rate of average accuracy while the DT algorithm obtained the minimum value for F_N.

The works mentioned above give a general idea about the type of dataset to use and the corresponding ML techniques to apply in order to increase accuracy of classification. Drawing from their conclusions and experimenting with other techniques, the DT classifier and NKD dataset is selected in this work. In this work, the built-in firewall, iptables, of Unix systems coupled with the proposed DT classifier (hybrid IDS) is utilized to block possible malicious traffic via whitelisting, inspired by [8], which was not implemented in the above-mentioned works. This work also deals with FS to filter irrelevant attributes and noise removal from the NKD dataset for obtaining an unbiased classifier as proposed by [2].

3 Dataset Description

This section describes the source of data from the KDD dataset [2] which is used to perform the experiments. It also describes the 42 parameters considered for class prediction. The NKD dataset [2] which is an improvised version of the KDD CUP 1999 dataset is employed in this work. Four classes are present in the dataset,

namely denial-of-service (DOS) like SYN flood attack, R2L (accessing in an unauthorized manner from a machine located remotely) attack like password guessing,
U2R (accessing the privileges of super users in an unauthorized manner) attack like
overflowing the buffers, and probing—surveillance, and other probing like scanning
of ports. NSL-KDD dataset with 42 attributes is used in this work. This dataset
represents an improvised version of the KDD CUP 1999 dataset in which redundant
records have been deleted for achieving unbiased outcomes. Each database record
represents a normal connection or an attack.

4 Proposed Work

This section describes the devised model for performing intrusion detection. Figure 1
demonstrates the big picture of the incoming network connection.

From Fig. 1, one can get to know the lifecycle of each incoming connection in the
network. During the life cycle end, a connection is going to be classified as either
malicious or normal. All incoming connections are subjected to the first and second
line of defense. External connections may or may not be subject to the first line of
defense (LOD).

Fig. 1 High-level design
framework

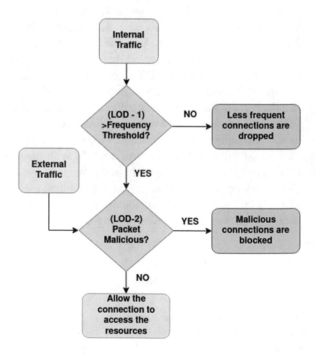

4.1 First LOD (LOD-1)

The LOD-1 for protecting from network attacks is whitelisting specific IP addresses and ports on a network with the default behavior of "DROP." The iptables are made use of as the firewall on a Linux machine. Whitelisting has been implemented in the following manner, concerning a single node: Initially, all incoming and outgoing traffic is allowed by setting default policy to "ALLOW" on the host node's iptables INPUT and OUTPUT chains. After getting a list of frequently used IP addresses and their respective ports, the default policy of iptables is modified back to "DROP" and the obtained IP addresses and ports are whitelisted by changing their policy to "ALLOW".

The threshold value of the frequency can be selected as the mean of the total packets captured by all the peer machines. This threshold is found to exhibit a positive correlation with the number of packets captured by each machine. Also, the history of averages can be considered to further improve the value of the threshold. To accommodate new traffic rules, it is recommended to apply whitelisting in regular intervals, so that any changes in traffic flow will be reflected in the firewall rules. Hence, any unusual connections and surge of connections will be prevented ensuring high availability.

4.2 Second LOD (LOD-2)

The LOD-2 involves blacklisting of IP addresses showing malicious behavior. Firstly, the model is trained and deployed on the cloud. The connections are then subjected to the action of the model and then classified as malicious or not. Figure 2 depicts the architecture of LOD-2 used in this work.

Given the dataset, meaningful insights from the data need to be obtained for creating a model that classifies the connections as malicious or not. Data preprocessing techniques like EDA, data scaling, data encoding, data sampling, FS, dataset partition, along with prediction for model training has been performed in this work. The data is split into test and train datasets, and on that data, then, EDA is done. Based on the EDA, Fig. 3 shows the insights on the distribution of classes.

The next step involves the feature scaling of numerical attributes. It essentially aids in data normalization, so that the range is not exceeded. Occasionally, it assists in faster computations in the algorithm. The standard scaler where the values are replaced by the z-scores are used.

$$\text{Standardization, } z = \frac{x - \mu}{\sigma} \tag{1}$$

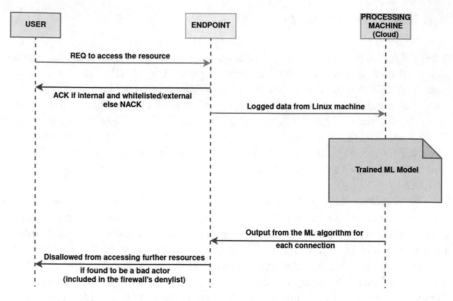

Fig. 2 Architecture of LOD-2

Fig. 3 Attack class distribution of the NSL-KDD dataset

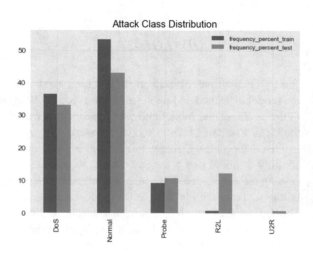

where μ is the mean given by Eq. (2), σ is the standard deviation given by Eq. (3)

$$\mu = \frac{1}{N} \sum_{i=1}^{N} (x_i) \tag{2}$$

$$\sigma = \sqrt{\frac{1}{N} \sum_{i=1}^{N} (x_i - \mu)^2} \tag{3}$$

The Min-Max scaling is given by Eq. (4)

$$X_{\text{norm}} = \frac{X - X_{\min}}{X_{\max} - X_{\min}} \tag{4}$$

The next step would be encoding of the categorical values. Label encoding is done in which the classes are coded between 0 and n. After this, random oversampling is done for reducing the bias of majority classes over the minority classes. Overfitting of models results due to oversampling the redundant records in a random manner from the minority class of the training dataset. The results of random over sampling are as follows:

- Original dataset shape counter ({1: 67,343, 0: 45,927, 2: 11,656, 3: 995, 4: 52}).
- Resampled dataset shape counter ({0: 67,343, 1: 67,343, 2: 67,343, 3: 67,343, 4: 67,343}).

Out of the 42 features, certain features are insignificant. Most of them may be very interrelated among them, and hence, it causes redundancy. Hence, the features that have the highest impact on the accuracy of the ML model is chosen. The matrix or a correlation heatmap for finding the correlation between various values and eliminating one out of every pair of most correlated values. However, the RF Classifier is utilized as a means of assessing the feature importance of the dataset.

RFE involves feature ranking with recursive feature elimination. Out of these features, the features which impact accuracy of the ML model is chosen. The dataset is then partitioned into equal samples, so that there are only two classes which are normal and malicious. This helps in generalizing the proposed model and adding more attack classes in the future. Moreover, it is better to abstract the classes into normal and malicious as it is one of the goals of the research. After the data preprocessing is done, the model building is done. The three ML models, namely NB, DT, and LR, are implemented and compared. The DT model demonstrated higher accuracy. An improvised version of the CART algorithm was implemented in this work. The scientific design is as follows:

Given training vectors $x_i \varepsilon R^n$, $i = 1, \ldots, n$ and a vector of class labels $y \varepsilon R^l$, the DT algorithm splits the space in a recursive fashion, so that the instances possessing the similar classes are classified together. Algorithm-1 shows the DT for detecting the network intrusions.

Algorithm-1: Decision Tree (DT) for intrusion detection
INPUT: S, the set of classified instances **OUTPUT:** Tree of Decisions Require: $S \neq \emptyset$, *num_attributes* > 0 1: **procedure** BUILDTREE 2: **repeat** 3: *max-G* ← *0 //Maximum-Gain* 4: *divideA* ← *null* 5: e ← *Entopy(Features) // Entropy* 6: **for all** *Features s in F* do 7: *gain* ← IG(s,e) 8: **if** *gain > max-G* then 9: *max-G* ← *gain* 10: *divideA* ← s 11: **end if** 12: **end for** 13: Partition(S,divideA) 14: **until** all splits are processed 15: **end procedure**

Suppose the node at m is denoted as Q. For every split of candidate $\theta = (j, t_m)$ comprising of an attribute and threshold, the data is divided into $Q_{left}(\theta)$ & $Q_{right}(\theta)$ subsets.

$$Q_{left}(\theta) = (x, y)|x_j \leq t_m \tag{5}$$

$$Q_{right}(\theta) = Q|Q_{left}(\theta) \tag{6}$$

The $H()$ computes the impurity at m. This selection relies on the job being performed (regression/classification).

$$G(Q, \theta) = \frac{n_{left}}{N_m} H\big(Q_{left}(\theta)\big) + \frac{n_{right}}{N_m} H\big(Q_{right}(\theta)\big) \tag{7}$$

Select the parameters that minimize the impurity as shown in Eq. (8).

$$\theta^* = argmin_\theta G(Q, \theta) \tag{8}$$

Recurse for subset $Q_{left}(\theta^*)$ & $Q_{right}(\theta^*)$ till the maximum permissible depth is obtained, $N_m < min_{samples} or N_m = 1$

The classification criterions are as follows. Suppose the target feature to be classified possesses the values 0, 1, ..., K-1, for node m, signifying a region R_m with N_m records. Equation (9) represents the distribution of class k records in nodes

$$p_{mk} = 1/N_m \sum_{x_i \in R^m} I(y_i = k) \tag{9}$$

Gini and entropy are the two standard impurity measures. Equation (10) gives the formula for entropy a node in DT.

$$H(X_m) = - \sum_k p_{mk} \log(p_{mk}) \tag{10}$$

The information gain (IG) with entropy measure of impurity is chosen as the classification criteria in this research as shown in Eq. (11).

$$IG = Entropy(before) - \sum_{j=1}^{K} Entropy(j, after) \tag{11}$$

5 Experimental Setup and Results

This segment presents the outcomes of the classifiers NB, LR, k-NN, and DT. The confusion matrix (CM) and metrics of performance namely accuracy, precision, F1-score, support, and recall were used to evaluate the classifiers.

Table 1 depicts the matrix of confusion used for evaluating the classifiers [7, 8]

5.1 NB Classifier

For NB classifier, the cross-validation mean score was 0.973776072139, and the final model accuracy was found to be 0.973768617377. Tables 2 and 3 show the CM and the classification report of the NB classifier.

Table 1 CM [7, 8]

Forecasted class	Actual class	
	Positive	Negative
Positive	True +ve (T_p)	False +ve (F_p)
Negative	False −ve(F_N)	True −ve (T_N)

Table 2 NB-CM

Predicted values	Actual values	
	Positive	Negative
Positive	65346	1997
Negative	1536	65807

Table 3 NB-classification report

	Precision	Recall	F1-score	Support
0.0	0.98	0.97	0.97	67343
1.0	0.97	0.98	0.97	67343
Avg/total	0.97	0.97	0.97	134686

Table 4 LR CM

Predicted values	Actual values	
	Positive	Negative
Positive	65527	1816
Negative	773	66570

Table 5 LR-classification report

	Precision	Recall	F1-score	Support
0.0	0.99	0.97	0.98	67343
1.0	0.97	0.99	0.98	67343
Avg/total	0.98	0.98	0.98	134686

5.2 LR Classifier

For the LR classifier, the cross-validation mean score was 0.980822083372, and the model accuracy was 0.980777512139. Tables 4 and 5 depict the CM and classification report of the LR classifier, respectively.

5.3 K-NN Classifier

For the k-NN classifier, the cross-validation mean score was 0.996569816347, and the model accuracy was 0.99775774765. Tables 6 and 7 illustrate the confusion matrix and classification report of the k-NN classifier, respectively.

Table 6: k-NN CM

Predicted values	Actual values	
	Positive	Negative
Positive	67287	56
Negative	246	67097

Table 7: k-NN classification report

	Precision	Recall	F1-score	Support
0.0	1.00	1.00	1.00	67343
1.0	1.00	1.00	1.00	67343
Avg/total	1.00	1.00	1.00	134686

Table 8 DT CM

Predicted values	Actual values	
	Positive	Negative
Positive	67343	0
Negative	7	67336

Table 9 DT-classification report

	Precision	Recall	F1-score	Support
0.0	1.00	1.00	1.00	67343
1.0	1.00	1.00	1.00	67343
Avg/total	1.00	1.00	1.00	134686

5.4 DT Classifier

For the DT, cross-validation mean score was 0.999769836897, and the model accuracy was 0.999948027263. Tables 8 and 9 show the CM and the classification report of the DT classifier, respectively.

From Fig. 4, it was observed that the DT classifier outperformed all other classifiers based on the metrics evaluated.

The trained ML model was deployed in the cloud after converting the trained model into pickle format as illustrated in Fig. 5.

The machines which must be protected will be continuously sniffing the network and will be storing the connection data in a pcap file format. The abro(zeek) script and C script were employed to convert the pcap data into the NKD format and hence to collectively extract more information about the connections. The collected data was then sent in the NKD format to the cloud, so that the trained ML algorithm acts on the data sent by the various machines and then acts upon the data to generate iptables rules specific to a machine. The whole process is serialized by using a message queue which will do the asynchronous processing where multiple nodes will send the collected network data in real time to one processing machine. Hence, each machine is protected by the dynamic firewall rules that will be generated by the ML model deployed in the cloud.

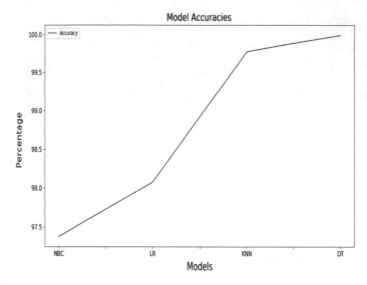

Fig. 4 Comparison of models

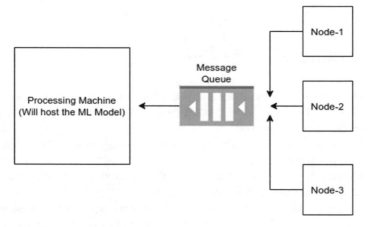

Fig. 5 Classification in real time

6   Conclusion and Future Scope

In this research, several methods to limit the damage caused by the unknown connections are explored. This was done by whitelisting the frequently used connections by calculating the threshold dynamically in LOD-1. This will account for the internal attacks as the frequent connections can be identified and whitelisted. Various ML algorithms like k-NN, DT, NB, and LR are explored, and DT was chosen as it obtained the most accuracy among the others. The data in a typical UNIX machine was collected using the packet sniffers, and the system logs were utilized to extract metadata of the connections. The ML algorithms were applied after storing the data in the database, so that the connections which are malicious will be blacklisted. In future, for LOD-1, a linear regression algorithm can be installed to calculate the value of the threshold instead of calculating the averages. Novel attacks can be detected by utilizing ML techniques. The data collected in the database can be used for mining and generating rules to prevent novel attacks.

References

1. M. Almseidin, M. Alzubi, S. Kovacs, M. Alkasassbeh, Evaluation of machine learning algorithms for intrusion detection system, in 15th International Symposium on Intelligent Systems and Informatics (SISY), pp. 277–282. IEEE (2018)
2. KDD CUP 1999 Dataset, https://kdd.ics.uci.edu/databases/kddcup99/task.html. Last accessed 3 July 2019
3. S. Kumar, A. Viinikainen, T. Hamalainen, Machine learning classification model for Network based Intrusion Detection System, in 11th International Conference for Internet Technology and Secured Transactions (ICITST), pp. 242–249. IEEE (2016)
4. Y. Xin, L. Kong, Z. Liu, Y. Chen, Y. Li, H. Zhu, M. Gao, H. Hou, C. Wang, Machine learning and deep learning methods for cybersecurity. IEEE Access **6**, 35365–35381 (2018)
5. R. Das, T. H. Morris, Machine learning and cyber security, in International Conference on Computer, Electrical & Communication Engineering (ICCECE), pp. 1–7. IEEE (2017)
6. N. Haq, Md. Avishek, F. Shah, A. Onik, M. Rafni, Md. Dewan, Application of machine learning approaches in intrusion detection system: A Survey. Int. J. Adv. Res. Artificial Intelligence **4**(3) (2015)
7. M. Almseidin, M. Alzubi, S. Kovacs, M. Alkasassbeh, Evaluation of machine learning algorithms for intrusion detection system, in 15th International Symposium on Intelligent Systems and Informatics (SISY), pp. 277–288. IEEE (2017)
8. R. Patgiri, U. Varshney, T. Akutota, R. Kunde, An investigation on intrusion detection system using machine learning, in Symposium Series on Computational Intelligence (SSCI), pp. 1684–1691. IEEE (2018)
9. Z. Ahmad, S. Khan, C. Shiang, F. Ahmad, Network intrusion detection system: A systematic study of machine learning and deep learning approaches. Trans. Emerging Telecommunications Technologies (2020). https://doi.org/10.1002/ett.4150

Machine Learning-Based Air Pollution Prediction

Sheethal Shivakumar, K. Aditya Shastry⊙, Simranjith Singh,
Salman Pasha, B. C. Vinay, and V. Sushma

Abstract Due to urbanization, air pollution (AP) has become a vital issue. AP is adversely impacting the humanity by causing asthma and other air-borne diseases. The accurate prediction of AP can aid in ensuring public health. Assessing and maintaining the quality of air has become one of the vital functions for the governments in several urban and industrial areas around the world today. There are a lot of parameters that play a significant role in increasing the air pollution like gases released from the vehicles, burning of remains of fuels, industrial gases, etc. So, with this increase in pollution, there is an increasing requirement of devising models which would record the information about the concentration of the pollutants. The deposition of the injurious air gases is adversely impacting people's life, especially in urban areas. Machine learning (ML) is a domain that has recently become popular in AP prediction due to its high accuracy. There are, however, many different ML approaches, and identifying the best one for the problem at hand is often challenging. In this work, different ML techniques such as linear regression (LR), decision tree regressor (DTR), and random forest regressor (RFR) algorithms are utilized to forecast the AP. Results revealed that the RFR performed better than LR and DTR on the given data set.

Keywords Air pollution · Prediction · Machine learning

1 Introduction

Globally, the rapid increase in population and economic upswing in cities has led to environmental problems such as AP, contamination of water, noise pollution, and many more. AP is a mix of solid elements and atmospheric gases. Emissions from vehicles, industrial chemicals, dirt, etc. are a major part of AP in metropolises. So,

S. Shivakumar · K. A. Shastry (✉) · S. Singh · S. Pasha · B. C. Vinay · V. Sushma
Nitte Meenakshi Institute of Technology, Bangalore 560064, India
e-mail: adityashastry.k@nmit.ac.in

V. Sushma
e-mail: sushma.v@nmit.ac.in

© The Author(s), under exclusive license to Springer Nature Singapore Pte Ltd. 2022   17
P. Shetty D. and S. Shetty (eds.), *Recent Advances in Artificial Intelligence and Data Engineering*, Advances in Intelligent Systems and Computing 1386,
https://doi.org/10.1007/978-981-16-3342-3_2

keeping a check on the quality of air has developed into a major issue in many of the cities, as the increase in air pollution is directly proportional to the risks of having airborne diseases and skin-related diseases. AP has direct impact on health of humans. There is an increased public awareness about the same worldwide. AP is found to cause global warming, and rains constituting of acids among other climatic disasters. Asthma patients are found to suffer a lot due to AP. Precise air quality forecasting can reduce the effect of maximal pollution on the humans and biosphere as well. Therefore, better and improved forecasting of AP forms one of the critical factors for preventing major climatic disasters from taking place [1].

Currently, three approaches used for forecasting AP are statistical models, numerical methods, and ML. The first approach is based on statistical models. Statistical models are built on LR possessing single variable. They have shown negative correlation among the atmospheric variables (wind, precipitation, and temperature). The second approach is based on numerical techniques like transport of chemicals and diffusion of atmospheric components. Even though these techniques are used for forecasting AP, the accuracy of their prediction is very much reliant on updated sources that are very hard to obtain or generate. Furthermore, the characteristics of geography vary from location to location along with terrain conditions that are very complex in nature. These factors pose serious challenges in developing the numerical models for AP prediction, mainly because of the air flow complexity such as the speed of wind and its direction about the topographic feature. In contrast to pure models of statistics, ML techniques are able to consider numerous parameters in a single model. Lately, ML algorithms are becoming popular for predicting air pollution. They have the potential to predict the pollution in air more accurately than the traditional methods [2].

This work deals with the application of three ML algorithms, namely LR, DTR, and RFR on the data set downloaded from Kaggle [3] for AP prediction. The dataset comprised of a total of 14 input attributes described later in the paper. The target attribute was the air humidity (AH). Exhaustive trials were performed for the prediction of AP. Results revealed that the RFR with grid search performed accurately with lowest root mean square error (RMSE) when compared to RFR without grid search, DTR, and LR.

The paper contents have been organized into five main sections. Section 2 discusses the related work in the area of air pollution prediction using ML techniques. The proposed work is deliberated in Sect. 3, and the results are discussed in Sect. 4. The conclusion and future scope are provided at the end.

2 Related Work

This segment deliberates certain relevant works done in the area of air pollution prediction using ML techniques.

In [4], urban pollution was analysed and mapped as per the terrestrial areas. Tehran data was considered for experimentation. Moreover, authors compared the accuracy of forecasts of LR and Naive Bayes (NB) algorithm. NB forecasts information more precisely than the other ML techniques for categorizing AQ classes which are not known. Methods related to regularization and optimization were used to forecast air pollutants' level (particulate matter, sulphur dioxide). Authors projected the values utilizing the data sets of two regions. One region predicted the O_3 and SO_2 values, while the second region forecasted O_3 and PM2.5 values. Data modelling was performed using similarity, and authors employed LR for classification. The assessment criteria utilized was RMSE. The major drawback of [4] is the inability of the LR techniques in forecasting unexpected events. Furthermore, the authors utilized data from only two regions that limited the model's generality.

In the work [5], a new deep learning (DL) technique was proposed for analysing IoT smart city data. Authors have proposed a new technique centred on long short-term memory (LSTM) networks to forecast future AQ values in an urban city. The authors demonstrated that the LSTM performed better than support vector regression (SVR). The work [6] proposes a method to predict the concentration of PM2.5 by utilizing the recurrent neural network (RNN) with LSTM. The results of [6] revealed that the devised approach was able to successfully estimate the PM2.5. During forecast in the long term, the value of PM2.5 was precisely forecasted to elevate. For predicting PM2.5 values that are distant via diverse climatic data related to flow of air, the progressing features of air contamination can be learnt more precisely. Precise PM2.5 forecast values can be provided by the model for lengthier periods of time. The LSTM is embedded in the RNN, where a directed graph is formed by the connections between units. This model has exhibited remarkable performance in several forecasting techniques that depend on historic data. Consequently, authors applied LSTM for solving such problem types. The work [7] mainly used the data gathered from IoT devices for the AQ forecast in smart cities. Data forms a major criterion for developing innovative solutions for smart cities. Regardless of the data prospects of smart cities, developing these solutions and applications forms a daunting task. The exposure to particulate matter which is a category of AP has triggered millions of deaths.

As observed from the above works, accurate AP prediction is the need of the hour as it can make the world healthier and safer. In this regard, ML techniques are being used effectively for AP prediction. Hence, this work focuses on accurate AP prediction using ML techniques.

3   Framework and System Design

This section discusses the framework and the system design of the proposed system for predicting the air pollution. The whole procedure is split into two segments, training and testing, as represented in Fig. 1.

In the first part, the training data (TD) was utilized to train the chosen algorithm. In the second part, the model is subject to testing. During testing, the model uses the previously generated results and predicts the value of target variable. The trained

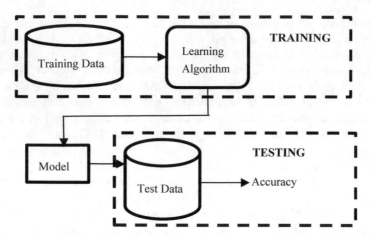

Fig. 1 Model design using testing and training data

model is employed to assess the accuracy of the trained model. The same procedure is repeated using all algorithms under consideration, namely LR, DTR, and RFR algorithms. Figure 2 illustrates the design of the devised technique.

The air pollution data set with 9358 records and 15 attributes was obtained from Kaggle [3]. The analysis of the data set involved estimating the types of attributes. Most of the attributes were numeric in nature. Preprocessing the data set involved estimating the missing values that were denoted as −200 in the data. These missing values (−200) were replaced by zero. Since the data set was small, deleting the instances whose values were missing would make the data set even smaller. Hence, we applied mean imputation method for predicting the missing values. In this method, the missing values were substituted by the mean of the column. This step was repeated for every attribute (column) in the data set.

The heatmap was generated for the data set. The heatmap gave the correlation between the features of the data set. The two features having negative values indicated negative correlation between them. The data was then partitioned into training and test sets. The TD comprised of 70% (6550 records) of the total records. The remaining data (30%) of 2808 records comprised the test data. In this work, three ML algorithms, LR, RFR, and DTR models, were applied on the air pollution data set for predicting the air pollution. In DTR model, the maximum depth (d) of the tree was varied from 2 to 5. Better result was obtained for DTR with $d = 5$. For the RF algorithm, the parameters were searched using grid search as well as without grid search (parameters were applied randomly). Better result was obtained for RFR with grid search.

As the data set consisted of only numerical values, the RMSE was used for measuring the error rate of the prediction. Mean of the target attribute (AH) was taken as the threshold value. If the predicted value of the model goes above this threshold, then we can infer that the air pollution is high, else it is less.

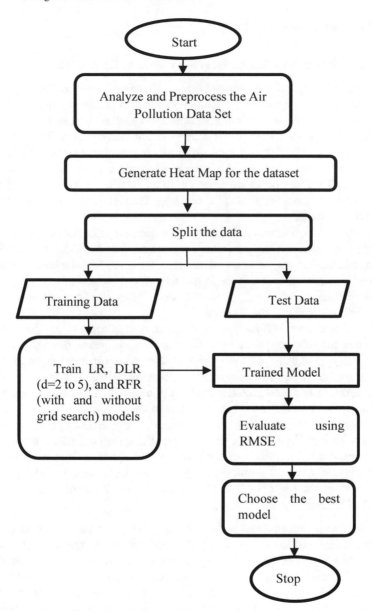

Fig. 2 Design of the proposed system

3.1 Implementation Details

Stepwise procedure of the developed system is summarized below:

(i)   Data set collection: A data set that is relevant to the problem statement is collected from a trusted source and cleaned to facilitate analysis. Cleaning involves dealing with missing values in any method best suited to the future steps. In our work, the data set of air pollution was collected from Kaggle [3].

(ii)   Feature selection: The selected attributes possess divergent significance levels to the problem statement. This technique is to modify the crucial feature which the confirmation technique relies on. In this work, the correlation among the variables was obtained using the heatmap. The features having negative values were removed.

(iii)   Exploratory data analysis (EDA): It represents a procedure in which preliminary examinations are performed on data set for identifying missing values, redundant records, and other data inconsistencies. This technique is utilized in the data preprocessing phase for preparing the data before the ML model is applied. In this work, EDA was done using the heatmap that gave the relationship among the features.

(iv)   Training phase: A TD is a set of data that is used for learning the model and fitting the relevant parameters. Most techniques search through the TD for identifying realistic relationships and tend to overfit the data, meaning they can recognize and exploit obvious relations in the TD that do not embrace in general. In this work, the training was done using three ML models, namely LR, DTR, and RFR on a TD of 6550 records.

(v)   Testing phase: A test data set is independent of the TD, but it possesses the same probability as that of the TD. Overfitting happens when the ML model delivers accurate predictions for TD but exhibits poor performance on test data. Therefore, a test set is utilized for performance assessment of ML models. In this work, the LR, DTR, and RFR models were evaluated on the test data of 2808 records.

The work was implemented in two phases. The first phase was data cleaning. The data set acquired from Kaggle [3] had 9358 entries. As the data set is already small, deleting rows consisting missing After applying various cleaning techniques as well as feature selection algorithms, we were left with a data set consisting of 10 attributes and 9358 records. Air Humidity represents the target feature. Missing values are tagged with -200 value. This set of data can be utilized solely for the purpose of research.

Table 1 Data set description

Attribute name	Attribute description
Date	Date of air quality measurement in DD/MM/YYY format
Time	Time of measuring the air quality in HH.MM.SS format
CO Conc	Concentration of carbon oxide on an hourly basis measured in mg/m^3
PT08.S1	Tin oxide measured hourly in $microg/m^3$
Benzene	Concentration of benzene on an hourly basis measured in mg/m^3
PT08.S2	Titania measured hourly
NOx	Hourly concentration of nitrogen oxide measured in ppb
PT08.S3	Tungsten oxide measured hourly
NO_2	Hourly measurement of nitrogen dioxide in $microg/m^3$
PT08.S4	Tungsten oxide measured hourly
PT08.S5	Indium oxide measured hourly
T	Temperature measured in degree per Celsius
RH	Relative humidity measured in percentage
AH	Target attribute representing absolute humidity

3.2 Dataset Details

The data set for experimentation was taken from the source: UCI data set type: xlsx with data set size: 9358 * 15 from [3]. Table 1 shows the description of the attributes of the data.

3.3 Algorithms

In this work, three ML algorithms were employed for the prediction of AP. They are discussed below.

(i) DTR: Decision tree (DT) learning represents a ML technique for modelling forecasts. The DT is employed as a predictive model in which the data records are represented in the intermediate nodes, while the class to be predicted will be in the leaf. The target attributes in DTR are continuous values representing real numbers. Due to their simplicity and fast learning, DT has become popular in recent times. A typical DTR used in our work is depicted in Fig. 3.

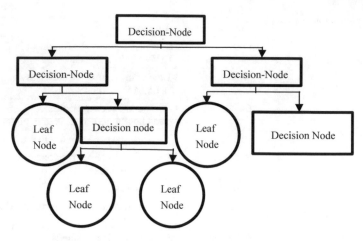

Fig. 3 DTR for AP prediction

(ii)   RFR: RFR represents a learning technique based on ensembles of regression trees. Here, several regression trees are constructed during training, and the mean prediction of each tree is taken to give the final output. A RFR used in this work is illustrated in Fig. 4.

(iii) LR model: LR models the association among scalar target and 1/more independent variables (explanatory variables). The LR model utilizes linear predictor functions for estimating unknown parameters from the data. LR is a type of supervised ML learning technique that does a regression task. The target feature is forecasted based on the independent features. Its primary usage is in determining the associations among the variables along with prediction of continuous values. In this effort, the target feature representing AH is predicted using the LR model.

4   Experimental Set-up and Results

The trials were performed on Windows 10 OS with 1 TB HDD and 8 GB RAM. Python 2.7 was used as the programming language along with the libraries Pandas, NumPy, Scikit-Learn, Matplotlib, and Seaborn.

The heatmap in Fig. 5 shows correlation between variables.

Figure 6 depicts the comparison between the LR, DTR, and RFR algorithms.

The RMSE value shows that the model can relatively predict the data accurately. After speaking to certain industry experts, it is concluded that the least RMSE value shows that the model is fit and predicts accurately. Considering the above statement, it is observed that RFR without grid search gives the least RMSE value of 0.1019.

Fig. 4 RFR for AP
prediction

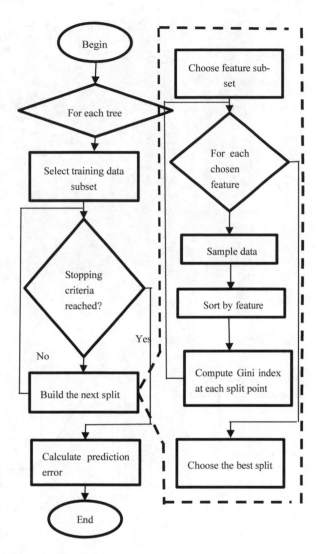

5 Conclusion and Future Scope

In this effort, three ML algorithms, viz. LR, DTR, and RFR, have been employed for
AP prediction. For modelling independent variables, we assess the interaction terms
to assess whether the outcome of one independent feature relies on the values of
other features. The RMSE value shows that the model can relatively predict the data
accurately. The RMSE values obtained for the ML algorithms were between 0.2 and
0.5 indicating higher accuracy in AP prediction. Exhaustive trials demonstrated that
RFR without grid search was the best-suited algorithm in comparison with LR and
DTR models. As the data was numerical in nature, RFR provided the best results.

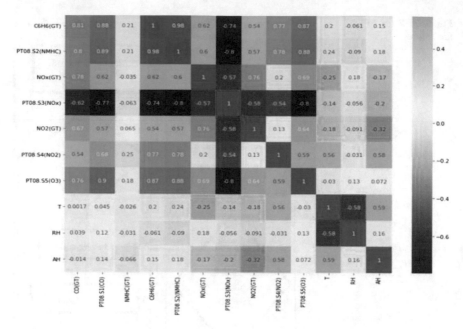

Fig. 5 Heatmap of the air quality dataset

Fig. 6 Comparison of algorithms for AP prediction

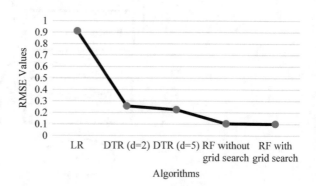

In future, this work can be extended to include real-time data to give more realistic prediction. A user interface can also be created to tell if the place is highly polluted or not. To improve the accuracy of the results, larger data set can be used. More robust ML algorithms can be utilized to perform better predictions.

References

1. V.R. Pasupuleti, P. Uhasri, S. Kalyan, H.K. Reddy, Air quality prediction of data log by machine learning, in 6th International Conference on Advanced Computing and Communication Systems

(ICACCS), Coimbatore, India, pp. 1395–1399 (2020)

2. G. Sakarka, S. Pillai, C.V. Rao, A. Peshkar, S. Malewar, Comparative study of ambient air quality prediction system using machine learning to predict air quality in smart city, in Proceedings of International Conference on IoT Inclusive Life (ICIIL 2019), NITTTR Chandigarh, India. Lecture Notes in Networks and Systems, ed. by M. Dutta, C. Krishna, R. Kumar, M. Kalra, vol 116. Springer (2020)

3. https://www.kaggle.com/sayakchakraborty/air-quality-prediction-of-relative-humidity. Accessed online on 24 December 2019

4. S. Ameer, M. Ali Shah, A. Khan, H. Song, C. Maple, S. Ul Islam, M. Nabeel Asghar, Comparative analysis of machine learning techniques for predicting air quality in smart cities. IEEE Access **7**, 128325–128338 (2019)

5. I. Kȯk, M. Ulvi Simsek, S. Ozdemir (2017) A deep learning model for air quality prediction in smart cities, in International Conference on Big Data (Big Data), pp. 1983–1990. IEEE (2017)

6. R.O. Sinnott, Z. Guan, Prediction of air pollution through machine learning approaches on the cloud. In IEEE/ACM 5th International Conference on Big Data Computing Applications and Technologies (BDCAT), pp. 51–60. IEEE (2018)

7. Y-T. Tsai, Y-R. Zeng, Y-S. Chang, Air pollution forecasting using rnn with lstm, in IEEE 16th International Conference on Dependable, Autonomic and Secure Computing, 16th International Conference on Pervasive Intelligence and Computing, 4th International Conference on Big Data Intelligence and Computing and Cyber Science and Technology Congress (DASC/PiCom/DataCom/CyberSciTech'), pp. 1074–1079. IEEE (2018)

Crop and Fertilizer Recommendation System Based on Soil Classification

Pruthviraj, G. C. Akshatha⬧, K. Aditya Shastry⬧, Nagaraj, and Nikhil

Abstract Agriculture forms a major occupation in countries like India. More than 75% people rely on farming for their daily wages. Hence, achieving good yield in the crops grown by farmers is the major concern. Various environmental factors have a significant impact on the crop yield. One such component that contributes majorly to the crop yield is soil. Due to urbanization and enhanced industrialization, the agricultural soil is getting contaminated, losing fertility, and hindering the crop yield. Machine Learning (ML) is employed for agricultural data analysis. The proposed ML based model aims at classifying the given soil sample datasets into four different classes, namely very high fertile, high fertile, moderately fertile, and low fertile soil utilizing support vector machine (SVM) technique. It also predicts the suitable crops that can be grown based on the class which the soil sample belongs to and suggests the fertilizers that can be used to further enhance the fertility of soil. Using proposed model, farmers can make decisions on which crop to grow based on the soil classification and decide upon the nitrogen–phosphorous–potassium (NPK) fertilizers ratio that can be used. Comparison of the SVM algorithm with k-nearest neighbor (k-NN), and decision tree (DT) has shown that SVM performed with a higher accuracy.

Keywords Machine learning · Soil classification · Support vector machine · Crop yield

1 Introduction

Data analysis is a procedure which involves data inspection, cleaning the data, and modeling the data with the aim of extracting useful information and conclusions. It is a technique of analyzing, extracting, and predicting meaningful information from vast data. Companies make use of this process to turn their client's raw data to

Pruthviraj · G. C. Akshatha (✉) · K. A. Shastry · Nagaraj · Nikhil
Nitte Meenakshi Institute of Technology, Banglore 560064, India
e-mail: akshatha.gc@nmit.ac.in

K. A. Shastry
e-mail: adityashastry.k@nmit.ac.in

useful information. This process of data analysis can also be used in the agriculture sector. Most of the farmers have been relying on their long-term field experience on specific crops to expect a higher yield during the next harvesting period. But they still do not get the crops' worth price. It is mostly because of selection of inadequate crops or improper irrigation. Hence, agricultural researchers stress on the need for an efficient mechanism to predict and improve the crop growth. Majority of research works carried out in agriculture field, focusing on biological mechanisms to identify crop growth and improve its yield. The crop yield primarily depends on various factors like diversity of crop, quality of seed, and various environmental factors like sunlight, temperature, soil (pH), water (pH), rainfall, humidity, and variety of soil nutrients. By analyzing soil of a particular region, best crop to obtain more yield can be predicted. This prediction will help farmers to identify the appropriate crops that is grown in their farm according to the soil classification. The estimation of yield harvest is a tedious task because it is affected by various factors such as soil productiveness, weather conditions, and practices followed for cultivation such as amount of irrigation and date of sowing [1–3].

Soil classification is one of the interesting areas of research in the engineering field. The traditional methods such as cone penetration test (CPT) consume lots of time and need experts for extracting accurate results. Latest approach to soil classification is to use ML techniques. Several ML algorithms such as k-NN, DT, and SVM can be used. SVM is suitable to solve classification problems [1–3]. Hence, in our proposed work, we are using SVM for classification of soil nutrition level into different classes like class 0(Low fertile), class 1 (moderately fertile), class 2 (high fertile), and class 3 (very high fertile). We have used SVM algorithm to classify the given soil data samples founded on numerous measurable attributes of soil. Similarly, based on these features, the soil can be classified as class 0 class 1, class 2, and class 3. And also, with the same information, suitable crop can be grown which will give more yield that will be predicted.

The remainder of this manuscript is comprised of four sections. Section 2 converses about the related work carried out in the field of soil classification. Section 3 deliberates the methodology followed for classifying the given soil sample datasets based on the various nutrients in the soil. Section 4 confers about the experimentation carried out on the sample soil dataset and validates the efficiency of the methodology used. The paper ends with conclusion section.

2 Related Work

Extensive amount of study is being done for classifying the soil samples using different dataset. Zaminur et al. [4] presented a hybrid model that can predict soil series based on land type, and as per the prediction, they will suggest suitable crops but do not specify the fertilizers that can be utilized to improve the soil fertility.

Ashwini et al. [5] proposed the classification technique centered on SVM. The soil samples are graded using different scientific features such as texture, drainage, color, and terrain. These soil samples are classified into red, black, clay, alluvial, etc., categories but do not specify about fertility. Jay Gholap et al. [6] proposed a classification approach using classification algorithms such as JRip, Naïve Bayes, J48, and regression algorithms like least median square regression and linear regression. The classification is not based on soil nutrients and does not recommend crop and fertilizer ratio. Ramesh et al. [7] have proposed a classification model which is based on Naïve Bayes classifier. In their work, the classification is not based on soil nutrients and do not propose a new approach.

Vibha et al. [8] designed a hybrid model based on clustering and classification. They mainly focus on soil nutrient mining. The different soil parameters are not considered for classification. The authors do not recommend how to improve the fertility of the soil. Sofianita Mutalib et al. [9] in their model use the techniques k-means and self-organizing map (SOM) for classifying the soil. It is basically comparing the performance of two existing algorithms, and the classification is not done based on the soil nutrients. It is mainly based on color, texture, etc.

The proposed model mainly focuses on classifying the soil records gathered from GKVK UAS, Bangalore, Karnataka, India. The dataset has 1550 soil samples obtained from Tarikere, Kadur, Sringeri, and Koppa, places of Chikamagaluru district, of Karnataka. Our proposed approach based on SVM is applied to the soil samples, and the classification is done based on various metal compositions such as calcium, magnesium, potassium, lime, and soil moisture content, which denote the soil parameters. It also aims at solving farmer problems by predicting the effective crops that can be grown in the given class of soil and also recommends NPK fertilizer ratio to increase the nutrients in soil based on the class to which it belongs, which will help farmers to get more crop yield.

3 Proposed Work

Motivated from the work done by various researchers in the domain of soil classification, we proposed an SVM based classifier for soil classification. The key idea behind the SVM algorithm is to generate the hyper plane that can partition n-dimensional space into classes such that a new data point can be placed in the appropriate class. SVM selects the extreme point or vectors called support vectors which aid in creating the hyper plane [8].

3.1 SVM Algorithm

SVM is the ML algorithm which determines the decision planes that creates decision boundaries. A decision boundary divides the class of objects from one another. Support vectors have highest prominence in ML field. The main objective of the Kernel function is to abstract the data and transform to the required form. SVM is an employed method both for classification and regression tasks and used widely for classification. SVM algorithm takes data as input and creates a hyper plane that splits the information into classes. Let us consider the subsequent case, let us assume we have two objects: circle and triangle, and our data has two key features let us denote it as x and y. What is that we need is a classifier that, which is given a pair of (x, y) coordinates, it outputs 1 if it is either circle or triangle. Labeled training data is plotted on a plane as shown in Fig. 1 [9].

SVM reads the data records and generates the hyper plane that best splits the labels. Multiple hyper planes may exist, among which it is necessary to identify the best hyper plane that splits the labels in many accurate ways shown in Fig. 2. In SVM, best hyper plane is the one that maximizes the margins from both objects [10].

In the above example, the data could be separated linearly; hence, we were able to draw a hyper plane to separate circles from triangles. Usually things are not that simple. Sometimes we must deal with nonlinear data as demonstrated in Fig. 3 [7].

From Fig. 3, it is clear that there is no linear decision boundary that divides both objects. To resolve it, a third dimension is added as shown in Fig. 4 [11].

In 3D, a hyper plane is parallel to x-axis at some z. Remapping it back to two dimension, a decision boundary is obtained which is a circumference of some radius that separates the objects triangle and circle using SVM as demonstrated in Fig. 5 [11].

Fig. 1 Illustration of two classes; blue dots depicts class 1 samples; red triangles depicts class 2 samples [9]

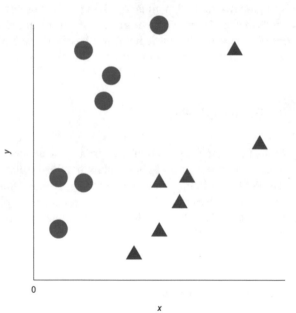

Fig. 2 Optimized hyper
plane separating classes of
two-dimensional data using
SVM [10]

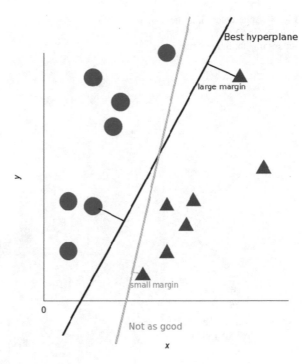

Fig. 3 Illustration on
nonlinear data [7]

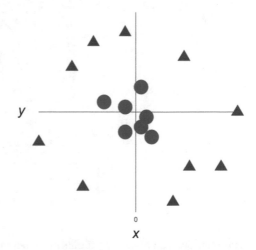

The proposed SVM based classifier is built by analyzing the soil sample training dataset. Class labels for given data are predicted during the classification step. The dataset records and their related categories of classes under analysis are split into training set and test set. 70% of the soil sample data are utilized for model training, and remaining 30% of the sample data are used for testing purposes. Individual tuples that constitute the training data are sampled randomly from the information under

Fig. 4 Application of kernel
function in SVM for
separating nonlinear data
[11]

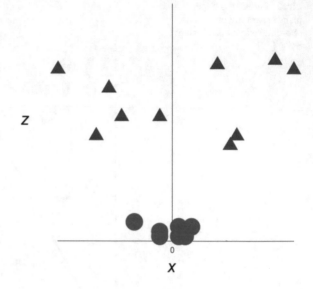

Fig. 5 Nonlinear data
separated by hyper plane
[11]

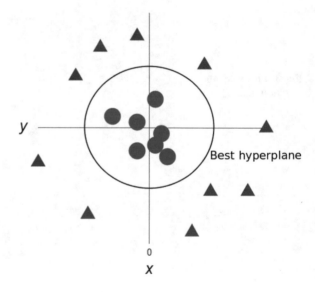

examination. The remaining tuples are independent of training tuples and constitute
the test data, which means that the test set will not be utilized to construct the classifier.
The test set is utilized to obtain the accuracy of classifier. The classifier accuracy is
based on the proportion of test records that are classified correctly by the classifier.
Based on the threshold values of the nutrients that we have used in our datasets, we
will be able to classify them into different classes like class 0 (low fertile), class 1
(moderately fertile), class 2 (high fertile), and class 3 (very high fertile), and based

on these classes, we will be able to predict the suitable crops and NPK fertilizers ratio. The proposed approach is represented in Fig. 6.

Description of the soil dataset [12]: The dataset used in our proposed work is obtained from GKVK UAS which include samples from different area of Chikkamagaluru district. It includes samples from various taluk of Chikkamagaluru district like Tarikere, Kadur, Sringeri, and Koppa. There are 1550 soil samples in the dataset, and the attributes of datasets are as follows: N, P, K, Ca, Mg, Lime, C, S, and moisture.

Crops' health depends on sufficient supply of soil nutrients and moisture. As the supply of moisture decreases, plant normal function and development are disturbed, and crop yields are reduced. Few soil sample data records used in our proposed approach is as demonstrated in Table 1.

The threshold nutrient values for the soil samples that we have used in our proposed approach is demonstrated in Table 2.

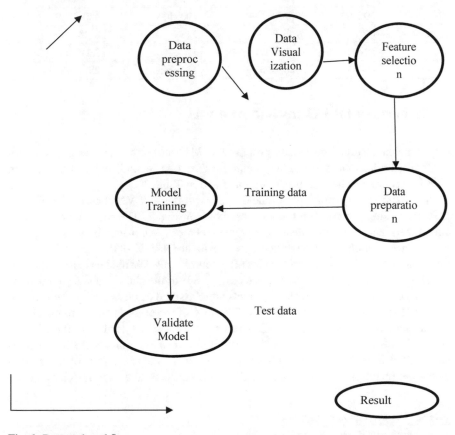

Fig. 6 Proposed workflow

Table 1 Soil sample dataset

Ca	Mg	K	S	N	Lime	C	P	Moisture
9.653	6.585	142	108	226.05	5.83	1.29	18	0.9
19.88	22.2	339.35	77	308.25	6.45	2	298	0.8
2.931	41.22	514.29	108	277.42	6.43	0.74	48	0.6

Table 2 Threshold values

	Low	Medium	High	Very high
Ca	0–1.5	1.5–11.2	11.2–33.6	33.6–45.1
Mg	0–18	18–36	36–54	54–72
K	0–150	150–300	300–450	450–620
S	0–35	35–80	80–120	120–170
N	0–200	200–300	300–400	400–1430
Lime	0–6	6–7	7–8	8–14
C	0–0.6	0.6–1.4	1.4–2.2	2.2–3.0
P	0–85	85–170	170–255	255–350

4   Experimental Results and Analysis

Experimental results show that the proposed SVM based classifier was able to classify 845 soil samples in to correct classes and misclassified 240 soil samples resulting in 77.85% accuracy. The results are demonstrated in Table 3.

Analysis: The overall performance of a proposed SVM based classifier is described using the confusion matrix as shown in Fig. 7. The rows indicate actual class labels, and columns indicate the predicted labels. According to the matrix, 58% of the class-0 labels are also predicted as class-0 labels. 27% of the class-0 labels are misclassified as class 1. 1% of the class-0 labels are misclassified as class 2. 14% of the class-0 labels are misclassified as class 3. 3.8% of the class-1 labels are misclassified as class 0. 91% of the class-1 labels are classified as class 1 only. 0.4% of the class-1 labels are misclassified as class 2. 0.46% of the class-1 labels are misclassified as class 3. 0.12% of the class-2 labels are misclassified as class 0. 0.8% of the class-2 labels are misclassified as class 1. 74% of the class-2 labels are classified as class 2. 5.9% of the class-2 labels are misclassified as class 3. 16% of the class-3 labels are misclassified as class 0. 0.02% of the class-3 labels are misclassified as

Table 3 Results of proposed approach

Correctly classified soil sample instances	Incorrectly classified soil sample instances	Total number of testing records	Accuracy of proposed approach
845	240	1085	77.85%

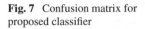 **Fig. 7** Confusion matrix for proposed classifier

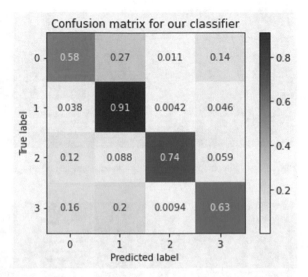

class 1. 0.9% of the class-3 labels are misclassified as class 2. 63% of the class-3 labels are classified as class 3.

Classification of soil is done based on the level of fertility. The values which are the features of the tested soil samples are given as input to the system. It is classified as to which class it belongs. There are four classes, namely class 0, class 1, class 2, and class 3. Class 0 refers to low fertile soil, class 1 refers to moderately fertile soil, class 2 refers to high fertile soil, and class 3 refers to highly fertile soil. Based on the classification, the crops that can be grown are suggested. This helps the farmers to grow the best-suited crop that is adaptable to their soil conditions. If the soil belongs to class 0, then crops like beans, green peas, carrot, and onion can be grown. If the soil belongs to class 1, then crops like radish, cow pea, cabbage, cucumber, and sweet potato can be grown. If the soil belongs to class 2, then crops like sugar cane, paddy, bajra, guava, acacia, etc., can be grown. If the soil belongs to class 3, then crops like barley, cotton, tobacco, sunflower, etc., can be grown.

The experimental results are depicted as shown in Figs.8 and 9.

The snapshot demonstrated in Fig. 8 shows the classification of given input soil sample dataset into class 0 and suggests the suitable crop such as beans and chickpea that can be grown, along with the fertilizer ratio.

The snapshot shown in Fig. 9 shows the classification of given input soil sample dataset into class 1 and suggests the suitable crop that can be grown, along with the fertilizer ratio.

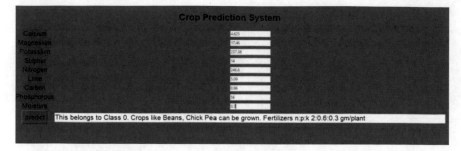

Fig. 8 User interface for classifying the soil sample and suggestion of crop and fertilizer ratio

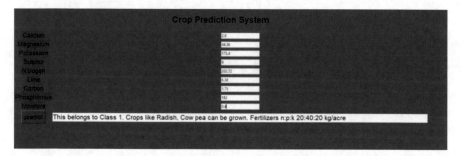

Fig. 9 User interface of the classified soil result of the proposed classifier and suggestion of crop and fertilizer ratio

4.1 Comparison of SVM with k-NN and Decision Tree

The soil sample dataset is verified against three classification algorithms like k-NN, DT, and SVM. Initially, the k-NN algorithm (K-nearest neighbor) is used for classification. For k-NN, the training and test data has been loaded. Then, we were supposed to choose the k value (which is nearest data point). K will be the new point. After that, for each point in test data, we computed the distance between data points and each row of training data with the help of Euclidian distance from new data point. We have included the test record for the given class label. But for the dataset chosen, k-NN gives less accuracy when compared to the other algorithms as it is susceptible to imbalanced dataset. Accuracy obtained using k-NN algorithm is 72.04%. Later DT algorithm has been used. In DT algorithm, soil parameters was classified based on the threshold value in each node. Based on the threshold value, soil samples are classified into different class level at the leaf or the last node. But it gave less accuracy when compared with SVM since it is prone to overfitting. Accuracy we got using DT algorithm is 69.46%. Finally, our proposed model was built using the SVM algorithm by using the training and testing dataset. Our dataset has 1550 records where we are using 70% data for training and remaining 30% of data for testing. SVM obtained higher accuracy because it captures nonlinearity in dataset (Table 4).

Table 4 Comparison of ML algorithms

Algorithm	Accuracy
SVM	77.85
k-NN	72.04
DT	69.46

Fig. 10 Accuracy comparison between SVM, k-NN, and DT

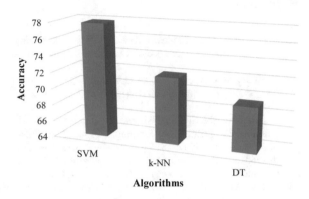

Figure 10 shows the comparison of accuracies between the SVM, k-NN, and DT algorithms.

As demonstrated in Fig. 10 for the SVM algorithm, we have got 5% to 8% of improvement in accuracy level when we compared with k-NN and DT algorithms. Accuracy we got using SVM algorithm is 77.85. From these results, we can infer that the SVM algorithm performed better classification when compared to k-NN and DT.

5 Conclusion

The proposed model aims at classifying the given soil sample datasets into four different classes, namely low fertile, moderately fertile, high fertile, and very high fertile soil, using SVM algorithm. It also predicts the suitable crops that can be grown based on the class which it belongs to and also suggests the NPK fertilizers ratio that could be employed to obtain better yield of the crops grown in case if the fertility of the soil is not up to the mark. Using this model, farmers can find out the fertility class of their field soil and can make decisions on which crops to grow, so that they will get better yield and fertilizer ratio to be used to further improve the fertility of the soil.

In the proposed classification model, linear SVM is used for classifying the given soil sample dataset. In future, the same work can be extended using kernel SVM on the same soil dataset to see how well it performs in terms of classifying soil type, given N, P, K, Ca, Mg, Lime, C, S, and moisture parameters as input. Hence, kernels like radial basis kernel, Gaussian kernel, etc., can be explored as a part of future work. Moreover, future work may also include incorporation of neural networks on the same training dataset to see if the classification accuracy improves any further.

References

1. K.A. Shastry, H.A. Sanjay, H. Kavya, A novel data mining approach for soil Classification, in 9th International Conference on Computer Science & Education, Vancouver, BC, pp. 93–98 (2014)
2. K. Aditya Shastry, H.A. Sanjay, G. Deexith, Quadratic-radial-basis-function-kernel for classifying multi-class agricultural datasets with continuous attributes. Appl. Soft Comp. **58**, 65–74 (2017)
3. S. Panchamurthi, Soil analysis and prediction of suitable crop for agriculture using Machine Learning. Int. J. Res. Appl. Sci. Eng. Technol. **7**(3), 2328–2335 (2019)
4. S. Al Zaminur Rahman, K. Chandra Mitra, S.M. Mohidul Islam, Soil classification using ML methods and crop suggestion based on soil series, in 21st International Conference of Computer and Information Technology (ICCIT), pp. 1–4. IEEE (2018)
5. A. Rao, A. Gowda, R. Beham, Machine Learning in soil Classification and crop detection. Int. J. Sci. Res. & Dev. **4**(1), 792–794 (2016)
6. J. Gholap, A. Ingole, J. Gohil, S. Gargade, V. Attar, Soil data analysis using classification techniques and soil attribute prediction. Int. J. Comput. Sci. Issues **9**(3), 1–4 (2012)
7. R. Vamanan, K. Ramar, Classification of agricultural land soils: A data mining approach. Agricultural J. **6**(3), 379–384 (2011). https://doi.org/10.3923/aj.2011.82.86
8. L. Vibha, G.M. HarshaVardhan, S.J. Prasanth, P. Deepa Shenoy, K.R.L. VenuGopal, M. Patnaik, A hybrid clustering and classification technique for soil data mining, in Proc. ICTES. IEEE, pp. 1090–1095 (2007)
9. S. Mutalib, S-N Fadhlun Jamian, S. Abdul-Rahman, A. Mohamed, Soil classification: an application of self organizing map and k-means, in: Proc 10th IEEE-ISDA. IEEE, pp 439–444 (2010)
10. G. Yi-Yang, Ren, Nan-ping, Data mining and analysis of our agriculture based on the DT, in Proc. IEEE-ISECS, pp 134–138 (2009). 10.1109 ICCCM. 5267962
11. N. Jain, A. Kumar, S. Garud, V. Pradhan, P. Kulkarni, Crop selection method based on various environmental factors using ML, February 2017.
12. K. Srunitha, S. Padmavathi, Performance of SVM classifier for image-based soil classification, in 2016 International Conference on Signal Processing, Communication, Power and Embedded System (SCOPES), pp. 411–415. IEEE (2016)

Human Activity Classification Using Deep Convolutional Neural Network

Aniket Verma, Amit Suman, Vidyadevi G. Biradar, and S. Brunda

Abstract The human activity classification is an identification of person's actions from an image or video frames captured by a camera. This helps in tracking person movements which are useful for various purposes. The human activity classification is a challenging task due to factors like complex background captured in the image, occlusion, and camera view angle. Therefore, there is a need for development of a robust human activity classification system. In this paper, a classification model based on deep convolutional neural network is proposed which extracts coordinates of human joints and parts affinity field from an image for predicting human pose and classification of activity. The model is designed to carry out classification of six major activities, and they are raising hand, pick-up an item, squats, namaste greeting, on-phone, and pointing a gun. These activities represent a class of suspicious and non-suspicious activities. The activity "namaste greeting" is included considering the COVID-19 issue. The proposed deep convolutional neural network comprises of two branches for joints coordinates estimation and parts affinity field estimation. The model is trained on MPII human pose dataset and tested on live videos captured from a webcam to evaluate its performance. The model gives highest training accuracy of 99.7 and 99.2% test accuracy.

Keywords Deep convolutional neural network · Pose estimation · Activity recognition · Parts affinity field

1 Introduction

In this era of online media, the amount of images and video databases are increasing having unbelievable progress. Using the data, there is a huge scope for computer vision community to come up with algorithms that can build smarter surveillance systems by validating the human activity, in order to prevent the occurring of criminal activities. Human activity recognition is a method in which one can predict the

---

A. Verma · A. Suman · V. G. Biradar (✉) · S. Brunda
Nitte Meenakshi Institute of Technology, Bangalore, India
e-mail: vidyadevi.g.biradar@nmit.ac.in

© The Author(s), under exclusive license to Springer Nature Singapore Pte Ltd. 2022   41
P. Shetty D. and S. Shetty (eds.), *Recent Advances in Artificial Intelligence and Data Engineering*, Advances in Intelligent Systems and Computing 1386,
https://doi.org/10.1007/978-981-16-3342-3_4

behavior of a person based on the trace of their movements. Basically the idea behind is that once the activity of the subject is identified and known, then a smart computer system model can provide assistance. In recent studies, human activity recognition has seen success by using deep learning techniques to solve problems.

Human activity classification aims to classify activities of a person from a still image or video frames. Image-based activity classification plays a significant role in various applications in developing healthcare systems, identification of suspicious activities, robotic applications, and event analysis. Human activity classification is a challenging task due to factors like complex background in the video or image, interaction of people leading to overlapping of body parts, occlusions, and more. A 2D human pose estimation is the core task in identification and classification of human activities. Human pose estimation techniques exploit the information on human body joints coordinates and their limbs associations.

Human activities are identified by observing movements in body joints and main joints of human skeleton. The traditional methods have limitations to accurately classify and identify human activity for scenarios where people in the scene are interacting in some way. Figure 1a and b show images with persons that are close to each other, and there is an overlap of limbs belonging to different person. It is evident from the stick figure as highlighted in Fig. 1a and b that their joints are overlapped. Deep learning models have been excelled in solving most of the computer vision problems due to their learning capability when subjected to large dataset. Deep convolutional neural networks are the major category of deep learning models, and these models are used in solving many problems like object detection, face detection, and image classification. This paper implements a model for human activity classification using deep convolutional network with two branches.

(a)   (b)

Fig. 1 Part affinity issue **a** persons hands joints are overlapping with body joints, **b** hand joints of middle person occluded

Human pose estimation is carried out using a CNN-based regression model for the extraction of body joints. This model uses multistage cascaded regressors, and results indicate that CNN regression model is efficient in localization of hidden joints. The research work in [1] presents a joint model consisting of CNN and graph model for prediction of human pose. A two stage is presented which comprises of two models, first is a multistage CNN model designed for extraction of required features from the image and the second is a prediction model that predicts the pose from the features extracted by CNN model [2].

The paper is organized into different sections, Section 2 gives a brief survey on human activity classification methods. Section 3 presents the methodology of proposed model for human activity classification. Section 4 demonstrates results, and conclusion and scope for future enhancement are given in Section 5.

2 Literature Survey

Literature review has been carried out to understand the methodology and limitation of various existing methods for human pose estimation. The advantages and disadvantages are summarized in Table 1.

The study of existing methods indicates that there are still challenges in human pose estimation, and there is a need for robust algorithm. This paper proposed a human activity classification model based on deep learning technique which is discussed in the next section.

3 Proposed Model

In this work, a double-branch CNN model is used for estimation of human pose and human activity classification. The upper branch is for detection of joints coordinates, and lower branch is for computation of parts affinity field as shown in Fig. 2. This work utilizes images from MPII human pose dataset for training the deep learning model. The images for activities like namaste greeting are locally generated as images in this category are not in plenty in MPII database, which are also used for training and testing. Human pose estimation requires estimation of joints coordinates and parts affinity field. This has been carried out using a double-branch CNN model as shown in Fig. 2.

The image is fed into double-branch CNN in which the upper branch estimates the coordinates of body joints and the lower branch estimates parts affinity fields. This information is combined to estimate human pose, and it is fed to dense network for classification. The model is tuned to accurate classification of six activities, and they are raising hand, pick-up an item, squats, namaste greeting, on-phone, and pointing a gun. These activities represent the categories of suspicious and non-suspicious activities.

Table 1 Methods for human activity classification

Paper	Techniques	Advantages	Disadvantages
Hao Du et al. [3]	This paper presents a method to classify human activity using transfer-learned residual network based on micro-doppler spectrograms. The performance of the model was evaluated on CMU Mocap dataset	Accuracy obtained 97% Jitter is less, results' deflection is less	Requires more computational capacity to train CMU Mocap dataset
Xiaoran Shi et al. [4]	Authors of this paper have used deep learning techniques on spectrograms data. The method uses convolutional generative adversarial networks and DCNN	It offers good scalability Requires less time to train	Accuracy is less than 82% Jitter is high, results deflect too often
Zhe Cao et al. [1]	This paper presents an effective method to detect 2D pose from images containing multiple people The method uses greedy bottom-up parsing strategy	It offers high accuracy, Jitter is very less, results rarely deflect Suitable for real-time application	Requires more computation power
Cheng Ning et al. [5]	This paper proposes a method to increase the accuracy of classification by improving Single Shot Detection (SSD) algorithm. The method is designed by incorporating architectural features of pretrained inception model	Algorithm is robust	Jitter is too high; results deflect too often Less accuracy for small objects

(continued)

Table 1 (continued)

Paper	Techniques	Advantages	Disadvantages
Fabian Caba et al. [6]	This paper introduces three possible applications where ActivityNet can be used. They are untrimmed video classification, trimmed activity classification, and activity detection	In terms of richness of taxonomy and activity diversity, more varieties are presented using ActivityNet	The obtained accuracy is less
Jaeyong Sung et al. [7]	In this paper, an inexpensive sensor called RGBD was used to take the input. To model different properties of human activities, a two-layered maximum entropy Markov model (MEMM) was used	The algorithm showed an accuracy of 84.3% on the training dataset (what activities)	The algorithm showed an accuracy of 64.2% on test dataset
Andrej Karpathy et al. [8]	In this paper, UCF-101 video dataset was used; a slow fusion network using CNN was developed	The slow fusion network out performed over early fusion and late fusion network; it showed an accuracy of 80% on sports data	The algorithm did not show much difference for human object interaction, body-motion, human–human interaction, and playing instruments
Ivan Laptev et al. [9]	This paper uses local space–time features, space–time pyramids, and multichannel nonlinear SVM to improve state-of-the-art results on standard KTH action dataset	The proposed algorithm achieves 91.8% accuracy on std. KTH action dataset	This algorithm need to be improved in the script-to-video alignment and extending the video collection to a much larger dataset

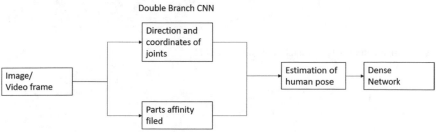

Fig. 2 Human pose estimation mode

The MPII human pose dataset is used in this work, and it has 25,000 images with annotated body joints. The dataset covers 410 human activities which is a benchmark dataset for validation of deep learning algorithms. This research work focuses on classification of six activities, and they are person raising hand, person picking up an item, person on-phone, and person pointing a gun. Therefore, to enhance the dataset and avoid model overfit, additional human pose dataset to represent these activities is created using local Web camera.

The architecture of a two-branch multistage CNN model shown in Fig. 3 is used for 2D human pose estimation. This model makes prediction approximations and computed the confidence map and affinity fields of points and associates each joint with another. Confidence maps are probability distribution that represent the confidence of the joint location at every pixel in given image and part affinity filed associates body joints of person. The images are resized to 224 × 224 and fed to the double-branch CNN. The input image is analyzed using VGG-19 pretrained model, and the resulting feature image is fed to multistage double-branch CNN. The predictions from multistage network are concatenated with image features which are fed to the next stage.

A set of 18 confidence map for each joint is generated by the upper branch, and the lower branch predicts a set of 38 parts affinity fields and represents the degree of association between joint parts. Part affinity fields help in pruning the graph that represents stick figure. The double-branch CNN model used for pose estimation gives the coordinates of all the body joints as its output. These coordinates are written to a text file. The text file is then converted into a csv file and that file is fed to the DNN.

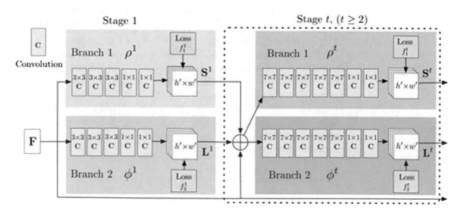

Fig. 3 Architecture of two-branch multistage CNN [1]

4   Results and Discussion

The DCNN model has been trained on MPII pose data and custom generated dataset which comprises of actions corresponding to namaste and point a gun. The dataset is further balanced by duplicating pose images for namaste. This model runs under Windows 10 working framework, and the PC setup is Intel Core i7 9750H CPU, Nvidia GTX 1650 4 GB GPU. The frame rate per second of estimating human pose in 2D is 6 FPS for VGG 16 model using GPU (GTX 1660).

The model is trained with batch size 20 and learning rate 0.001 on TensorFlow. The testing of the model is carried out on downloaded videos and live videos captured from webcam. The human joints coordinates are extracted from these video frames which are stored in a csv file. Sample images are shown in Fig. 4.

The research work focus on classification of six activities in one-person image or video frame. The motivation behind choosing raising hand, person picking up an item, person on-phone, and person pointing a gun is that these represent suspicious and non-suspicious activities. The activity 'namaste' is chosen keeping present COVID-19 pandemic in view where instead of handshake greeting, namaste is recommended. The activity 'pointing-gun' is for suspicious activity, and other activities represent social behavior.

Figure 5a–f shows the results of activity classification and stick figure of human pose for videos downloaded from YouTube. The activity is indicated by a caption on input images.

a

b

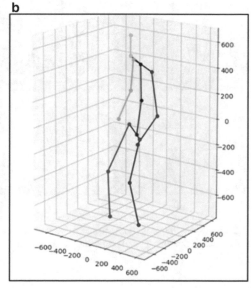

Fig. 4 Human pose and skeleton **a** person **b** skeleton of person in **a**

Fig. 5 Results showing body joints of a person and the actions performed videos collected from YouTube **a** Namaste, **b** pointing gun, **c** on-phone, **d** raising hands, **e** squats, and **f** picking an item

Figure 6a–f shows the results for videos captured on fly using Web camera. The model shows robust performance for both the cases.

The model is trained by increasing number of epochs with adam optimizer and categorical cross entropy loss function. Figure 6 shows the performance of the model for six activities. Figure 6a shows increase in the model accuracy with increase in number of epochs, and Fig. 6b shows decease in the model accuracy with increase in number of epochs. The model achieves accuracy of 99.7% training accuracy and 99.2% of test accuracy. The model shows satisfactory performance (Fig. 7).

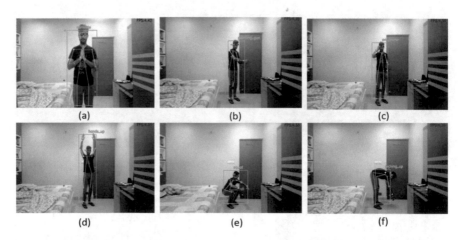

Fig. 6 Results showing body joints of a person and the actions performed from live videos **a** Namaste, **b** pointing gun, **c** on phone, **d** raising hands, **e** squats, and **f** picking an item

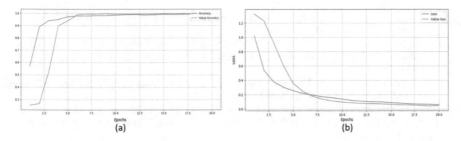

Fig. 7 Model performance **a** graph of accuracy vs epochs and **b** graph of loss vs epochs

5 Conclusion

In this paper, a deep convolutional neural network model is proposed for human activity classification for six activities, they are raising hand, pick-up an item, squats, namaste greeting, on-phone, and pointing a gun. The MPII human pose dataset is used for evaluation of model. The model has given encouraging results for still images, YouTube videos, and for the videos captured using webcam on fly. The model gives training accuracy of 99.7% and 99.2% test accuracy which is satisfactory. The future work may be to classify images with multiple persons in the image.

References

1. Z. Cao, T. Simon, S. Wei and Y. Sheikh, Realtime multi-person 2D pose estimation using part affinity fields, in Computer Vision and Pattern Recognition (CVPR), IEEE Conference, pp. 1302–1310, 2017
2. H.D. Mehr, H. Polat, Human activity recognition in smart home with deep learning approach, in 7th International Istanbul Smart Grids and Cities Congress and Fair (ICSG). Istanbul, Turkey **2019**, 149–153 (2019)
3. H. Du, Y. He, T. Jin, Transfer learning for human activities classification using micro-doppler spectrograms, in IEEE International Conference on Computational Electromagnetics (ICCEM), Chengdu, pp. 1–3, 2018
4. X. Shi, Y. Li, F. Zhou, L. Liu, Human activity recognition based on deep learning method, in International Conference on Radar (RADAR), Brisbane, pp. 1–5, 2018
5. C. Ning, H. Zhou, Y. Song, J. Tang, Inception single shot multiBox detector for object detection, in International Conference on Multimedia & Expo Workshops (ICMEW), pp. 549–554, 2017
6. F. Caba Heilbron, V. Escorcia, B. Ghanem, J. Carlos Niebles, Activitynet: A large-scale video benchmark for human activity understanding, in Proceedings of the IEEE conference on computer vision and pattern recognition, pp. 961–970
7. J. Sung, C. Ponce, B. Selman, A. Saxena, Human activity detection from RGBD images, plan, activity, and intent recognition, 64 (2011)
8. A. Karpathy, G. Toderici, S. Shetty, T. Leung, R. Sukthankar, L. Fei-Fei, Large-scale video classification with convolutional neural networks, in Proceedings of the IEEE Conference on Computer Vision and Pattern Recognition, pp. 1725–1732 (2014)
9. I. Laptev, M. Marszalek, C. Schmid, B. Rozenfeld, Learning realistic human actions from movies, in 2008 IEEE Conference on Computer Vision and Pattern Recognition (pp. 1–8). IEEE, June 2008

10. W. Liu, D. Anguelov, D. Erhan, C. Szegedy, S. Reed, SSD: Single shot multibox detector, in Computer Vision and Pattern recognition, pp.1–7, 2015
11. P.F. Felzenszwalb, R.B. Girshick, D. McAllester, D. Ramanan, Object detection with discriminatively trained part-based models. IEEE Trans. Pattern Anal. Mach. Intell. **32**(9), 1627–1645 (2010)
12. Y. Zhang, J. Liu, K. Huang, Dilated Hourglass Networks for Human Pose Estimation, in Chinese Automation Congress (CAC), pp. 2597–2602, 2018
13. G. Gkioxari, B. Hariharan, R. Girshick, J. Malik, Using k-Poselets for Detecting People and Localizing Their Keypoints, in IEEE Conference on Computer Vision and Pattern Recognition, pp. 3582–3589, 2014
14. K. He, X. Zhang, S. Ren, J. Sun, Deep residual learning for image recognition, in IEEE Conference on Computer Vision and Pattern Recognition (CVPR), pp. 770–778, 2016
15. E. Insafutdinov, L. Pishchulin, B. Andres, M. Andriluka, DeeperCut: A deeper, stronger, and faster multi-person pose estimation model, IEEE transactions of Pattern Analysis and Machine Learning, Vol. No.9910 pp. 34–50, 2016
16. S. Johnson, M. Everingham, Clustered pose and nonlinear appearance models for human pose estimation, in British Machine Vision Conference, pp.1–11, 2010
17. W. Ouyang, X. Chu, X. Wang, Multi-source Deep Learning for Human Pose Estimation, in IEEE Conference on Computer Vision and Pattern Recognition, Columbus, pp. 2337–2344, 2014
18. Z. Cao, G. Hidalgo Martinez, T. Simon, S. Wei, Y. A. Sheikh, OpenPose: Realtime multi-person 2D pose estimation using part affinity fields. In IEEE transactions on pattern analysis and machine intelligence, pp.1–14, 2019
19. V. Badrinarayanan, A. Kendall, R. Cipolla, SegNet: A deep convolutional encoder-decoder architecture for image segmentation. IEEE Trans. Pattern Anal. Mach. Intell. **39**(12), 2481–2495 (2017)
20. Z. Zhigang, D. Guangxue, L. Huan, Z. Guangbing, W. Nan, Y. Wenjie, Human behavior recognition method based on double-branch deep convolution neural network, in Chinese Control And Decision Conference (CCDC), Shenyang, pp. 5520–5524, 2018

Intelligent IoT-Based Product Dispenser and Billing System

Roshan Fernandes, Anisha P. Rodrigues, Anusha Rani, Rachana Pandit, Vijaya Padmanabha, and B. A. Mohan

Abstract The rise of wireless technology and other communication techniques has made electronic commerce extremely famous. Buying at broad malls is becoming a regular occurrence in metro towns. For these areas, there is a massive rush during weekends and holidays. People buy different things and place them in a trolley. After purchases are completed, one has to go to the payments billing counter. The cashier checks the bill at the billing counter using a barcode reader which is a very time-consuming procedure which results in a long queue at the billing counter. A smart product is the one that helps make daily life easy, convenient, and secure. In this paper, we present an innovative shelf-checkout concept that helps smart shopping and billing. The main concept here is to support a person shopping on a regular basis in terms of the reduced amount of time spent when buying a product. The proposed work also implements an analysis of the products to gain a best-buy option. To achieve this feat, we have incorporated a recommendation system along with the general product display. The main objective of this work is to provide a technology-oriented, cost-effective, easily scalable, and robust system for personal shopping.

Keywords RFID · Smart shopping · E-commerce · Cosine similarity · Smart recommendation system · Arduino microcontroller

R. Fernandes (✉) · A. P. Rodrigues · A. Rani · R. Pandit
Department of Computer Science and Engineering, NMAM Institute of Technology, Nitte, Udupi, India
e-mail: roshan_nmamit@nitte.edu.in

A. P. Rodrigues
e-mail: anishapr@nitte.edu.in

V. Padmanabha
Department of Mathematics and Computer Science, Modern College of Business and Science, Baushar, Muscat, Sultanate of Oman

B. A. Mohan
Department of Computer Science and Engineering, NMIT, Bengaluru, India
e-mail: mohan.ba@nmit.ac.in

© The Author(s), under exclusive license to Springer Nature Singapore Pte Ltd. 2022
P. Shetty D. and S. Shetty (eds.), *Recent Advances in Artificial Intelligence and Data Engineering*, Advances in Intelligent Systems and Computing 1386,
https://doi.org/10.1007/978-981-16-3342-3_5

1   Introduction

Due to the advent of wireless technology and other networking innovations, e-commerce has become more and more trendy. Shopping has become a regular occurrence at large mall in major cities. People buy something special and put it in their cart. A billing counter for payment should be installed upon completion of the order. The cashier at the billing counter prepares a bill employing the barcode reader which is a very time-consuming operation and leads to long billing counter queue. The key concept in the proposed work is to help a customer, when they have less time to purchase a product in day-to-day shopping. The customer can sprint away from the shop once he or she has done shopping without waiting in the queue for billing. The idea behind the device is based on RFID technology, where RFID [1] tags are inserted into items and user cards and RFID readers are accustomed to reading this tag for the store's effective, efficient, and theft-controlled service. In using the new method, most of the flaws of the barcode technique are solved. In the proposed system, GSM technology is used to alert the customer for the best goods. Building a small prototype of a smart shopping network that can make the process of buying goods and handling payments makes easier for the customer. The proposed work also keeps track of information about the customer. Smart cards and "shelf checkout" can capture customer buying preferences, provide useful data to help the service providers understand consumer dynamics, maximize inventory rates, and minimize congestion in grocery stores. This is accomplished by using an analytics component in the proposed work.

The proposed work is based on RFID technology, where RFID tags are inserted in the access cards and RFID readers are used to read these tags for safe, secure, and theft-controlled shop service. In the proposed method, GSM technology is used to warn consumers with the products taken. Because of advanced warning systems, it is automatically added to the cart whenever a user picks up an object. If the package is not correctly checked out or the payment fails, the exit is blocked for the object till in possession.

Content-based solution involves a fair deal of knowledge on the correct aspects of the objects, instead of using experiences and input from the consumers. For instance, it may be object attributes such as category, year, manufacturer, price, or other important aspects using natural language processing. The natural language processing involves analyzing the feedbacks given by the customers. Collaborative filtering, on the contrary, requires little more than the historical preference of users over a collection of objects. Since it is based on historical evidence, the main assumption here is that in the future the people who have agreed to purchase the items in the past will continue to do the same in the future too. The proposed work also provides an intelligent recommendation system for the customers, based on the purchase history of the same customer or other customers having similar interests. This helps the customer to make a better choice in terms of the item to be purchased, choice of the brand, price savings, and many more. For this to accomplish, the proposed method uses cosine similarity using matrix factorization. Equation (1) gives the cosine similarity.

$$\text{Sim}(u_i, u_k) = \frac{r_i \cdot r_k}{|r_i||r_k|} = \frac{\sum\limits_{j=1}^{m} r_{ij} r_{kj}}{\sqrt{\sum\limits_{j=1}^{m} r_{ij}^2 \sum\limits_{j=1}^{m} r_{kj}^2}} \tag{1}$$

The reason for using this algorithm despite its shortcomings is because of the limited size of the dataset. Cosine similarity works well in limited size dataset. Also, the assumption is that both the vectors having non-null values. With the growth in size of the dataset, the processing power progressively worsens. Also, the algorithm does not provide accurate results if no nearby projected items are found. Cosine similarity method works with matrix projections of given data objects. The similarity measures ranges between -1 and $+1$, where -1 is completely dissimilar and $+1$ is completely similar. The closer the items are in the projection, the similar they are. The next section discusses the related work on intelligent product dispenser system.

2 Related Work

Eriyeti Murena et al. [2] proposed a control system design for a vending machine. The proposed system handled coin or mobile payment methods. The control system is used for dispensing snacks, newspapers, beverages, and smoking cigarettes. It uses a digital touchscreen–user interface. It also provides refrigeration system by continuously controlling the speed of compressor and fails in recovering after the failure. No intelligent suggestions provided. Wejdan Alsaeed et al. [3] proposed a mobile application to manage the various services in an organization, namely restaurants. It uses the geographical information system to handle the services in a better way. It computes the arrival time more accurately. The main drawback of this system is that it cannot be generalized for any organization or shopping malls. Hariharan Ramalingam et al. [4] have discussed the Internet of things-based application usage in the banking domain. The paper discusses the various challenges involved in developing automated, intelligent shopping, or payment system. The challenges include data privacy, data security risk, and data density. Any automated, intelligent IoT-based system must take care about these three parameters. Kaveh Azadeh et al. [5] presented an overview of the recent trends in automated warehousing by using robotic techniques and intelligent systems. The study shown that the item pick process in a shopping mall with intelligent suggestions have a major effect on the growth of these shopping malls. The proposed system, hence, come up with a study on an intelligent suggestion system. Katerina Berezina et al. [6] have reviewed various applications of robots, artificial intelligence, and service automation in various service organizations, namely in restaurants. The study shows that artificial intelligence-powered devices in the shopping malls increase the personalization of customers and thereby increasing the business. Radoslaw Klimek [7] proposed an intelligent queue management system to serve individual customers. It uses IoT sensor camera network device system which classifies the customers as elders, people in a hurry, and many more. Accordingly, the system handles the queue. The system is tested using a simulator and not in a real-time scenario. Adilakshmi Satya Sri et al. [8] designed an automated product dispenser system for five specific products. It also enables the customer for the payment. The proposed method does not provide the intelligent recommendations. Monika V N et al. [9] proposed an automated medicine box vending machine. It uses ARM microcontroller, relay, GSM, and RFID technology to accomplish the same. No intelligent recommendation is provided. Chandrasekaran N. et al. [10] discuss how blockchain, artificial intelligence, machine learning and deep learning have helped in increasing the business in various fields. It also helps in performing better predictive analysis. Jinlong Wang et al. [11] proposed a new warehouse integration technology to improve automation in tobacco industry. Authors used IoT on cloud platform. This improved business in tobacco industry and reduced the overall cost. This work is restricted to only the cigarettes and tobacco products. Abhaya Kumar Sahoo et al. [12] proposed a recommendation system based on individual's health profile. This increases the sale of items there by increasing the business. The drawback is to collect the health profile of users as most of hospitals do not share the patient details.

By analyzing the various research articles, we have proposed a system which combines both automatic billing of items and intelligent recommendation for the customers. The proposed system is discussed in the next section.

3   Proposed System

In attempt to connect the consumers with the right options and exclusive discounts based on their previous shopping orders, the proposed system presented users with the latest and highly requested items using the collaborative filtering algorithm, which returns the most often ordered goods depending on the users' purchase history. The proposed system uses various technology to provide the end users with a quick and secure solution.

3.1   System Control Flow

To start with, the system has built a very simple and important interface for the consumers for quick access and to enjoy all the items at the comfort of their homes. To encourage administrators to view both data and related material, a dedicated platform to handle the users' records and documentation has been deployed. Usage of artificial intelligence has been introduced in order to include real-time suggestion framework. Components such as microcontrollers, RFID, LCD, keypad, motors, and several more are at the hardware end to create the shelf. The system also provides a

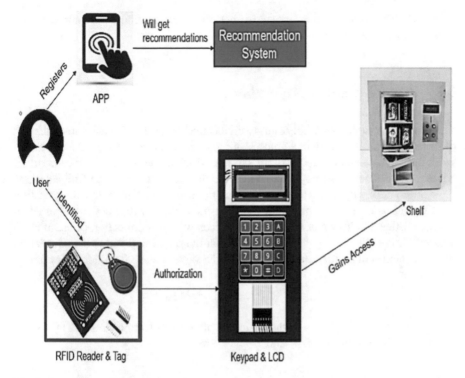

Fig. 1 Proposed system

personalized IoT-based system in order to allow users access from anywhere. Figure 1 shows the overall flow of the proposed system.

The accompanying procedure explains how the proposed system process operates using various methods. It is a machine bootstrapped with multiple procedures. The various steps are explained below:

Step 1: The required client registers with the Play Store application.

Step 2: The user will be provided with an RFID after the registration process, which allows the system to authenticate the users at any shopping outlet.

Step 3: The consumer can visit the nearest outlet when he / she wants to go and buy goods.

Step 4: Consumers use LCD and keypad to insert their secondary authorization key in order to obtain access to the shelf at the shelf panel in a shopping mall.

Step 5: The customer picks his/her items of their choosing and then clicks on purchase.

Step 6: The user will be asked for the amount on his chosen UPI Payment App until transaction is completed.

Step 7: After conclusion of payment, all goods should be collected by the customer.

Step 8: All items ordered will now be added to the Users App under the group My Orders.

Step 9: The user can also review his/her other information in the app for even richer device experience.

3.2 Recommendation Algorithm

The proposed system essentially aims to predict the "preference" that a consumer will offer to an object. It is seen in a number of industries. The suggestion method in this work indicates when the customer buys peach, which belongs to the fruit category; it advises that the customer buy watermelon which belongs to the same fruit category. Also, customers will score the goods on recommendation. The matrix factorization finally gives us the details about how much a person aligns with a collection of latent characteristics and how much the interest blends with this collection of latent characteristics. The difference of the closest regular neighborhood is that the correlation between them will be identified if they have the same underlying interests or not.

$$min_{q,p} \sum (r_{ui} - p_u^T q_i)^2 + \lambda (p_u^2 + q_i^2) \tag{2}$$

Given, the characteristics of the system, matrix factorization (in Eq. 2) provides the output with minimum displacement measure.

3.3 System Architecture

A description of the architecture shows how the concept works. Figure 2 shows the high-level architecture. The user is logging in through the application. The customer will only have access to one account. The app has a page to log in. Users are defined via RFID reader and name. The tag transfers information which is stored in the memory of the device. All tags have their own unique identification number which reuses these tags. The tag creates electromagnetic field in the reader. As a tag moves into the area, the reader interprets the information that is contained in the tag and sends it to the storage server, which in turn stores or retrieves label return information. The approved key which contains four digits is given to the user. The client or the owner should be able to enter the shelf and order the items using approved key. The shelf contains helical spring or coil spring that holds the product in the shelf. When removed this spiral spring returns to its original position. The keypad is used here to enter the quantity as well as the type of products that can be selected. There is also a part of the advice in this. It recommends the product in recommendation part. For example, when the customer buys orange it belongs to the category of fruit so it suggests another fruit in the same category. After scanning the customer card-id tag and product tag, it shows customer and product details taken. It can then perform take/retake operations.

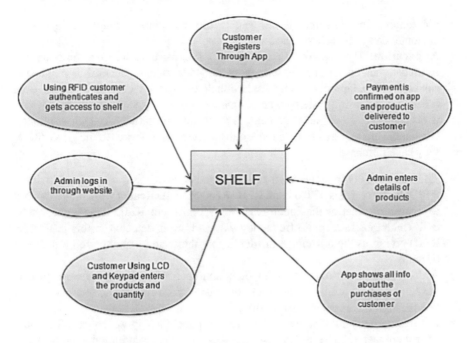

Fig. 2 High-level architecture

The project consists of two modules in it. Figure 3 gives the concept design of the system. They are software and hardware part. In software part, it includes mobile application/Web application, recommendation system, and in hardware, it includes RFID, Arduino micro-controller, LCD, and keypad. The software part includes:

Fig. 3 Concept design

- **Customer:** The customer is allowed to buy the items through the app. The customer can only have one account.
- **Application:** The app contains a login tab. It consists of information about the user/customer, such as name, sex, phone no, address, and gender. It is planned to run on desktop, laptop, or computer handheld devices. It can be downloaded from the mobile operating system owner, namely Google Play Store. On many devices, users can access Web sites such as desktops, laptops, tablets, and smart phones. Customers are required to give the feedback in terms of scores to the goods/items they have purchased.

- The hardware components used are discussed below:
- RFID: RFID consists of TAG and RFID Reader. RFID Reader works with the tags to receive information that identifies them. Within them, RFID Tag has a magnetic coil which generates radio frequency waves. User is detected in this project by RFID giving us the advantage of identifying thefts and unauthorized movement of items.
- Arduino microcontroller: It is an open-source hardware and software project developing and manufacturing single-board microcontrollers and microcontroller kits for digital computer construction.
- LCD: Liquid crystal display (LCD) is a flat panel display which uses the liquid crystal characteristics combined with polarizers. It is programmable with ease.

3.4 Recommendation System

The recommendation system consists of two techniques, namely content-based recommendation and the second one is collaborative filtering recommendation system. Content-based solution involves a fair deal of knowledge on the correct aspects of the objects, instead of using experiences and input from the consumers. For instance, it may be object attributes such as category, year, manufacturer, price, or other important aspects using natural language processing. It gains knowledge from the user feedback written in natural language, namely English. Collaborative filtering, on the contrary, requires little more than the historical preference of users over a collection of objects. It uses historical evidence and more information gathered from the users. Since it is based on historical evidence, the main assumption here is that in the future the people who have agreed in the past will continue to agree. The pseudocode for the recommendation system is given below.

Recommendation System Pseudocode:

Step1:
    indices = pd.Series(metadata.index,index = metdata['prod_name']).drop_duplicates().
    Here, indices represent how the system links to the cosine similarity matrix.

idx = indices[name].
    Here, idx will have indices only of product name.

Step 2:

Get the pairwise similarity scores of all products with that product.
sim_scores = list(enumerate(cosine_sim([idx])).

Here, we have enumerated the cosine similarity scores along with the indices. So, we will link the indices 0,1,2,3 that is our products along with the cosine similarity scores. So that we know that this product has these number of scores. And we store these as a single list.

Step 3:

Now sort the products based on the similarity scores. This step is needed because whatever scores are generated, they need not always be in the same order.
sim_scores = sorted(sim_scores, key = lambda x:x[1], reverse = True.

Here, we have given reverse = True, which means, we want to sort the score in the descending order, that is, to fetch the top highest similarity scores.

Step 4:

Get the scores of the five most similar products.
sim_scores = sim_scores[1:6].

Get the product indices.
prod_indices = [i[0] for i in sim_scores].

Return the top 10 most similar product.
return metadata['prod_name'].iloc[prod_indices].

Now if the system prints the similarity scores, the customer will not understand anything. So, the proposed method links these scores to the product name. Metadata is the original dataset that the proposed system has imported. Iloc will identify the row of index and print the product name.

The experiments were conducted by considering both the approaches, and the results obtained are discussed in the next section.

4 Results and Discussion

The proposed system has been designed to provide easy and fast checkout methods which incorporate cheaper processes and components. Based on the outcomes recorded, the system functions are expected with financial efficiency of up to 20% less than that of conventional dispensing systems. Also, the proposed system involves recommendation algorithms, which have proven to reduce shopping time by 7% on an average and at the same time increase sales of related items. Using one such algorithm "cosine similarity," we see that the recommendations always remain on the same Euclidean level hence proving to be accurate. Figures 4 and 5 show the results of collaborative-based filtering recommendation and content-based filtering recommendation, respectively.

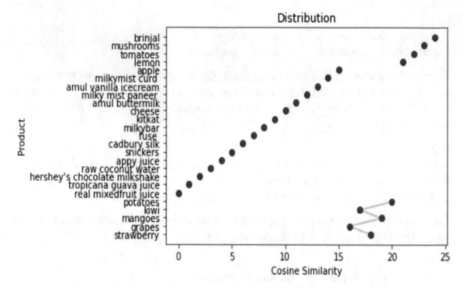

Fig. 4 Collaborative-based filtering recommendation

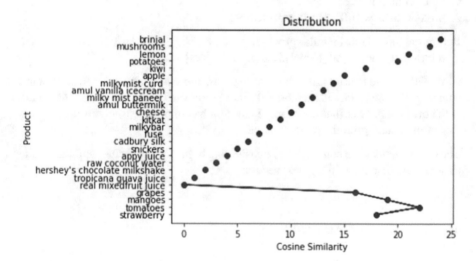

Fig. 5 Content-based filtering recommendation

The red dots indicate the cosine similarity scores. In collaborative filtering (Fig. 4), the points are joined together which are very close to each other. They are on the same level between the range 15 and 20. But if we observe the graph in Fig. 5, that is, content-based filtering, the cosine similarity scores range from 0 to 23. So, the recommendation system can choose any point in the given slot. The results are analyzed by only considering the features from the dataset. Collaborative filtering predicts more accurately because it has a lot of data.

Figure 6 shows the results of probability distribution of metadata vs features. The X-axis represents the number of products, and the Y-axis represents the probability. The red dots represent the probability of "features" being matched. The blue dots represent the probability of "metadata," respectively. The overlapping of red and blue dots represents the matching of both metadata and features. The graph is obtained when calculating probabilities, the product features appearing through the product names one after the other in the results. The basic content filtering dataset is indicated by the red dots that uses only the features column from the dataset and as you see they are more scattered. The scattering thus leads to anomalies in the output. The blue dots indicate the metadata used for collaborative filtering which not only includes features but also other factors such as cost, weight, category, and contents. The blue dots are continuous and have consecutive data points in the state space. The more data you feed, more accurate will be your prediction.

Talking about the accuracy of the model, used for recommendations, the easiest algorithm that comes to mind is content-based recommendation/filtering. As we have seen in the previous distribution, only using features/tags or names do not suffice in fast and accurate recommendation of items. Hence, an updated and more accurate algorithm is used, namely collaborative filtering. Collaborative filtering uses past data of a single data producer and generates a hypothesis that this attribute is true for all instances of the data point. Hence, the next time the same data point produces some data, and it is highly likely to choose from its previous dataset or to choose something near to it. Collaborative accuracy (Fig. 4) depicted by yellow increases gradually with the continuous generation of new data objects. The more the

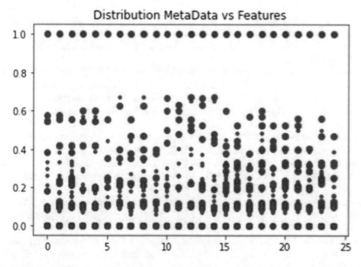

Fig. 6 Probability distribution of metadata vs. features

availability of data, more accurate the prediction. On the other hand, content-based filtering provides a short-time spurt of accuracy, but clearly dies down soon enough.

5 Conclusion and Future Scope

The proposed system enables customers to bypass the queues and confusion by converting their mobile into a point of service (PoS). This reduces the burden of customers to wait in queues and checkouts by turning their mobile into a point of service (PoS). The proposed system saves time and human labor, which if well applied, has the power to transform the lifestyle of the people forever. It is an evolution from the self-checkout version, which is being seen today in several supermarkets. The cost of introducing, upgrading, and improving the program will be balanced by raising checkout clerks to surpass cost-savings. The proposed system can install these devices at a cheaper rate for those that are unable to buy specific things, such as food or clothes or other items purchased on a regular basis. This helps the people who can buy it and do not wish to use it. They can deposit into this system and obtain other incentives that they can use instead. Such items/goods can be offered to disadvantaged or needy citizens to support them. More intelligent and sophisticated customer feedback analysis program, to increase product service that would help provide customers with a better user experience, is an open research problem.

References

1. M.I. Younis, SLMS: a smart library management system based on an RFID technology. Int. J. Reason.-Based Intell. Syst. **4**(4), 186–191 (2012)
2. E. Murena, et al., Design of a control system for a vending machine, in Procedia CIRP 91, 758–763 (2020)
3. W. Alsaeed, K. Alhazmi, An intelligent spatial-based queue management system, 2019 IEEE Asia-Pacific Conference on Computer Science and Data Engineering (CSDE). IEEE, 2019
4. H. Ramalingam, V. Prasanna Venkatesan, Conceptual analysis of Internet of Things use cases in Banking domain, TENCON 2019–2019 IEEE Region 10 Conference (TENCON). IEEE, 2019
5. K. Azadeh, R. De Koster, D. Roy, Robotized and automated warehouse systems: Review and recent developments. Transportation Sci. 53(4), 917–945 (2019)
6. K. Berezina, O. Ciftci, C. Cobanoglu, Robots, artificial intelligence, and service automation in restaurants, in *Robots, Artificial Intelligence, and Service Automation in Travel, Tourism and Hospitality* (Emerald Publishing Limited, 2019)
7. R. Klimek, Context-aware and pro-active queue management systems in intelligent environments, in 2017 Federated Conference on Computer Science and Information Systems (FedCSIS). IEEE, 2017
8. U. Sri, A. Lakshmi Satya, et al., Design of Automatic 5× 5 Dispenser with three kinds of payment modes using Verilog HDL, in 2020 IEEE International Students' Conference on Electrical, Electronics and Computer Science (SCEECS). IEEE, 2020
9. K.M., Hema, M.M. Savitha MM, An IOT based intelligent medicine box using vending machine-medical ATM, Institute of Scholars (InSc) (2019)

10. N. Chandrasekaran, N., et al., Digital transformation from leveraging blockchain technology, artificial intelligence, machine learning and deep learning. İn *Information Systems Design and Intelligent Applications* (Springer, Singapore, 2019), pp. 271–283
11. J. Wang, Z. Li, R. Higgs, The application of dense storage integration technology in Tobacco logistics centers based on the internet of things, 2017 IEEE International Conference on Computational Science and Engineering (CSE) and IEEE International Conference on Embedded and Ubiquitous Computing (EUC). Vol. 1. IEEE, 2017
12. A.K. Sahoo, et al., Intelligence-based health recommendation system using big data analytics, in *Big Data Analytics for İntelligent Healthcare Management* (Academic Press, 2019), pp. 227–246

Models Predicting PM 2.5 Concentrations—A Review

Anusha Anchan, B. Shabari Shedthi, and G. R. Manasa

Abstract With accelerated seriousness of air contamination, particulate matter with diameter 2.5 μm (PM2.5) is considered as the most significant air pollution. It has adversely impacted on individual's health, life and work and has made damage to individuals' well-being. Quick urban development is causing PM 2.5 contamination. Logical and viable forecast of PM2.5 can empower individuals to play it safe ahead of time to stay away from or diminish damage to the creatures on earth. In this manner, the prediction of PM2.5 has become a subject of extraordinary pragmatic essentialness. This examination assessed the forecast of fine particulate matters and summed up the aftereffects of foreseeing models. This review lights up two categories of models, i.e., machine learning and statistical/mathematical models. In addition, forecasting can essentially be enhanced by joining meteorological factors with ground measured parameters of every locale, alongside aerosol optical depth (AOD) rather than just considering meteorological factors. This work suggests the use of high-level AOD information, and future further works could be centered around creating satellite-related foreseeing models for forecasting PM 2.5 and other air contamination.

Keywords Autoregression · Linear support vector machine · Wavelet transform · XGBoost

1 Introduction

In this era of raising technology and globalization, air contamination has become a genuine danger for personage well-being and communal or public government assistance in ongoing many years as well as it creates negative impact on individual's health life [1]. With the improvement of financial aspects on the planet, the air contamination issue has become increasingly more genuine lately, particularly the particles of diameter under 2.5 μm (PM2.5) [2]. The estimation unit of PM 2.5 is

A. Anchan (✉) · B. S. Shedthi · G. R. Manasa
NMAM Institute of Technology (Visvesvaraya Technological University), Nitte, Karnataka, India
e-mail: anusha@nitte.edu.in

© The Author(s), under exclusive license to Springer Nature Singapore Pte Ltd. 2022   65
P. Shetty D. and S. Shetty (eds.), *Recent Advances in Artificial Intelligence and Data Engineering*, Advances in Intelligent Systems and Computing 1386,
https://doi.org/10.1007/978-981-16-3342-3_6

by micrograms per cubic meter (μg/m 3). Increased concentrations of PM 2.5 have become a worldwide issue. Different investigations have shown that fine airborne particles have the most grounded well-being impacts. Also, airborne particles are vital in influencing barometrically radiation, cloud development, climatic photo-chemical responses and light elimination impacts that impact worldwide climate changes [3, 4, 41]. The impacts of PM 2.5 on general well-being have been all around reported in the writing.

The current patterns for particulate observing procedures will, in general, screen PM 2.5 in view of the immediate relationship with well-being impacts and the avoidance of regular particulate obstruction. Industrial exercises with high essential particulate outflows, for example, coal, concrete, cement or mining, significantly affect air quality because of their concentrated particulate discharges in 2.5–10 μm extend. Utilization of firewood and charcoal for warming and cooking is unavoidable issue, which is likewise adding to PM 2.5 outflows.

This gives the inspiration to exploring the strategies for assessment of particulate issue in territories having less screens at a unique scale, a methodology presently drawing in extensive scholarly intrigue. The focus of this review is as follows:

- Classify and stipulate frequently used methodologies which assesses the PM 2.5 concentration.
- Study the impact of input features considered with various algorithms.
- Find what are the possible performance metrics on which various models' performances are compared.
- Overview of various prediction modeling method recommendation.

2   Objectives of Various Research Articles

This review covers the studies which were carried out with following objectives: prediction of PM2.5 concentration (hourly/day a head), feature importance for PM 2.5 prediction, assessment probability of PM2.5 surpassing a predefined secure limit and sway of meteorological features up on PM 2.5. Most of study had several algorithms selected for comparison as to decide which among them performed best in predicting the specified contamination. This study has carried out by considering papers published 2010 onwards. In this paper, all the study regarding concerned topic has considered data acquired between 2010 and 2020. Table 1 targets featuring the basic standards of machine leaning methods and about their part in improving the expectation execution. Table 2 shows the consolidated study of statistical models/time series model/chemical models.

Table 1 ML models used for PM2.5 prediction

	Objective w.r.t to PM2.5	Region	Data collection duration	Features/parameters	Algorithms	Performance metrics	Best Model as result of comparison
[5]	Prediction (hourly)	Tianjin city, China	Dec 2016	Hourly PM10, SO_2, NO_2, CO, O_3	XGBoost RF MLR DTR SVM	RMSE, MAE R2	XGBoost RMSE = 17.298 MAE = 11.774 R2 = 0.9520
[6]	Daily estimation	USA	2000–2015	Land-use variables Satellite data. Meteorological variables	NN RF Gradient boosting Ensemble model NN + RF_XGBoost	R2, RMSE	Ensemble model NN + RF_XGBoost R 2 = 0.89 RMSE = 1.26
[8]	Feature importance for PM 2.5 prediction	Tehran, the capital city of Iran	2015–2018	Satellite data, Meteorological data, Ground-measured PM 2.5, Geographical data (23 features)	RF XGBoost Deep learning	R2, MAE RMSE	Ranking of features XGBoost gives the leading performance with 0.81(R2), 0.894(R), 10.0(MAE), 3.62(RMSE)
[9]	Assessment probability of surpassing a pre-defined secure limit	Bangkok, Thailand, Kasetsart University	2015–2019	Meteorological data Air pollutants	Naïve Bays Logistic regression, RF NN	F1 score, confusion matrix	Random forest with 0.82 F1-score

(continued)

Table 1 (continued)

	Objective w.r.t to PM2.5	Region	Data collection duration	Features/parameters	Algorithms	Performance metrics	Best Model as result of comparison
[10]	Predicting and Classification	Quito, Ecuador, Cotocollao Belisario	2007 - 2013	Aerosol data, fine particle concentrations, meteorological data	Boosted Trees (Bts), L-SVM, NN, CGM	TPR/FNR MSE, ROC curves	AUC-BT(0.72) is larger than AUC-L-SVM(0.66) MSE of 22.1% for NN and 15.6% for CGM
[12]	(i)Level detection (ii)Prediction	UCI repository	-	Temperature, dew point, wind speed, PM2.5, pressure,	Logistic regression, autoregression (AR)	Mean accuracy Standard deviation, Mean Squared Error(MSE)	Mean Accuracy 0.998859 (LR) Standard deviation Accuracy 0.000612(LR) MSE 27.00(AR)
[19]	Short period prediction	Beijing	2015	AOD, Meteorological factors Gaseous pollutants	MLR	R2, RMSE	Regression model with yearly data R2(0.766) goodness-of-fit and R2 (0.875)
[21]	Prediction	Delhi	2016–2018	Meteorological data, pollutant parameters	SVM	Prediction graph	Close prediction to real value

(continued)

Table 1 (continued)

	Objective w.r.t to PM2.5	Region	Data collection duration	Features/parameters	Algorithms	Performance metrics	Best Model as result of comparison
[22]	Forecast	Pingtung station, Taiwan	2012–2017	Meteorological data—speed, wind, temperature air contamination data—CO_2, PM2.5, PM10, CO, NO, SO_2, HUI	LR, RF, XGBoost KNR, MLP, DT-CART	RMSE R2	XGBoost outperforms other models R2 = 0.9336 RMSE = 0.1302
[29]	Forecasted	Romania Stations Munchen,(Germany) Ploieste(Romania)	M	PM2.5	ANN, ANFIS	MAE, RMSE IA, R	ANN MAE = 1.9278, RMSE = 3.1931, IA = 0.9804, R = 0.9619
[30]	Estimation	55 countries	1997–2014	Remote sensing observations, meteorological observations, CO NO_2	Multivariate, non linear, nonparametric machine learning	R	R for each of the five training data sets is 0.96
[31]	Prediction	25 stations, Seoul, South Korea	2015–2018	Hourly pm2.5, pm10, meteorological features	LSTM, DAE	RMSE	LSTM with RMSE = 12.174
[33]	Forecast	Jiayuguan, Datong, Fushun, Qiqihar, Weinan, and Xuchang, China	2014–2019	PM2.5	WT, SAE, LSTM SAE based on LSTM	MAE	MAE based on SAE-LSTM is least compared to other models

(continued)

Table 1 (continued)

	Objective w.r.t to PM2.5	Region	Data collection duration	Features/parameters	Algorithms	Performance metrics	Best Model as result of comparison
[34]	Prediction	CPCB online station	–	CO, NO, NO$_2$, SO$_2$, O$_3$, VOC, PM2.5, meteorological parameters	SVM, ANN	RMSE, MAE,R	ANN RMSE = 1.51(compared to SVM)

Methods XGBoost, random forest(RF), multiple linear regression(MLR), decision tree regression(DTR), support vector machine(SVM), neural network (NN), linear support vector machine(L-SVM), convolutional generalization model (CGM), true positive rate(TPR), false negative rate(FNR), Area under curve (AUC), mean-squared error (MSE), autoregression (AR), linear regression (LR), K-neighbor regressor (KNR), multilayer perception (MLP), decision tree (DT), index of agreement (IA), artificial neural networks (ANNs), adaptive neuro-fuzzy inference DSystem (ANFIS), long short-term model (LSTM), deep auto-encoders (DAEs), wavelet transform(WT), stacked auto-encoder(SAE).

2.1 Machine Learning Models

Current examinations in the field of climate science and designing show that deterministic models scuffle to catch the connection between the convergence of environmental contamination and their emanation sources. The ongoing advances in factual displaying dependent on AI/ML approaches have developed an answer to tackle these issues. Input variable sort to a great extent influences the execution of an algorithm, in any case, it is yet to be known why an algorithm calculation is favored over the other for a specific undertaking.

2.2 Statistical Models/ Time Series Model/Chemical Models

See Table 2.

3 Region and Data

Almost 90 percent of study reviewed here used air pollution contamination data and meteorological features as their prime feature for the study they were intended to. Along with that, some studies have got benefit of having access to AOD data [8], geographical data[24] and satellite data. In paper [5], chemical transport model (CTM) expectations of PM 2.5 vertical profiles would help adjust AOD. Along with ground information, this examination additionally gained AOD item information from Aqua-Moderate Resolution Imaging Spectrometer (MODIS) 550 nm collection in paper [19]. In article [39], IRMS was used to even detect the fire emission.

4 Data Processing Method

In paper [8], standard normalization was applied on data before training. The resampling method such as synthetic minority over-examining technique (SMOTE) is utilized in [9], in order deal with an imbalanced data set. Linear interpolation technique was used in study [21], and data were normalized using min–max normalization. Authors in paper [22] used Fourier arrangement and spline multinomial approaches to deal with missing values. In papers [23, 25], spatial and temporal distribution is used. To choose the vital boundaries out of an enormous number of boundaries, the Akaike information criterion (AIC) was assessed in the paper [26]. Summer-time air quality overview, spatial variability and temporal variability were carried out in the work [34].

Table 2 Statistical/time series models/chemical models

	Objective	Region	Duration	Features	Algorithms/Methods	Performance metric	Outcome
[7]	Prediction	Beijing	2018	Meteorological data, Air pollutants	Time series models AR, AR-Kalman filter, IMM	MAE, MAPE, RMSE	IMM with best performance
[11]	Trend prediction, value prediction, hidden factors prediction	Shanghai Beijing Guangzhou,Chengdu	2012–2014	Department of US state air quality files	ARMA Model, SV Model SW Model	MSE	Better prediction related to PM 2.5 concentration(six hours)
[13]	Determination	Tehran, the capital of Iran	2006–2016	Spatial data, temporal data	SVR, GWR, ANN, NARX—external input	R2, RMSE	NARX gives best performance
[14]	Analyzing the impact of features	Fuling District, Chongqing	2016–2017	Air pollutants, AOD data, meteorological data	Mathematical model of BP neural network	RMSE	Effective correlation between AOD and PM2.5 is established by BP NN
[15]	Prediction	China	–	Hourly CO, NO_2, SO_2, PM10	ANN BP NN, genetic algorithm BP NN optimized by genetic algorithm	Average relative error and maximum relative error	NN optimized by genetic algorithm gives best performance
[16]	Forecast	Ningbo, Qingdao, Guiyang, Yinchuan	–	–	Novel multi-resolution ensemble model	MAE, MAPE, IA	Proposed ensemble model gives better performance

(continued)

Table 2 (continued)

	Objective	Region	Duration	Features	Algorithms/Methods	Performance metric	Outcome
[17]	Estimation (daily)	USA consisting of 48 adjoining states and Washington DC	2011	AOD data, meteorological data, land-use data, ground-truth PM 2:5 measurement data	Convolutional neural network (CNN)		CNN shows competitive accuracy with contrast to the lately built models Feature rank
[18]	Forecast		2010–2014	PM 2.5, meterological data and snowfall (Snow)	AE, Bi-LSTM	RMSE	Combination of AE AND Bi-LSTM reduced training time, improved prediction accuracy RMSE-3.091
[20]	Prediction	Wuhan, Hubei Province	-	SO2, NO2, O3, PM10, CO, meterological data	Modified Gaussian smoke plume model	RMSE	Successful simulation
[23]	Impact of meteorological conditions	Nagasaki Japan	2013	Meteorological data: temperature, humidity, wind speed, precipitation, PM 2.5	Linear analysis, Spearman corrlation analysis	R2	Precipitation negative correlation, temp positive correlation, wind speed and humidity threshold

(continued)

Table 2 (continued)

	Objective	Region	Duration	Features	Algorithms/Methods	Performance metric	Outcome
[24]	Contribution	Central Chile, Santiago	2012	PM10, PM 2.5, SO_2, NH_3, NO_2, CO, VOC(volatile organic compounds)	MATCH CTM	Average concentrations of pollutants in Unit: $\mu g\ m^{-3}$	CTM simulates higher concentration inert components, compared to the winter period, elemental carbon, dust
[25]	Prediction [Chem v/s machine learning]	Shanghai, China	2015–2018	Meteorological measurement, air pollutants	GBDT/XGBoost, WRF-Chem model, Lasso regression model	MB, ME RMSE R3	XGBoost out performed with MB = 3.6 ME = 0.3 RMSE = 26.1 R = 0.6
[26]	Prediction	Chengdu Sichuan Basin	2013–2017	Meteorological data, Radiosonde meteorological data	GAM	AIC R2 F-value P-Value	Strongest predictors are **T,** wind speed, precipitation, trajectory direction, distance GAM achieved an R 2 of 0.73
[27]	Determine leading nodes	Taiwan	2015 -2017	PM 2.5, Synoptic weather patterns	BD-oriented SNA, PCA	Z-value , daily average concentration of PM2.5	BD-oriented SNA is reasonable means to determine PM2.5 leading nodes

(continued)

Table 2 (continued)

	Objective	Region	Duration	Features	Algorithms/Methods	Performance metric	Outcome
[28]	Hourly concentration study	Gansu and Lanzhou China	2015	PM 2.5, PM 10, NO_2, SO_2, O_3 CO, meteorological factors	Pearson's correlation coefficient (relationship between factors) multivariate analysis	Average concentration of pollutants, correlation coefficient	Wind speeds—negatively correlated, temperature—negatively correlated, humidity—positively correlated, vaporous, pollutants showed no big effect
[32]	Forecast	Shenzhen, China	2013	Meteorological factors, SO_2, NO_2, CO, O_3	ARIMA, SVM, Ensemble ARIMA + SVM, The hybrid-Garch forecasting model	MAE, RMSE, DA, IA	Precision level of half breed Garch model was steadily higher than that of ARIMA or SVM
[35]	Day-ahead forecast	Wuhan and Tianjin	2014–2016	PM 2.5	WT, DE, BP, DE + BP, Hybrid WT-VMD-DE-BP,VMD,	MAE MAPE, TIC, RMSE,	Hybrid WT-VMD-DE-BP decomposition technique is effective with MAE = 4.05 RMSE = 6.25 MAPE = 8.88 PIC = 0.05

(continued)

76   A. Anchan et al.

Table 2 (continued)

	Objective	Region	Duration	Features	Algorithms/Methods	Performance metric	Outcome
[36]	Forecast	Tehran	2013–2016	T, PBL, P, RH, PM 2.5	MART, DFNN, hybrid model based on LSTM	RMSE MAE	LSTM with lowest RMSE = 8.91 MAE = 6.21 (µg m −3)
[37]	Forecast [ML VS Geographical Model]	Tianjin Hebei, Beijing	–	Meteorological parameters, air pollutants	Air mass trajectory-based geographic model, wavelet transformation ANN, hybrid model trajectory-based geographic parameter—input to ANN	MAE, RMSE, IA	Accuracy of hybrid model was high
[38]	Forecast	Beijing, China	2016		ESN, PSO, PSR, Hybrid IPSO-PSR-ESN	RMSE, MAE, SMAPE	Hybrid IPSO-PSR-ESN has better performance
[39]	Estimation	China	1999–2002	Meteorological, fire AOD	GWR	RMSE APE	RMSE = 17.2 APE = 18.5

Methods Autoregressive moving average (ARMA) model, stochastic volatility (SV), model and Stock–Watson (SW) model, artificial neural network (ANN), auto-encoders (AE), bidirectional long short-term memory (Bi-LSTM) chemical transport model(CTM), multi-scale atmospheric transport and chemistry model (MATCH), volatile organic compounds(VOC), gradient boosting decision tree (GBDT), weather research and forecasting (WRF) model coupled with chemistry, generalized additive models (GAMs), social network analysis (SNA), big data (BD), wavelet transform (WT), variational mode decomposition (VMD), differential evolution-back propagation(DE-BP), planetary boundary layer height (PBL), multiple additive regression trees (MART), deep feedforward neural network (DFNN), particle swarm optimization (PSO), echo state network (ESN).

Metrics Mean bias (MB), mean error (ME), mean absolute error (MAE) µg/m 3, root mean square error (RMSE) µg/m 3, correlation coefficient, Akaike information criterion (AIC), Direction accuracy (DA), mean absolute percentage error (MAPE),Theil's inequality coefficient (TIC), symmetric mean absolute percentage error (SMAPE), geographically weighted regression (GWR), absolute percentage prediction error(APE), temperature (T), surface-level pressure (P).

5 Methodology/Algorithms

5.1 Neural Networks

ANNs are the apparatuses used for applying AI which numerically model the working of a human cerebrum, and it has been utilized to construct answers to various issues from different areas. These organizations are dependent on connected designs, preparing components which can be called as nodes. These nodes are regularly stacked in a completely associated arrangement of at least three layers specifically, input layer, hidden layer(s) and the output layer. In papers [5, 6, 10, 21, 22, 34, 37, 39], authors have considered ANN as one of the models. In paper [8], author has used variant of ANN, i.e., deep learning. The **ANFIS** uses fuzzy inference system (FIS)which is a hybrid architecture enriched with ANN features proposed by author in paper [29]. Deep or stacked auto-encoder (DAE) is one among the incredible neural organization designs, used in papers [31, 33]. In paper [17], CNN works by stacking a convolution three layer and a common completely associated layer. The convolution layer comprises nonlinear activation function and different weight-shared kernels. Two layers fundamentally perform nonlinear changes of the input so the portrayal of information indicators is more sensible to get PM2.5 estimated. Authors of paper [14, 15, 35] have used BP NN including at least one shrouded layers including one or more hidden layers which was used to connect AOD and PM2.5, and it reduces overfitting.

5.2 XGBoost and Random Forest

Extreme gradient boosting (XGBoost) is a triumphant ML library emanated from gradient boosting algorithm. It uses more regularized model formalization to have good hold on overfitting. Presented works in papers [5, 6, 8, 22, 25] utilize this model, and out of these five studies, four studies proved that this model is better compared to considered benchmark models. Random forest is emanated from decision trees which fall under the category of supervised ensemble learning methods. Works carried in papers [5, 6, 8, 9, 22] have utilized this model.

5.3 Support Vector Machines

SVM converts the input space into a multi-dimensional in character space; hence, it is used as a discriminative classifier technique. This model can be used for application of classification and regression. In papers [5, 10, 21, 32, 34], SVM was utilized for regressing required pollutant.

5.4   Decision Tree Regression and MLP Regression

Decision tree (DT) algorithm comes under umbrella of ML supervised learning
models. In paper [23], DT was used to choose the property for the root node in each
level. Multilayer Perception (MLP) belongs to a family of ANNs. The nodes in this
model are named as neurons. MLP was utilized in paper presented [24].

5.5   LSTM

Recurrent neural networks (RNN) are emanated from DNN which had cyclic connection between nodes. This helps in persisting the information. Generally, inclination
of gradient in the RNN results in gradient explode or shrink exponentially. One solution is to utilize the long short-term memory (LSTM). In paper [18], the presented
forecast model comprises two layers: the auto-encoder layer and the Bi-LSTM layer.
Mix of auto-encoder and Bi-LSTM models reduces the planning time and improves
conjecture accuracy. In articles [31, 33], RNN is along with LSTM to establish a
climate for the calculation cycle, weighted lattice, get include and make yield.

5.6   Time Series Model

It is a statistical model. This arrangement incorporates autoregressive (AR) model,
moving normal (MA) model and autoregressive and moving normal (ARMA) model.
In paper [12], autoregression is applied on time series information so as to anticipate
the PM 2.5 which estimates seven days preceding the current date.

Autoregressive–Kalman Filter Method: In article [7], AR model was set up;
afterward, AR model is changed into state condition for Kalman filtering. This
work presents three models: stochastic volatility (SV) model, autoregressive moving
average (ARMA) model and Stock–Watson (SW) Model. Brand new model is
obtained joining SW model with time series architecture.

ARIMA: In article [32], ARIMA model and ensemble of ARIMA and SVM were
experimented. These were compared with **hybrid Garch forecasting model**.

Interactive Multiple Model Approach: The intuitive numerous model (IMM),
which utilizes at least two models, i.e., soft switching algorithm and effective
weighted fusion was used in work carried out in [7]. The examination demonstrates
that the intuitive numerous model (IMM) be up to anticipate more precisely than
alone AR model or alone AR-Kalman model.

5.7 NAXRX, GWR, EZN

In paper [13], NARX strategy had least error plus the best assurance coefficient, trailed by SVR along with preferable execution over GWR and trailed by ANN. Paper [36] introduces air mass trajectory-based geographic model which is been a useful tool to find direction if pollutants as well as source location of them. The paper [38] proposed a new a hybrid model combining echo state network (ESN) and an improved particle swarm optimization (IPSO) algorithm. Phase space reconstruction (PSR), a time series model, constructs the group of coordinate components that characterized the behaviors of original system. In this study, IPSO-PSR-ESN hybrid method was introduced.

5.8 Weather Research and Forecasting Model with Chemistry (WRF-Chem)

The WRF-Chem is the model novel regional atmospheric dynamic chemical coupling model. Benefits are coupling of chemical transmission mode and meteorological mode in time resolution and spatial resolution [40]. This model was used by authors in the presented work in [25].

5.9 Wavelet Transform

WT is based on the idea of short-time Fourier transform localization. This method surpasses a problem of constant window with frequency. Furthermore, it acts as best apparatus to analyze the system and to process. Mallat algorithm is used for wavelet decomposition which is used in experiment in papers [33, 35].

5.10 Differential Evolution (DE)

Experiments in paper [35] used DE algorithm which is a stochastic, population-based algorithm and direct search algorithm, which has the characteristics of simple structure, less control, fast convergence, parameters along with significant advantages of dealing with convergence, and strong robustness and non-differentiable. The DE-BP model is introduced to improve the function approximation capacity of BP neural network, especially on the catastrophe points. The study describes the purpose of DE algorithm to optimize the weight matrices and thresholds. Hybrid WT-VMD-DE-BP forecasting model is established for daily PM 2.5 concentration.

6   Conclusion

The PM 2.5 accumulation of a specific point can be estimated by suitable methods with exact information; nevertheless, it is difficult to get values constituting to the restrictions over a full examination area where setting up the essential tools is not attainable. Hence, an expanding number of techniques for assessing PM 2.5 focuses in zones were introduced by the scientists/researchers. The primary methodologies are partitioned into two classes in this study, to be specific are (a) statistical/mathematical methods (b) ML techniques. Be that as it may, they normally overlap as far as source data and preprocessing are considered. Nonetheless, ML techniques have risen as the most famous procedure during late years, and it is worth an extraordinary notice. Presented work audited pertinent as of late distributed papers and just as old style papers in the field of techniques and utilization of PM 2.5 prediction and assessment. The primary discoveries include the following:

1. In light of the ascent in computing power, quick improvement in ML has become a significant examination center territory for assessing and anticipating issues through persistent procurement, joining and investigation of an assortment of heterogeneous and huge information in urban areas. Consolidating physical models with ML holds evident potential for assessing spatial-transient powerful dissemination of metropolitan PM 2.5 focuses.
2. The conventional methodologies for assessment of PM 2.5 focuses are through obsolete in light of the fact that they give insights about future possible options and development in study area. Moreover, there is an expanding number of integrated ensemble models and strategies applied to different restrictive applications.
3. The most advantageous and time-effective technique is using ML techniques. Accuracy of these strategies is generally high. The more intricate of the above strategies consistently identifies with ML which can anticipate the obscure information of PM 2.5 on spatiotemporal scales. There are numerous enhancements and reconciliations of various strategies which can give more precise outcomes under explicit conditions.
4. There are different degrees of exactness/accuracy, which are controlled by an enormous components including conditions explicit to exploration territory, goal of the source information, boundaries picked by explicit models, and the details all the while, utilized by various techniques to appraise the PM 2.5.

References

1. Y.F Xing, Y.H. Xu, M.H. Shi, Y.X. Lian, "The impact of PM2.5 on the human respiratory system". https://doi.org/10.3978/j.issn.2072-1439.2016.01.19, https://doi.org/10.3978/j.issn.2072-1439.2016.01.19.
2. Health effects of particulate matter, Policy implications for countries in eastern Europe, Caucasus and central Asia, World Health organization, PublicationsWHO Regional Office for EuropeUN City, Marmorvej 51DK-2100 Copenhagen Ø, Denmark, ISBN 978 92 890 00017
3. L. Miller, X. Xu, "Ambient PM 2.5 Human Health Effects—Findings in China and Research Directions", Atmosphere 2018, **9**, 424. https://doi.org/10.3390/atmos9110424m,MDPI
4. D. Zhu, C. Cai, T. Yang, X. Zhou, "A Machine Learning Approach for Air Quality Prediction: Model Regularization and Optimization", big data and cognitive computing, MDPI, 24 Feb 2018.
5. B. Pan, "Application of XGBoost algorithm in hourly PM2.5 concentration prediction", ICAESEE 2017, IOP Publishing. https://doi.org/10.1088/1755-1315/113/1/012127
6. Q. Di , H. Amini, L. Shi, I. Kloog, R. Silvern, J. Kelly, MB. Sabath, C. Choirat, P. Koutrakis, A. Lyapustin, Y. Wang, LJ. Mickley, J. Schwartz, "An ensemble-based model of PM 2.5 concentration across the contiguous United States with high spatiotemporal resolution", Environment International. **130**, 104909 (2019), ELSEVIER
7. J. Li, X. Li, K. Wang, "Atmospheric PM 2.5 Concentration Prediction Based on Time Series and Interactive Multiple Model Approach", Hindawi Advances in Meteorology Volume 2019, Article ID 1279565, 11. https://doi.org/10.1155/2019/1279565
8. M. Zamani Joharestani, C. Cao, X. Ni, B. Bashir, S. Talebiesfandarani, "PM 2.5 Prediction Based on Random Forest, XGBoost, and Deep Learning Using Multisource Remote Sensing Data", atmosphere, MDPI, 4 July 2019.
9. J. Boonphun, C. Kaisornsawad, P. Wongchaisuwat, "Machine learning algorithms for predicting air pollutants", E3S Web of Conferences **120**, 0 30 0 4 (2019) CGEEE 2019, https://doi.org/10.1051/e3sconf/20191 200 3004
10. J.K. Deters, R. Zalakeviciute, M. Gonzalez, Y. Rybarczyk, "Modeling PM 2.5 Urban pollution using machine learning and selected meteorological parameters", Hindawi J. Electric. Comput. Eng. **2017**, Article ID 5106045, 14 https://doi.org/10.1155/2017/5106045
11. J. Shen, "PM 2.5 concentration prediction using times series based data mining"
12. C.R. Aditya, C.R. Deshmukh, D.K. Nayana, P.G. Vidyavastu, "Detection and prediction of air pollution using machine learning models", Int. J. Eng. Trends Technol. (IJETT). **59**(4) (May 2018)
13. M.R. Delavar, A. Gholami, G.R. Shiran, Y. Rashidi, G.R. Nakhaeizadeh, K. Fedra, S.H. Afshar, "A novel method for improving air pollution prediction based on machine learning approaches: a case study applied to the capital city of Tehran", Int. J. Geo-Inf. MDPI (2019)
14. Y. Chen, "Prediction algorithm of PM2.5 mass concentration based on adaptive BP neural network", Computing **100**, 825–838 (2018). https://doi.org/10.1007/s00607-018-0628-3. Crossmark
15. X. Wang, B. Wang, "Research on prediction of environmental aerosol and PM2.5 based on artificial neural network". Neural Comput. Appl. **31**, 8217–8227 (2019). https://doi.org/10.1007/s00521-018-3861-y. Crossmark
16. H. Liu, Z. Duan, C. Chen, "A hybrid multi-resolution multi-objective ensemble model and its application for forecasting of daily PM2.5 concentrations", Inf. Sci. Elsevier (2019)
17. Y. Park, B. Kwon, J. Heo, X. Hu, Y. Liu, T. Moon, "Estimating PM2.5 concentration of the conterminous United States via interpretable convolutional neural networks", Environmental Pollution, journal homepage: www.elsevier.com/locate/envpol, Elsevier (2019)
18. B. Zhang, H. Zhang, G. Zhao, J. Lian, "Constructing a PM 2.5 concentration prediction model by combining auto-encoder with Bi-LSTM neural networks", Environ. Model. Softw. J. Homepage: http://www.elsevier.com/locate/envsoft (2020)

19. R. Zhao, X. Gu, B. Xue, J. Zhang, W. Ren, "Short period PM 2.5 prediction based on multivariate linear regression model", PLOS ONE I July 26, 2018. https://doi.org/10.1371/journal.pone.020 1011

20. P. He, B. Zheng, J. Zheng, "Urban PM 2.5 diffusion analysis based on the improved gaussian smoke plume model and support vector machine", Aerosol Air Qual. Res. **18**, 3177–3186 (2018), ISSN: 1680–8584 print / 2071–1409 online. https://doi.org/10.4209/aaqr.2017.06.0223

21. A. Masood, K. Ahmad, "A model for particulate matter (PM 2.5) prediction for Delhi based on machine learning approaches", Science Direct, Proc. Comput. Sci. **167**, 2101–2110 (2020), Elsevier

22. KS Harishkumar, KM Yogesh, I. Gad "Forecasting air pollution particulate matter (PM 2.5) using machine learning regression model", Procedia Comput. Sci. **00**, 000–000 (2019). ScienceDirect, Elsevier

23. J. Wang, S. Ogawa, "Effects of meteorological conditions on PM 2.5 concentrations in Nagasaki, Japan", Int. J. Environ. Res. Public Health, ISSN 1660–4601 www.mdpi.com/jou rnal/ijerph, **12**, 9089–9101 (2015). https://doi.org/10.3390/ijerph120809089

24. J. Langner, L. Gidhagen, R. Bergström, E. Gramsch, P. Oyola, F. Reyes, D. Segersson, C. Aguilera, "Model-simulated source contributions to PM 2.5 in Santiago and the central region of Chile", Aerosol Air Qual. Res. **20**, 1111–1126 (2020), ISSN: 1680–8584 print / 2071–1409 online. https://doi.org/10.4209/aaqr.2019.08.0374

25. J. Ma, Z. Yu, Y. Qu, J. Xu, Y. Cao, "Application of the XGBoost machine learning method in PM 2.5 prediction: a case study of Shanghai", Aerosol Air Qual. Res. **20**, 128–138 (2020), ISSN: 1680–8584 print / 2071–1409 online. https://doi.org/10.4209/aaqr.2019.08.0408

26. Y. Zeng, D.A. Jaffe, X. Qiao, Y. Miao, Y. Tang, "Prediction of potentially high PM 2.5 concentrations in Chengdu, China", Aerosol Air Qual. Res. **20**, 956–965 (2020), ISSN: 1680–8584 print/2071–1409 online. https://doi.org/10.4209/aaqr.2019.11.0586

27. I. Cheng Chang, "Identifying leading nodes of PM 2.5 monitoring network in Taiwan with Big Data-oriented social network analysis", Aerosol Air Qual. Res. **19**, 2844–2864 (2019), ISSN: 1680–8584 print/2071–1409 online, https://doi.org/10.4209/aaqr.2019.11.0554

28. M. Filonchyk, V. Hurynovich, "A study of PM2.5 and PM10 concentrations in the atmosphere of large cities in Gansu Province, China, in summer period", Article in J. Earth Syst. Sci. (August 2016). https://www.researchgate.net/publication/306052138

29. M. Opera, Sanda Florentina Mihalache ad Maraian Popescu, "Computational intelligence-based PM2.5 air pollution forecasting", Int. J. Comput. Commun. Control, ISSN 1841–9836, **12**(3), 365–380, June 2017, Research Gate.

30. D.J. Lary, T. Lary, B. Sattler, "Using machine learning to estimate global PM2.5 for environmental health studies", Environ. Health Insights, **9**(s1), SAGE Publishing, https://doi.org/10.1177/EHI.S15664

31. T. Xayasouk, HwaMin Lee, G. Lee, "Air pollution prediction using long short-term memory (LSTM) and deep autoencoder (DAE) models." Sustainability **12**, 2570 (2020). https://doi.org/10.3390/su12062570. MDPI

32. P. Wang, H. Zhang, Z. Qin, G. Zhang, "A novel hybrid-Garch model based on ARIMA and SVM for PM 2.5 concentrations forecasting", http://www.journals.elsevier.com/locate/apr, Atmospheric Pollut. Res. **8**, 850e860 (2017)

33. W. Qiao, W. Tian, J. Zhang, "The forecasting of PM2.5 using a hybrid model based on wavelet transform and an improved deep learning algorithm", http://creativecommons.org/licenses/by/4.0/, **7** (2019)

34. J. Shah, B. Mishra, "Analytical equations based prediction approach for PM2.5 using artificial neural network", Springer Nature Switzerland AG 2020, SN Appl. Sci. **2**, 1516 (2020). https://doi.org/10.1007/s42452-020-03294-w

35. D. Wang, Y. Liu, H. Luo, C. Yue, S. Cheng, "Day-Ahead PM 2.5 concentration forecasting using WT-VMD based decomposition method and back propagation neural network improved by differential evolution". Int. J. Environ. Res. Public Health, **14**, 764 (2017). https://doi.org/10.3390/ijerph14070764, MDPI

36. H. Karimian, Q. Li, C. Wu, Y. Qi, Y. Mo, G. Chen, X. Zhang, S. Sachdeva, "Evaluation of different machine learning approaches to forecasting PM 2.5 mass concentrations", Aerosol Air Qual. Res. **19**, 1400–1410 (2019), ISSN: 1680–8584 print / 2071–1409 online. https://doi.org/10.4209/aaqr.2018.12.0450

37. X. Feng, Q. Li, Y. Zhu, J. Hou, L. Jin, J. Wang, "Artificial neural networks forecasting of PM 2.5 pollution using air massct rajectory based geographic model and wavelet transformation". https://doi.org/10.1016/j.atmosenv.2015.02.030 1352–2310/© 2015 The Authors. Published by Elsevier Ltd

38. X. Xu, W. Ren, "Application of a hybrid model based on echo state network and improved particle swarm optimization in PM 2.5 concentration forecasting: a case study of Beijing, China", Sustainability, **11**, 3096 (2019). https://doi.org/10.3390/su11113096 www.mdpi.com/journal/sustainability.

39. W. You, Z. Zang, L. Zhang, Y. Li, X. Pan, W. Wang, "National-scale estimates of ground-level PM2.5 concentration in China using geographically weighted regression based on 3 km resolution MODIS AOD", www.mdpi.com/journal/remotesensing, Remote Sens. **8**, 184 (2016) https://doi.org/10.3390/rs8030184

40. G. Zhang, X. Rui, Y. Fan, "Critical review of methods to estimate PM 2.5 concentrations within specified research region", Int. J. Geo-Inf., MDPI (2018)

41. N. Zhang, H. Huang, X. Duansd, J. Zhao, B. Su, "Quantitative association analysis between PM 2.5 concentration and factors on industry, energy, agriculture, and transportation", www.nature.com/scientificreports (2018)

Performance Analysis of Modified TCP New Reno for MANETs

Sharada U. Shenoy, Udaya Kumar K. Shenoy, and M. Sharmila Kumari

Abstract The transmission control protocol (TCP) controls packet loss due to congestion by additive increment (AI) and multiplicative decrement (MD) algorithm on congestion window size. This works fine in case of wired networks. In wireless ad hoc networks, the packet loss may not always be due to congestion, whereas most of the time the reason could be the problems in wireless medium. In case of path break, the routing algorithm will reestablish the route and packet loss may last for short duration. The MD phase of TCP algorithm reduces the congestion window by 50% on detection of packet loss. This reduces the transmission rate drastically. So instead, the congestion window can be reduced by a small fraction until the path gets resumed by a routing algorithm. So in this paper, a comparative study has been carried out by reducing the congestion window only by 20% instead of existing 50% during congestion in the MD phase of TCP New Reno for MANETs. This resulted in improved performance in terms PDR, throughput, and packet delivery rate compared to existing method.

Keywords MANET · TCP · Transmission control protocol · Congestion control · Congestion window · Modified TCP New Reno

1 Introduction

The transmission control protocol (TCP) congestion control algorithms work based on adjustment of congestion windows to cope with the congestion and packet losses. The reduction in congestion window size to 50% in case of packet loss decreases the data rate until the congestion situation is resolved. This is carried out irrespective of

S. U. Shenoy (✉) · U. K. K. Shenoy
NMAM Institute of Technology, Nitte, India
e-mail: sharadauday@nitte.edu.in

U. K. K. Shenoy
e-mail: ukshenoy@nitte.edu.in

M. S. Kumari
PA College of Engineering, Mangaluru, India

causes for packet loss. Especially in wireless networks, the packet loss is not always due to congestion [1]. The packet loss in wireless networks is usually due to the error-prone medium of transmission. Possible reasons may be path break, scattering, interference, multipath propagation, and weak signals unlike wired networks. In wired networks, packet losses are mostly due to congestion. Hence, it is possible to get fair results in terms of throughput, PDR, and packet delivery by modifying the congestion control mechanism used in traditional TCP. Especially in ad hoc networks where there is no infrastructure, reduction of congestion window by 50% causes severe degrading in data transmission. At the same time, the dynamic routing algorithm used in wireless networks will successfully find the alternative paths and regain the connection quite fast. Thus, the performance of TCP can be increased if we do not reduce the congestion window by a factor of 0.5 (50%) but reduce with a smaller fraction because packet loss in wireless networks is due to temporary problems while rerouting will be done by the routing algorithm quickly. This paper discusses the result of reducing the congestion by only 20% during packet loss and continuing with existing additive increment phase later on. This paper also provides a comparative study on the existing method of 50% window reduction.

The transmission delay and packet loss rate are determined by data link layer (DLL) parameters also, and they affect the TCP performance. Hence, the link errors and congestion need to be managed in a different way. The worse performance at the DLL and physical layer will primarily cause transmission errors like packet loss in the TCP at transport layer [2, 3, 4].

The traditional TCP which is designed for wired networks is the reliable protocol at large demand. But the situation in wireless networks is entirely different, and packet loss may happen because of reasons other than congestion, like mobility, path loss, signal fading, reflection, or weakening of the signal [1]. Traditional TCP senses any loss as congestion and shrinks the congestion window by usually 50% with an intention to decrease the network load. This continues if congestion persists or until timeouts. According to the additive increment and multiplicative decrement (AIMD) approach [1, 5], TCP at the connection setup initializes the congestion window (cwnd) to 1. Then, it starts with a slow start algorithm which does additive increment (AI) of the window size. Then, it continues in the AI phase until the slow start threshold (ssthresh) is reached. Then, increments the window linearly up to maximum limit on window size. The congestion window size gets decremented by 50% and in case of congestion till the lower limit being cwnd = 1. This technique is called as multiplicative decrement (MD) and works well with wired networks where the congestion is because of overflow in the router.

Most of the real-time traffic also uses TCP protocols in wireless. Considering the comparison of TCP New Reno and CUBIC, the study shows that New Reno recovers fast from single packet loss and suffers from slow start when multiple packets are lost. CUBIC is fair and stable, but data flow is lesser compared to New Reno [6]. The problem is with the window decrement strategy when congestion is detected and slow start during recovery. So few algorithms, receiver-assisted slow start (RAAS) and feedback-assisted recovery (FAR), proposed in [7] modify the initial window size according to the acceleration factor that is calculated based on the receiver advertised window, and it follows slow growth of congestion window. It works based on three sizes of initial windows. This gives additional throughput of 37% and 52% against TCPNCE and New Jersey.

In another approach, a Markov chain model is defined for real-time delay distribution in TCP. It checks for real media data, voice and video streams and finds that there will be lesser TCP delays when RTT is 100 ms with error rate is less than 3% [8]. It is not only enough to change the window size by taking only the packet loss and need to consider other factors too in the wireless network. So feedback on the cause of loss can be taken from the underlying layers to enhance the performance of TCP. The queue length value is written to the advertised window (awnd) field of TCP in which the active window mechanism (AWM) in the access gateway [9]. The gateway informs the source to send controlled data with the help of this optimum value.

TCP should get a fair amount of bandwidth and also need to be stable while coexisting with other TCP protocols. Hence, the performance of TCP also needs to be examined when multiple flows coexist. The increment congestion window size in the additive increment (AI) algorithm phase of TCP can take feedback from the bottleneck node, while the multiplicative decrement (MD) is calculated separately by each source [9]. The AIMD window adjustment is carried by the window controller component at the source node. Similar work shows scaling of the window size for maintaining the traffic shape, and working of MD phase based on the RTT value is done by the traffic controller at bottleneck node [10]. Another approach takes the feedback from the bottleneck node, based on contention ratio (CR) and channel utilization (CU) [11]. This feedback regulates transmission rate of source nodes. In case there exist fluctuations in the channel due to signal fading or mobility, the cross layer (CL) approaches too can lead to enhanced results. The delay information-based fairness congestion control (DFCC) protocol that estimates congestion through the contention in the medium uses CL approach [12]. The packet delay is used by each source to determine average congestion window size. Similarly, modification is done to TCP Vegas where the transmission rate is based on the calculation of RTT [13]. The node increases the congestion window by 1, if the number of packets transmitted is greater than the limit, alpha. It decrements the congestion window by 1, if the limit is in between alpha and beta, else it remains constant. An active window that can interact with TCP source is also found to be useful [9] and implemented on a gateway to interact with TCP source. The router uses "awnd" field of TCP header to regulate the transmission rate of the source.

The protocols mentioned so far focus on window increase strategies and not on window decrease. The protocols that pay importance to window decrease strategies are mentioned below. Multiplicative increment multiplicative decrement (MIMD) [14] proposes a method where a constant time is taken to recover from error and it still depends on RTT, leading to overflow of bottleneck buffer causing severe packet loss. This affects stability and fairness. Alternatively, binary increase congestion control (BIC)-TCP [15], uses linear and logarithmic functions at MD phase. It decreases the window by β on loss. The maximum congestion window size, Wmax, will be retained to the previous size which was just before the start of packet loss. Minimum window size, Wmin, will be set to the size that was reduced after packet loss detection. Later it sets the window size to the middle value of Wmax and Wmin, for every positive ACK. In case the distance of midpoint and Wmin is greater than the predefined maximum increment, Smax, it uses logarithmic increment. This works fine for the droptail queue in the routers. The compound TCP (CTCP)[16] uses a delay-based approach for AIMD. BIC-TCP and CTCP reduce the window by a factor of $\beta = 0.125$ because reducing it to 50% is not suitable. A quite different method is used in Westwood TCP [17], and it considers BDP (bandwidth delay product) instead of 50% reduction and is known as additive increment adaptive decrement (AIAD). İt adjusts the window to available BDP upon packet loss. In TCP Westwood + [18], which is advanced Westwood, the filtering approach is modified since the ACKs that are compressed and delayed were not allowing the Westwood to accurately measure the available BDP.

Few solutions are proposed to improve the transmission rates in case the packet loss is not due to congestion [19]. A TCP protocol introduced the explicit link notification (ELN) [20]. Here, the reason for the transmission error is conveyed by the receiver to the sender through ELN bit.

The issues of channel fading, transmission delay, and high BER cause the route rescheduling and retransmissions. Increased retransmission also causes congestion. So heaps of cross layers were introduced and are capable of identifying issues like channel attenuation, transmission delay, BER, or channel interference [3]. A forward error correction (FEC) that passes the information of error in DLL along with the ELN [21] is shown to enhance the performance by almost 100%. A revision of existing mechanisms in [21], ARQ along with FEC is proposed in [22]. Mapping based on SNR value is proposed in [22], and it conveys the errors in the physical layer / DLL to higher layers, and AMC is also used that has improved performance and QoS.

In MANETs, each node must work as a node as well as router. The existing congestion control method is not suitable since the frequent packet drop is a common scenario in wireless networks. In the existing approach, the rate of packets sent decreases and the available bandwidth is underutilized, thus resulting in reduced throughput. MANETs are popular, and it demands high data rates for data transfer like audio, video, live streaming, etc. Conventional TCP is not suitable because of low throughput due to its multiplicative decrement congestion management [15, 16]. If the window decrement mechanism of conventional TCP can be adapted for wireless scenarios, then this can provide better performance. So this paper has made an attempt to modify the MD stage of TCP New Reno and is able to get better performance for the given scenario.

2 Result Analysis for TCP New Reno

2.1 Performance Parameters for Existing TCP New Reno

This section provides analysis of TCP New Reno with existing method of congestion control.

Tables 1 and 2 show the performance parameters for the existing cwnd reduction of 50% for varying time over 50 and 100 nodes MANET. Table 3 shows the performance parameters for the existing cwnd of 50% by varying the speed over 50 nodes and 100 nodes mobile ad hoc network. In Tables 1 and 2, the average jitter is around 649 ms and ranges from 414.18 ms to 786.93 ms. The e2e delay ranges from 0.87 ms to 1.1217 ms. The average throughput varies from 545 to 555kbps. The PDR is in the range of 93.139% to 96.83%, throughput range of 465.3kbps to 596.96 kbps as seen in Tables 1 and 2.

With varying speed and constant simulation time, the packets sent is between 34,000 to 36,000 and average of 33,364 packets are sent as shown in Table 3.

Table 1 TCP New Reno performance metrics with existing approach over varying time for 50 nodes

TCP New Reno—Performance metrics for existing cwnd = 50%, 50 nodes

Time (s)	Jitter (ms)	e2e delay (ms)	Throughput (kbps)	Number of packets sent	Number of packets received	PDR (%)
100	414.18	0.880579	510.74	5101	4598	90.1392
150	464.24	0.87843	485.93	10,597	10,086	95.1779
200	566.66	0.946554	516.89	14,159	13,518	95.4728
250	650.7	1.01921	554.01	17,930	17,296	96.464
300	729.55	1.11597	561.46	19,846	19,072	96.11
350	645.09	1.01547	564.02	23,097	22,095	95.662
400	631.36	1.04358	529.53	23,605	22,462	95.1578
450	730.33	1.12182	567	32,298	31,106	96.3094
500	710.67	1.03279	548	36,514	35,099	96.1248
750	769.74	1.11213	547.24	53,403	51,448	96.3392
1000	786.93	1.00143	570.79	68,001	65,820	96.233
1500	755.34	1.09226	590.43	111,725	107,762	96.4529

Table 2 TCP New Reno performance metrics with existing approach over varying time, 100 nodes

TCP New Reno—Performance metrics for existing cwnd = 50%, 100 nodes

Time (s)	Jitter (ms)	e2e delay(ms)	Throughput (kbps)	Number of packets sent	Number of packets Received	PDR (%)
100	625.33	0.942108	465.3	5622	5390	95.8734
150	683.89	0.950464	519.36	9473	9173	96.8331
200	709.41	1.10076	515.03	13,068	12,026	92.0263
250	595.78	0.909859	550	17,133	16,431	95.9026
300	627.82	0.982212	547.16	20,692	19,687	95.1431
350	564.04	0.915938	554	24,782	23,785	95.9769
400	692.84	1.06943	572	28,559	27,594	96.621
450	646.53	0.980932	585.7	32,945	35,895	96.9009
500	607.51	0.952863	584.14	37,247	35,895	96.3702
750	664.23	0.942374	590.18	53,883	51,929	96.3736
1000	637.16	0.952118	596.95	73,251	71,119	96.178
1500	743.42	0.919559	584.78	100,521	96,730	96.2286

2.2 Result Analysis for Congestion Window Set to 80%—TCP New Reno Modified Window (MW

For the cwnd = 80%, the performance metrics have been plotted in Tables 4, 5 and 6. The e2e delay has been in the range of 0.637–1.081 ms as seen from the above-mentioned tables. This is less compared to existing approach and also the modified 60 and 70% approach. The jitter value starts from 285.42 ms to 779.24 ms and is less than the existing method as well as the 60% and 70% congestion window sizes.

The throughput has values ranging from 532.2kbps to 608.2kbps for varying simulation times from 100 to 1000 s and 570.87kbps to 614.75kbps for varying speed from 10 to 30 mph. This is high compared to the existing approach. The PDR is 95.31–97.4%, whereas the existing method of cwnd = 50% has 90.14–96.9% PDR. There is slight increase in PDR. If the packet count is considered, the number of sent and received packets is more than the existing approach as seen in Tables 4, 5 and 6 than seen in Tables 1, 2, and 3.

3 Improvement Over Existing Method

This section discusses an improvement seen by setting cwnd to 80% over the existing method.

Table 3 TCP New Reno—performance metrics existing approach for varying speed

TCP New Reno—Performance metrics for cwnd = 50% (existing) for varying speed

Nodes	Throughput (kbps)		PDR (%)		e2e delay (ms)		Jitter (ms)		Packets sent		Packets received	
speed (mps)	50	100	50	100	50	100	50	100	50	100	50	100
10	541.73	567.75	94.4	95.22	1.07971	1.00382	676.99	704.4	33,941	35,945	32,040	34,226
15	562.2	579.67	95.91	96.09	1.00568	1.00783	715.84	639.02	35,031	34,620	33,596	34,006
20	569.87	575.52	95.9	96.2	0.99029	0.971939	656.53	649.92	32,927	35,353	31,578	34,017
25	575.54	573.96	95.13	95.03	0.970074	0.878402	633.68	573.96	35,100	35,248	33,391	33,497
30	571.59	572.58	95.18	91.91	0.874423	0.95093	566	644.41	34,842	36,053	33,163	33,135

Table 4 TCP New Reno-MW performance metrics with cwnd set to 80% over varying time for 50 nodes

TCP New Reno MW—Performance metrics for cwnd = 80%, 50 nodes

Time (s)	Jitter (ms)	e2e delay (ms)	Throughput (kbps)	Number of packets sent	Number of packets received	PDR (%)
100	587.51	0.73292	529.62	5418	5182	95.64
150	555.45	0.83152	542.6	7743	7306	94.36
200	511.33	0.87593	532.2	13,143	12,664	96.36
250	703.22	0.89869	564.2	17,280	16,578	95.94
300	701.29	0.94256	577.6	21,441	20,804	97.03
350	692.36	0.95408	534.6	24,737	23,765	96.07
400′	698.29	1.06347	592.2	29,515	28,568	96.80
450	733.16	1.05839	587	36,802	35,487	96.43
500	701.36	1.08121	597	37,365	36,128	96.69
750	741.97	1.00301	566.64	56,113	54,191	96.58
1000	710.23	1.00111	608.96	73,301	70,540	96.233
1500	721.26	1.00056	616.67	116,893	112,973	96.5

Table 5 TCP New Reno–MW performance metrics with cwnd set to 80% for varying time, 100 nodes

TCP New Reno MW—Performance metrics for cwnd = 80%, 100 nodes

Time (s)	Jitter (ms)	E2e delay (ms)	Throughput (kbps)	Number of packets sent	Number of packets received	PDR (%)
100	285.42	0.67204	445.9	5323	5045	94.7774
150	497.85	0.78004	533.25	9035	8728	96.6021
200	505.06	0.79457	548.67	13,159	12,574	95.5544
250	586.2	0.97345	569	17,461	16,705	95.6704
300	541.06	0.91449	577.8	21,330	20,774	97.3933
350	615.37	0.96779	573.23	25,063	24,161	96.4011
400	633.53	1.01205	589	29,778	28,454	95.5538
450	708.43	1.08887	580.4	37,469	36,267	96.792
500	708.1	0.9516	567.9	57,006	54,874	96.26
750	636.83	0.63701	589.68	55,309	53,644	96.9896
1000	657.34	0.76981	585.49	76,133	73,477	97.0895
1500	733.12	0.78125	628.13	115,172	110,907	96.5234

Table 6 TCP New Reno MW—performance metrics with cwnd set to 80% for varying speed

TCP New Reno MW—Performance metrics for cwnd = 80%

Nodes / speed (mps)	Throughput (kbps)		PDR (%)		e2e delay (ms)		Jitter (ms)		Packets sent		Packets received	
	50	100	50	100	50	100	50	100	50	100	50	100
10	614.75	587.68	96.64	96.48	1.04439	1.04077	751.33	635.45	38,447	36,059	37,154	34,787
15	595.75	570.87	95.91	96.21	1.06672	1.05404	727.81	721.49	37,512	36,432	35,106	33,885
20	599.9	602.94	96.46	96.33	1.00497	1.00969	680.71	681.47	37,926	37,715	36,582	36,328
25	571.32	596.79	95.62	96.1	1.02546	0.92486	746.23	644.9	37,854	37,773	36,193	36,243
30	592.45	594.22	95.31	95.98	1.01495	1.00614	779.24	755.07	37,002	37,395	35,263	35,888

Table 7 TCP New Reno MW—% improvement in packet transmission (%) for various simulation time

TCP New Reno MW—% improvement in packet transmission for varying time

Time (s)	50 nodes		100 nodes	
	Improvement packets sent (%)	Improvement packets received (%)	Improvement packets sent (%)	Improvement packets received (%)
100	6.22	2.57	24.48	24.48
250	0.56	10.02	1.92	1.67
300	8.04	4.16	3.09	5.53
350	2.88	25.7	1.97	2.3
400	25.04	19.51	4.27	3.12
450	13.95	3.31	13.74	1.04
500	2.34	8.04	53.05	52.88
750	5. 3	9.18	6.68	6.48
1000	7.8	7.92	4.92	4.31
1500	4.63	2.56	14.58	14.66
Average	7.67	9.3	12.87	11.65

If Tables 1, 2, 4, and 1, 2, and 5 are considered, for different execution time, the throughput increase is seen in both cases of 50 and 100 node MANETs except at few case. There is an increase in average PDR of 1.214% for smaller and 0.51% for larger MANETs. Average of 5019 more packets are sent for 50 node and 4541 for 100 node MANETs. Jitter improvement 69.95 ms is seen for larger networks, and there is e2e delay improvement of 0.08 ms and 0.1 ms (Table 7).

From Tables 3 and 6 for varying speeds, there is no improvement in average jitter and e2e delay. The throughput has seen increment by 12.6%, and PDR has increased by 0.68% to 1.3%. There is 10.62% to 16% improvement which is seen in number of sent, 4.7% and 6.3% improvement in received packets when cwnd = 80% for 50 nodes and 100 node scenarios than the existing method as given in Table 8.

4 Conclusion

The work carried out shows that the performance can be improved by varying the congestion window of TCP New Reno to 80% will yield good performance in MANETs, and it is advantageous over traditional method. 80% congestion window size gives time to recover from problems like path breaks, temporary signal fading, or effect of mobility which will last for a short while. Further, it can continue to increment the window exponentially rather than again reverting back to a slow start phase.

Table 8 TCP New Reno MW—% improvement in packet transmission for various speeds

TCP New Reno MW—% Improvement in packets sent/received for varying speed

Speed (mps)	50 nodes		100 nodes	
	Improvement in packets sent (%)	Improvement in packets received (%)	Improvement in packets sent (%)	Improvement in packets received (%)
10	13.28	15.96	0.32	1.64
15	7.09	6.54	5.24	6.37
20	15.18	15.85	6.69	6.83
25	7.85	8.4	7.17	8.2
30	6.2	6.4	3.73	8.31
Average	9.92	10.62	4.7	6.3

References

1. C. Siva Ram Murthym, B.S. Manoj, Adhoc wireless networks architectures and protocols, Pearson Education, 2004
2. S. Shakkottai, T.S. Rappaport, P.C. Karlsson, Cross-Layer Design for Wireless Networks. IEEE Commun. Mag. **41**(10), 74–80 (2003)
3. F. Foukalas, V. Gazis, N. Alonistioti, Cross-layer design proposals for wireless mobile networks: a survey and taxonomy. IEEE Commun. Surv. Tutorials **10**(1), 70–85 (2008)
4. P. Bender, CDMA/HDR: a bandwidth efficient high-speed wireless data service for nomadic users. IEEE Commun. Mag. **38**(7), 70–77 (2000)
5. Fu. Zhenghua, X. Meng, Lu. Songwu, How bad TCP can perform in mobile ad hoc networks. Proceedings ISCC Seventh International Symposium on Computers and Communications **2001**, 298–303 (2002)
6. M.C. Dias, L.A. Caldeira, A. Perkusich, Traditional TCP congestion control algorithms evaluation in wired-cum-wireless networks. International Wireless Communications and Mobile Computing Conference (IWCMC), Dubrovnik, 805–810 (2015)
7. M.J.A. Jude, S. Kuppuswami, Enhanced window increment and adaptive recovery TCP for multi-hop wireless networks. Electron. Lett. **53**(6), 438–440 (2017)
8. E. Brosh, S.A. Baset, V. Misra, D. Rubenstein, H. Schulzrinne, The delay friendliness of TCP for real-time traffic. IEEE/ACM Trans. Netw. **18**(5), 1478–1491 (October 2010)
9. M. Barbera, A. Lombardo, C. Panarello, G. Schembra, Queue stability analysis and performance evaluation of a TCP compliant window management mechanism. IEEE/ACM Trans. Netw. **18**(4), 1275–1288 (August 2010)
10. H. Nishiyama, N. Ansari, N. Kato, Wireless loss-tolerant congestion control protocol based on dynamic aimd theory. IEEE Wirel. Commun. **17**(2), 7–14 (April 2010)
11. X. Zhang, N. Li, W. Zhu, D.K. Sung, TCP transmission rate control mechanism based on channel utilization and contention ratio in AD hoc networks. IEEE Commun. Lett. **13**(4), 280–282 (April 2009)
12. H.J. Lee, J.T. Lim, Fair congestion control over wireless multihop networks. IET Commun. **6**(11), 1475–2148 (July 2012)
13. R.S. Cheng, D.J. Deng, H.C. Chao, Congestion control with dynamic threshold adaptation and cross-layer response for TCP over IEEE 80.11 wireless networks. Proceedings of the International Conference on Wireless Information Networks and Systems, Spain, 95–100 (2011)
14. T. Kelly, Scalable TCP: improving performance in highspeed wide area networks. ACM Comput. Commun. Rev. **33**, 83–88 (2003)

15. H. Bullot, R.L. Cottrell, R. Hughes-Jones, Evaluation of advanced TCP stacks on fast long-distance production networks. J. Grid Computing. **1**, 345–414 (2004)
16. K. Tan, J. Song, Q. Zhang, M. Sridharan, *A Compound TCP Approach for High-Speed and Long Distance Networks* (Microsoft Press, MSR-TR, 2005)
17. S. Mascolo, C. Casetti, M. Gerla, M. Sanadidi, R. Wang, TCP westwood: end-to-end bandwidth estimation of efficient transport over wired and wireless networks. Proceedings of the 7th annual international conference on Mobile computing and networking (MobiCom '01) ACM. USA, 287–297 (2001)
18. L.A. Grieco, S. Mascolo, Performance evaluation and comparison of Westwood+, New Reno, and Vegas tcp congestion control. ACM Comput. Commun. Rev. **34**, 25–13 (2004)
19. G. Xylomenos, G.C. Polyzos, P. Mahonen, M. Saaranen, TCP performance issues over wireless links. IEEE Commun. Mag. **39**(4), 52–58 (2001)
20. H. Balakrishnan, R. H. Katz, Explicit loss notification and wireless web performance. Proc. GLOBECOM Internet Mini-Conference. 1998
21. M. Miyoshi, M. Sugano, M. Murata, Improving TCP performance for wireless cellular networks by adaptive FEC combined with explicit loss notification. IEICE Communications Society: Transactions on Commun. Vol. E85-B(10), 2208–2214 (2002)
22. H. Zheng, H. Viswanathan, Optimizing TCP performance with hybrid ARQ and scheduler. Proc. 14th International Symposium on Personal, Indoor and Mobile Radio Commun. **2**, 1785–1789 (2003)

Smart Health and Pattern Recognition

Identification of Helmets on Motorcyclists and Seatbelt on Four-Wheeler Drivers

Divyansh Saini, Vedashree Arundekar, K. V. Priya, and Divya Jennifer D'Souza

Abstract The main objective of this research is to detect helmets and seatbelts on drivers in traffic scene. This will help people follow the rules and also make sure they drive safely. Its will also help the police monitor traffic from a safe place and away from elements such as heat and rain. Moreover, this project will help increase safety for both the commuters and the police. Using this data, police may also fine the individuals breaking the law. We intend to reprimand violators by making them aware of their violations by detecting features of vehicles driven by people who not only endanger themselves but also others on the road by their actions.

Keywords Helmet detection · Seatbelt detection · Machine learning · Image processing · Computer vision

1 Introduction

Traffic police personnel are always dependant on manual methods for intercepting traffic rule violators. Traditional methods are costly and often risky. Drivers especially Indians are obstinate toward wearing helmets and following traffic rules and the traffic police apply rudimentary methods to stop them like standing in front of the vehicle to forcefully stop them. Most of the drivers are adamant and apathetic, which leads the police to take impromptu and risky actions. Inclement weather is also one of the major challenges for manual traffic inspection by personnel. In this era of automation, an arrangement which requires less or no manual intervention would be really benevolent for governing authorities.

Traditional methods of identifying traffic violators include traffic police officers posted on various check posts or designated spots staying continuously alert and observing all incoming vehicles for presence of helmets/seatbelts on the drivers. In recent observations, traffic departments of various developed countries are automating this process by the use of CCTV or other medium to automate this

D. Saini (✉) · V. Arundekar · K. V. Priya · D. J. D'Souza
Department of Computer Science and Engineering, NMAM Institute of Technology, Nitte, Karnataka, India

P. Shetty D. and S. Shetty (eds.), *Recent Advances in Artificial Intelligence and Data Engineering*, Advances in Intelligent Systems and Computing 1386,
https://doi.org/10.1007/978-981-16-3342-3_8

process and directly levy fines on offenders with an attached image of the time and location of capture serving as a proof to the people who violate traffic safety norms. It is undeniable that it is physically tiring and demanding for the police officers to stay vigilant at all times. Our project aims to perish the stereotypes of being physically present for surveillance and panopticon purposes. With our suggested approach, one can easily identify people who are driving recklessly by not wearing helmets or seatbelts. In the coming times with advent in technology, automation of various processes would be benevolent for ease and avoidance of complications related to human intervene.

2 Literature Survey

Kunal Dahiya et al. proposed a framework for automatic detection of traffic rule violators in real time [1] who ride bike without using helmet. Their structure additionally helps the traffic police for distinguishing such violators in odd conditions such as heat and other such factors. Their experiment results exhibit the correctness of 98.88 and 93.80% for detecting bike riders with helmets and of violators, respectively. Likewise, their proposed structure naturally adjusts to new situations whenever required, with slight tuning. Their approach can be extended to identify and report number plates of violators.

In [2], Huiwen Guo et al. proposed a method for identifying seatbelts on drivers in traffic. To achieve the same, they have used gradient orientation method. To begin with, they detect the front window area and human face identification is used to limit the recognition area. If their face detection process fails, they use a straight-line fitting method and detect lines of 45°slope. Their accuracy is generally reliant on the consequence of human face recognition.

A system to give an idea about the number of traffic offenders was developed by Ramesh Babu et al. in [3]. They produced a database of all motorcycle riders driving helmetless with a proof image generated. They used open source software including tensorflow, OpenCV for reduction in costs. Provided adequate lighting, their system gave full proof and accurate results.

Vishnu proposed an efficient method for detecting people not wearing helmets, while riding bikes. In [4], they specified an algorithm for first segmenting and classifying vehicles, and then detection of offenders on motor bikes. For vehicle classification, they used various approaches for background calculation and object tracking, and achieved an accuracy of 0.9778. Their proposed system significantly low miss rates.

The papers [5] and [6] present novel methods for the detection of seat belt in a monitoring image which contains the full scene information of the moving car. First, the driver area is located based on the vehicle outline, then they use gradient orientation Image Processing Techniques.

3   Proposed Methodology

Below mentioned is the architecture we propose as a solution to the aforementioned problem statement.

A software solution to the incessant anguish of policeman from fatiguing manual surveillance of vehicles. The software we are developing is an application that enables one to detect traffic offenders who violate safety rules by not wearing helmets or seatbelts, while driving vehicles.

The proposed system contains two modules: one for helmet detection on motorcyclists and the other for seatbelt detection in four-wheeler drivers. There is an easily operable GUI which enables the user to choose from either of the options. The sequence diagram for the functioning of the project has been depicted in Fig. 1.

Once user has made a choice, control will be directed to the desired module and user will be asked to upload an image of the traffic scene. The image will be given as input to the selected module, which will further be processed and classification will be performed. Post classification, image will be resulted as a positive case, i.e., image with the presence of helmet/seatbelt, or as a negative class, i.e., absence of helmet/seatbelt.

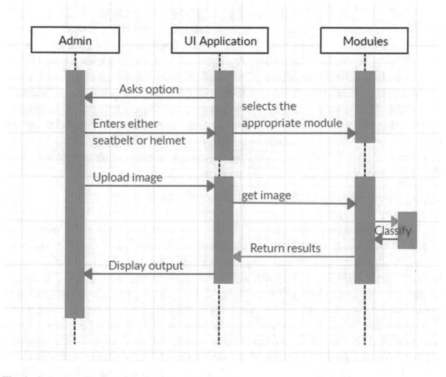

Fig. 1 Sequence diagram for working of our project

Fig. 2 Helmet detected on
subject

The user can then decide to notify the offenders once he has obtained results from our application.

(A)   *Helmet Detection Using YOLO*

The model for detecting helmets on motorcyclists was made on YOLOv3 model [7]. YOLOv3 classifies images in one go and has short inference time. Thus, it was suitable to fit for our model. Additionally, Indian traffic contains a greater number of motorcycle riders than four-wheeler drivers. Hence, the model must be fast for identifying helmet and non-helmet wearers on motorcycle. A bounding box is made on the subject's head area if the confidence of detecting the presence of helmet is above 50%. This ensures that scarfs, masks and other objects resembling the helmet are not mistakenly classified as a helmet by the model. Figure 2 shows the helmet detected case with a bounding box depicting the accuracy of recognition.

Whereas Fig. 3 contains a subject not wearing a helmet and hence no bounding box drawn.

(B)   *Seatbelt Detection Using Convolutional Neural Network*

The seatbelt detection is done using CNN (convolution neural network) initially by preparing image dataset of positive (pictures with seatbelt) and negative (pictures without seatbelt) images, which are pre-processed by subjecting it to various edge detection and thresholding algorithms; post which the image is loaded and resized to square image of 224*224 pixels to train the CNN using Tensorflow as backend. The model is created by developing four convolution layers with four max pooling layers in between. The last fully connected layer with software activation function is added to it in order to get probabilities of prediction. Lastly, the model is trained to evaluate and verify its accuracy and test result by giving the 'fit' and 'predict' command.

For detecting the seatbelt, we applied Canny edge detection algorithm first. Then, the same image was passed through Hough Transform. Now as specified by Huiwen

Fig. 3 No bounding box as helmet is not present on the subject

et al. in [2], seatbelt is usually at a specific angle. So, lines with slopes between 30 to 60 degrees are considered for being distinguished as seatbelts.

As shown in Fig. 4, the subject is wearing a seatbelt, and the same upon detection by the convolutional neural network is highlighted in red. A message indicating the presence of seatbelt is also printed in the console.

Fig. 4 Seatbelt correctly detected by our algorithm

Fig. 5 Seatbelt not present on the subject

Similarly, in Fig. 5 as there in the absence of seatbelt an appropriate message is printed in the console.

4 Results and Discussions

The purpose of our project was to automate the process of detection of traffic rule violators who jeopardize the safety of themselves as well as others by not wearing helmets or seatbelts as a mandatory precaution. Traffic police controllers risk their lives in an attempt to stop the speedy vehicles for interrogation and imposing fine on safety violators. Our project mainly aims at making the process computerized as feeds from CCTV or other sources in the form of video or photos can be processed and drivers not following rules can be identified. The UI of the said application is depicted in Fig. 6.

For this purpose, we have developed an application wherein we have added features to process images/frames and a bounding box is formed if a helmet is detected, whereas on detection of seatbelts highlighted lines are formed. For the same, we have implemented trained neural networks with a significant accuracy for detection. This project work will also help the authorities to minimize man power in order to observe the ongoing traffic.

Features from images were extracted and subjected to KNN Algorithm for comparison with our approach and the following results were obtained.

Our YOLO model shows an accuracy of 92.4% for helmet detection whereas KNN algorithm showed an accuracy of 87.7% (shown in Fig. 7). Similarly, for

Fig. 6 User interface of the application

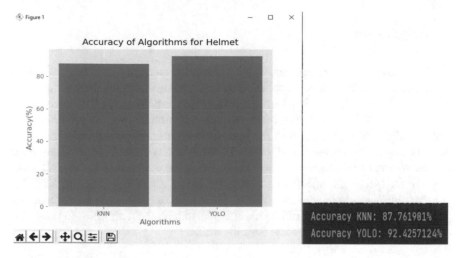

Fig. 7 Comparison of YOLO and KNN for helmet detection

seatbelt detection our trained CNN showed 91.8% efficiency and KNN had 85.7% accuracy (Fig. 8).

The test images were manually clicked images for helmet detection whereas for seatbelt detection there was a combination of manually clicked photos and images obtained from Google. It proves that model is fit for real world application.

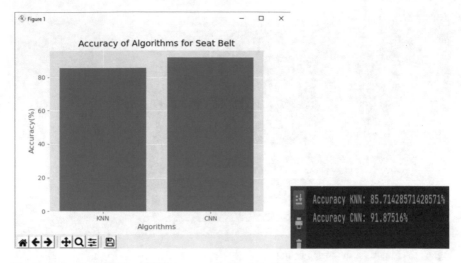

Fig. 8 Comparing CNN and KNN accuracy for seatbelt detection

5 Conclusion

To detect helmet on motorcyclists, we have proposed an approach which with given input image of a traffic scene, applies image processing and machine learning techniques. We have used techniques like Haar Cascade Classifier and LBP transform for face detection, YOLO Architecture for Helmet Detection [8], and Hough Transform and convolutional neural network for Seatbelt Detection. We analysed KNN classifier for comparative purposes and found that our method of operation provided better results. Wearing seatbelts and helmet, while driving car or motorcycles is an effective measure for safety and adequate attention should be given to it. The project helps in improving ways to have watch over traffic for police personnel and introduces a refined way for detection of offenders. It will also help the authorities to minimize man power in order to observe the ongoing traffic.

6 Future Work

The system that we have currently developed has multiple features that benefit the traffic police personnel in apprehending violators not following traffic safety rules. Further, we want to focus more on making the application more user-friendly and host the application on a server so that it can be accessed by authorities via a website too. The above project currently works on images and it needs to be transformed for operation on videos similar to the approach in [9] and [10]. We plan to focus on improving the overall performance of the system. License plate detection of offenders' vehicles [11] in front view is a task that we want to accomplish. The creation of a database that links personal information of a vehicle owner with the vehicle registration number is also a feature we look forward to add and maintain. A system to notify the traffic violators along with photographic proof of location and time of their negligence in not using safety precautions such as helmets and seatbelts would also be a part of our development. We also intend on developing an android application for our model which is a better platform for interaction and each traffic personnel can have it installed in their phones for flexible use.

Acknowledgements We thank our mentor Ms. Divya Jennifer D'Souza for her valuable inputs and guidance. We also extend gratitude to Dr. Niranjan N Chiplunkar, Principal, NMAM Institute of technology and Dr. Uday Kumar Reddy, Head of department, NMAM Institute of technology for their support and giving us an opportunity to conduct this research project in college.

References

1. K. Dahiya, D. Singh, C.K. Mohan, "Automatic detection of bike-riders without helmet using surveillance videos in real-time," Visual Learning and Intelligence Group (VIGIL)
2. H. Guo, H. Lin, S. Zhang, S. Li, "Image-based seatbelt detection," Speech Commun., 978–1–4577–0577–9IEEE, pp. 161–164 (2011)
3. R. Babu, D.R. Amandeep Rathee, K. Kalita, M. Singh Deo, "Helmet detection on two-wheeler riders using machine learning," ARSSS Int. Conf. **41**(4), 603–623 (Sept 2018)
4. C. Vishnu, D. Singh, C. Krishna Mohan, S. Babu, "Detection of motorcyclists without helmet in videos using convolutional neural network", Elsevier, Visual Intelligence and Learning Group (VIGIL), Department of Computer Science and Engineering, Indian Institute of Technology Hyderabad
5. B. Zhou, L. Chen, J. Tian, Z. Peng, "Learning-based seat belt detection in image using salient gradient". Ind. Electron. Appl. (ICIEA) 2017 12th IEEE Conference on, pp. 547–550 (2017)
6. D. Yu, H. Zheng, C. Liu, "Driver's seat belt detection in crossroad based on gradient orientation". Inf. Sci. Cloud Comput. Companion (ISCC-C) 2013 International Conference on, pp. 618–622 (2013)
7. J. Redmon, A. Farhadi, "YOLOv3: an incremental improvement", University of Washington, https://arxiv.org/abs/1804.02767
8. J. Wang, G. Zhu, S. Wu, C. Luo, Worker's helmet recognition and identity recognition based on deep learning. Open J. Model. Simul. **9**, 135–145. https://doi.org/10.4236/ojmsi.2021.92009 (2021)
9. L. Shine, C.V. J., Automated detection of helmet on motorcyclists from traffic surveillance videos: a comparative analysis using hand-crafted features and CNN. Multimed Tools Appl. **79**, 14179–14199 (2020). https://doi.org/10.1007/s11042-020-08627-w
10. X.H. Qin, C. Cheng, G. Li, X. Zhou, "Efficient seat belt detection in a vehicle surveillance application", Ind. Electron. Appl. (ICIEA) 2014 IEEE 9th Conference on, pp. 1247–1250 (2014)
11. Y. Kulkarni, S. Bodkhe, A. Kamthe, A. Patil, "Automatic number plate recognition for motorcyclists riding without helmet", Current Trends Towards Converging Technologies (ICCTCT) 2018 International Conference on, pp. 1–6 (2018)

Prediction of Autism in Children with Down's Syndrome Using Machine Learning Algorithms

D. N. Disha, S Seema, and K. Aditya Shastry

Abstract Many human diseases that have a causative association with genetic components are defined as inherited diseases. It is a very difficult task to identify the different variants linked with inherited diseases using only lab-based technologies. One of the case studies of inherited diseases considered here is autism spectrum disorder. There are various methods are available to detect autism such as expensive screening methods that are used in well-equipped hospitals. But due to cost and also the screening timings, it is not able to efficiently detect the autism disease well in advance. Hence, it is best to utilize machine learning techniques to diagnose the disease well in advance and classify based on the severity using classification methods. One of the best ways to diagnose autism is through questionnaire method. Using machine learning methods, we concentrate on efficient detection of autism disease by using a real-time questionnaire.

Keywords Autism spectrum disorder · Machine learning · K-means clustering · Logistic regression · Support vector machine

1 Introduction

Inherited diseases are the type of disorders that need to be treated with the utmost care, and precision has to be maintained when it comes to its diagnosis. They directly affect the individual's behavior, the way they interact with the outside world, and the way they respond to situations. Autism is one among the several disorders that

D. N. Disha (✉) · K. A. Shastry
Department of ISE, NMIT, Bangalore, India
e-mail: disha.dn@nmit.ac.in

K. A. Shastry
e-mail: adityashastry.k@nmit.ac.in

S. Seema
Department of CSE, MSRIT, Bangalore, India
e-mail: seema.s@msrit.edu

© The Author(s), under exclusive license to Springer Nature Singapore Pte Ltd. 2022
P. Shetty D. and S. Shetty (eds.), *Recent Advances in Artificial Intelligence and Data Engineering*, Advances in Intelligent Systems and Computing 1386,
https://doi.org/10.1007/978-981-16-3342-3_9

an individual can be diagnosed with and can often be confused with Asperger's syndrome and attention-deficit/hyperactivity disorder (ADHD).

There was an incident that took place in Bangalore, Karnataka, India, where an individual, who was actually autistic and was exhibiting the symptoms that fell into the spectrum. But the doctor could not diagnose him in a short span of time, and eventually, was misdiagnosed. The treatment toward the wrongly diagnosed disorder, therefore, did not help the individual in any way, and no improvement was seen in the individual. After this observation, another doctor was approached, where the individual was properly diagnosed with autism through expensive screening techniques. To make sure that this suffering should not be faced by anyone else and provide a cost-efficient, user-friendly solution to the issue, this was taken as a motive for the research.

1.1 Issues and Challenges

The major issues are that the symptoms of autism are not specific. ASD stands for autism spectrum disorder. The term 'spectrum' is used for the disorder as the symptoms vary along a very broad range; hence, prediction of autism becomes difficult. Also, as autism is an inherited disorder, the acquiring of the information needed for the dataset to train the algorithm is extremely difficult as the information is kept confidential by doctors and chooses to not disclose the information about the patient as they consider the data as highly sensitive. Selecting of symptoms in order to yield efficient diagnosis is completely analysis-based, and hence, it is time-consuming.

1.2 Objectives

- Identify behavioral patterns of the inherited diseases by collecting useful inferences from behavioral data.
- To create a budget-friendly, efficient method to detect autism and differentiate from other disorders with maximum accuracy using machine learning.
- To build a simple interface that can be used by anyone to collect data for initial screening purposes.
- To provide a budget-friendly, self-assessment method to parents to detect autism in an individual at initial stages before consulting a neurologist or a pediatrician.

1.3 Architecture of the Proposed Study

Schematic architecture is as follows. Initially, the behavioral data is collected and is tested with several machine learning algorithms. Based on the behavioral patterns detected, efficient prediction can be achieved (Fig. 1).

2 Literature Survey

Najwani Razali et al. [1] proposed that autism is identified by using motor movements of the child; the motor movements considered here are: imitating a finger tapping, clinching a hand. Dataset: Dataset ids are collected using EEG device (a device which is mainly used to record a brain signal) and multilayer perception (MLP) is used as classifier to discriminate between autistic and normal kid. Previous studies: Many researchers concentrated on finding the relationship between motor functions, cognition language development, and social communication disorder. Some concentrated on building a brief report about autism and impairment of motor function. Proposed method: Feature extraction is done by using Gaussian method and Fourier series.

Joaquin Rapela et al. [2] proposed that EOG technology is used as a therapy. Here, computer game is designed, and from the game, eye training will be given to the child. Based on the eye movement, speed and accuracy of eye movement and concentration, a disease can be detected.

Shyam Sundar Rajagopalan et al. [3], in this paper, algorithm is developed for identifying the self-stimulatory behaviors from the public dataset, which contains the video of a child which is tracked by selecting poslet bounding box prediction using the algorithm K-nearest neighbor.

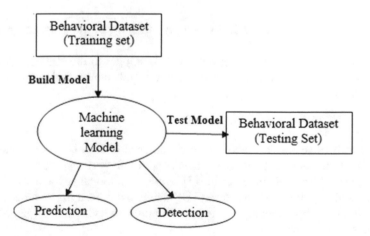

Fig. 1 Schematic architecture

Histogram of dominant motion can be computed using dominant motion flow by using detected body regions. The motion model built using this descriptor can be used for detecting self-stimulatory behavior.

The proposed study uses postlets and is used to identify bounding box predictions. In related studies, two broad categories have been identified in self-stimulatory behaviors.

In the first category, they use wearable sensors either in the hands or the body which help in tracking of the sensor data for fixed time period. From the results of tracked body parts, we can analyze repetitive self-stimulatory behaviors. Wetyn et.al used wearable wristband in order to track the hand motion, and they built Markov model using for classifying the behaviors.

a.   Pdoetz et.al with the help of sensors fixed to the legs, acceleration data can be obtained. Obtained data can be segmented to generate the behavior episodes which can be used for extracting the features in order to train the model which can be used for identifying the aggression, disruption, and self-injury behavior.

Second category includes tracking the child's body part motion in a video and then analyzing the track to detect the behaviors.

a.   Marwa et.al identified a new descriptor to find and localize the rhythmic body movements by using color and depth images.

Anna Sudar [4] used Sulinet (an educational webpage). Web page usability and face perception patterns in the population of people with autism spectrum disorders are noted.

a.   Face recognition in typical development: Usually, infants with normal condition recognize the human face and gaze at them, but this would not be in autism-affected children.

Several researches have been done by using eye-tracking movements. Latest is based on stimuli and stimulus dataset.

Some other researchers have given a movie snippet to the children. A normal child could able to recognize but autism child was only concentrating on mouth and surroundings.

b.   Eye tracking and website ergonomics

Allowed to browse for free, and how children scan different faces are examined. Autistic child treats everything as an object, and it fails to detect face.

Alejandra Ornelas Banrajas et al. [5]. Here, serious game for the children is prepared with tangible user interface (TUI) and geographical user interface (GUI). Later, effects of using serious games in play therapy are noted. Common game therapy is 'building block toys.'

Arodami Choriannopoulou et al. [6]. Here, the degree of children interaction with their parents is identified. All the features that are obtained from the participants are identified such as acoustic and dialogs. Visual investigations are also performed by observing the recorded video sessions consisting of the tasks performed by the children. From the video, patterns related to autism can be derived.

Mohana et al., a checklist called M-CHAT is used for the children of age between 16 to 30 months. By using the checklist, best classifier with reduced features can be found and which is used for autism risk level prediction. At last performance, evaluation for all the classifiers is evaluated.

Aim is to find toddler autism level by using minimum attributes.

Mythili et.al., discussed various symptoms of autism spectrum disorder and classification models are used for the detection of levels of ASD. They considered autism student database and considered the factors with respect to language, behavior, social interaction, and levels of autism. With the help of an open-source tool called WEKA, analysis, and interpretation of the classification is done, and performance analysis of the students affected with autism are calculated.

Wenbo Liu et al., here, autism is predicted by continuous observation of face scanning and eye movement data ASD prediction is done based on the face scanning eye movement data. Eye movements were recorded using Tobii T60 eye tracker. A framework for prediction is proposed.

Jianbo Yuan et al., in this paper, autism detection is done with the help of NLP based on the information obtained from medical forms of autistic patients. Here, they converted unstructured and semi-structured medical forms into a readable digital format, preprocessed it and finally done with classification.

3   Study Area and Methodology

The area of concentration is on the health care, and, in particular, the prediction of inherited diseases called autism spectrum disorder well in advance. Early prediction of the inherited disease is important as it helps the child to get the treatment and therapy.

Steps are as follows.

3.1   Collection of Data

i.   Dataset to train the model

It is one of the important steps as it involves collection of behavioral patterns of autistic patients. Neurologists, pediatricians and Head of Autistic schools were consulted to help us in order to collect the dataset. Due to the confidentiality issue, none of the doctors and medical institutions provided the data. Hence, a dataset for autism has been taken from UCI repository. https://archive.ics.uci.edu, which is created by Fadi Fayez Thabtah, was sufficient to train the machine learning algorithms.

ii.   **Real-time data collection for testing model**

After the data was collected from the UCI repository, in order to make the algorithm usable with the data that exists in the real world, data was collected from peers and was circulated to the consulted doctors, and also to the autistic centers. This was made possible by hosting a web page online, which consisted of a questionnaire, that is, a set of questions that had exactly two answers, yes and no. The questionnaire was formed in such a way that it included all the attributes that could determine if the individual is autistic or not. The responses were recorded onto an Excel sheet, where the final score was calculated and fed into the algorithm for testing purposes.

The data which is recorded is based on the responses provided by people who have either been diagnosed for autism or show behavioral symptoms of autism. The main intention of this paper is to make sure that autism is diagnosed accurately, and not be mistaken for any other disorder. The treatment for a misdiagnosed disorder will be of no significance and will be rendered useless. In order to accurately achieve results, machine learning has been incorporated. With the usage of algorithms, the aim is to achieve an accurate diagnosis based on the answers received. The most accurate algorithm among all the algorithms used will be selected as the optimum solution to detect autism.

3.2   *Data Cleaning*

Data cleaning is very important as it makes it easier to feed the algorithm with data and achieve the expected results. When any form of a dataset is created or retrieved, the data is not perfect. It contains values that are not essential for the computation of results, and there are values that are most needed for the computation of results but do not exist in the dataset. The dataset had to be restructured. In the dataset, all the missing values were filled with the mean of all the values in the attribute's column and all the duplicate values were merged together. Outlier values were also eliminated from the dataset, and the attributes that did not make a difference to the computational process were removed, leaving with a limited number of attributes that had complete records, thus making it convenient for the algorithm to train with.

Among the 21 attributes, the unnecessary attributes were removed as they only took up space and did not contribute to the computation of the data. The scores of 10 questions, age, gender, ethnicity, jaundice, autism history, country and result, were retained. All the spelling errors were corrected, and the missing values were filled with the mean of all the data in that column. All the outliers were discarded. In this manner, the crude dataset was refined and transformed into a usable state. Figure shows all the missing data have been substituted and all the irrelevant features or attributes have been removed (Fig. 2).

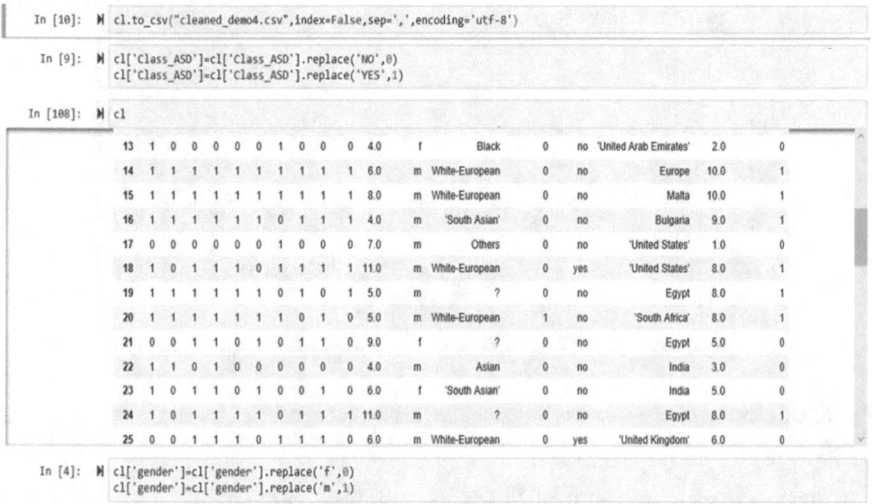

```
In [10]:  ▶  cl.to_csv("cleaned_demo4.csv",index=False,sep=',',encoding='utf-8')

In [9]:   ▶  cl['Class_ASD']=cl['Class_ASD'].replace('NO',0)
             cl['Class_ASD']=cl['Class_ASD'].replace('YES',1)

In [108]: ▶  cl
```

13	1	0	0	0	0	0	1	0	0	0	4.0	f	Black	0	no	'United Arab Emirates'	2.0	0
14	1	1	1	1	1	1	1	1	1	1	6.0	m	White-European	0	no	Europe	10.0	1
15	1	1	1	1	1	1	1	1	1	1	8.0	m	White-European	0	no	Malta	10.0	1
16	1	1	1	1	1	1	0	1	1	1	4.0	m	'South Asian'	0	no	Bulgaria	9.0	1
17	0	0	0	0	0	0	1	0	0	0	7.0	m	Others	0	no	'United States'	1.0	0
18	1	0	1	1	1	0	1	1	1	1	11.0	m	White-European	0	yes	'United States'	8.0	1
19	1	1	1	1	1	1	0	1	0	1	5.0	m	?	0	no	Egypt	8.0	1
20	1	1	1	1	1	1	1	0	1	0	5.0	m	White-European	1	no	'South Africa'	8.0	1
21	0	0	1	1	0	1	0	1	1	0	9.0	f	?	0	no	Egypt	5.0	0
22	1	1	0	1	0	0	0	0	0	0	4.0	m	Asian	0	no	India	3.0	0
23	1	0	1	1	0	1	0	0	1	0	6.0	f	'South Asian'	0	no	India	5.0	0
24	1	0	1	1	1	1	0	1	1	1	11.0	m	?	0	no	Egypt	8.0	1
25	0	0	1	1	1	0	1	1	1	0	6.0	m	White-European	0	yes	'United Kingdom'	6.0	0

```
In [4]:   ▶  cl['gender']=cl['gender'].replace('f',0)
             cl['gender']=cl['gender'].replace('m',1)
```

Fig. 2 UCI Dataset after cleaning

3.3 Implementation Process to Build a Model (Module 1 Dataset Using UCI Repository)

With the help of UCI repository data, the following algorithms were executed on the training dataset.

A. k-Means Clustering

K-means clustering is one of the most important algorithms that is used to solve clustering problems. The way the algorithm works is by classifying a given dataset into a certain number of clusters. The number of clusters is denoted by k, which means that the entire dataset is divided into k clusters with k centers. The clusters are being placed far away from each other in order to obtain maximum accuracy based on the dataset fed into the algorithm. Each point is placed to the nearest center. New centers have to be recalculated based on the clusters in the previous iteration. The k centers change their location until no more changes are possible. For the dataset, the attributes 'age,' 'gender,' 'jaundice,' and 'result' have been used to divide the dataset into two major clusters, namely autistic and non-autistic (Fig. 3).

Observation: The algorithm was successful in classifying into two main classes, namely autistic and non-autistic. The red dots represent the autistic cluster and the green dots represent the non-autistic cluster. There is a graphical representation of the outcome of the K-means Clustering.

B. Logistic Regression

It is the most widely used technique for performing binary classification problems. It identifies the correlation between the dependent variable and independent variables.

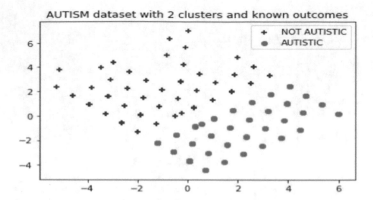

Fig. 3 UCI dataset after K-means clustering (autistic and non-autistic clusters)

It computes a weighted sum of the input variables and runs the result through a nonlinear function that is the sigmoid function in order to produce the result of the computation. With the help of logistic regression, the output of the algorithm can be converted into a class variable, that is 0, which is 'no', and 1, which is 'yes,' and for this, the sigmoid function is used.

The UCI repository dataset was split into a training set and a testing set. The features that were used to train the algorithm are age, gender, jaundice, and result.

On training the algorithm and using the training the set and feeding the testing set, an accuracy of 94.52% was obtained.

C.   Support Vector Machine

It is a very fast and a dependable classification algorithm that is efficient, even with a concise dataset. In the dataset, there are two major tags, autistic and non-autistic. From the data-set, two main features are being considered, that is 'jaundice' and 'result' attributes. An SVM takes these points and produces a hyperplane, which is a decision boundary that helps in classifying data points, thus dividing the dataset into two classifications, autistic and non-autistic. In SVM, the linear function's output is considered, and if it exceeds 1, it is identified as one class, and if it is −1, it is identified as the other class, thus giving a range of ($[−1,1]$) which becomes the margin. A margin is the separation of the line to the closest class points. A margin is said to be considerable when the separation is large for both of the classes. The points remain in the respective classes without merging into the other class. A bad margin is when the separation is small and where there are points, which are supposed to be a part of one class, exists in the other.

Observation: A hyperplane is produced, which separates both the classes. The features that were selected for this was 'Jaundice' and a combination of results gathered from UCI dataset parameters. Any score that ranged from 0 to 6 are classified as non-autistic individuals and any score that exists beyond 6 are classified as autistic. A function was created, that allows an individual to enter the values for the features

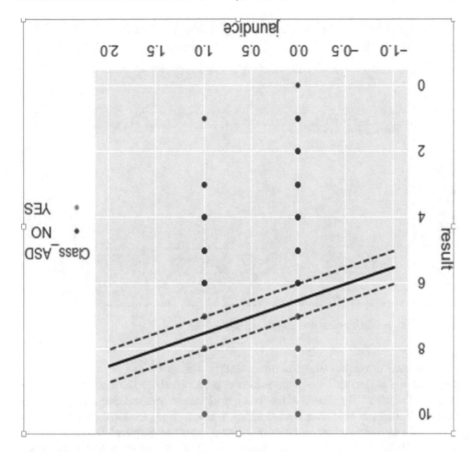

Fig. 4 SVM graph with hyper plane and intermediate values

Jaundice and the combined result parameters in order to check if the algorithm is classifying it accurately or not. Finally, the accuracy with UCI dataset came up to 83.3% (Figs. 4 and 5).

3.4 Implementation Process to Build a Model (Module 2 Real-Time Dataset)

A.   **Module 2 Dataset**.

This phase involves the collection of data from the public, which includes parents of autistic and non-autistic children, students and faculty members of autistic schools belonging to various parts of the country, and also doctors of some hospitals. The data

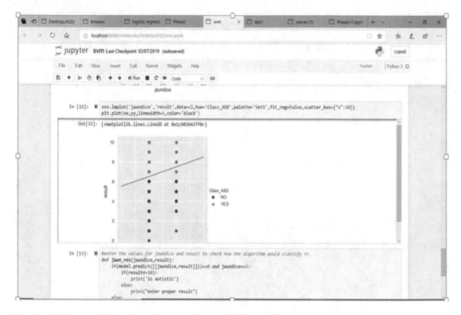

Fig. 5 SVM graph classifying into autistic and non-autistic

was collected through a Google Form questionnaire that was integrated into a web page and was hosted online. All the entries were recorded onto an Excel sheet, and on cleaning the data, they were fed into the algorithms for computation. Along with the existing features, more features were added. These features feed in data related to the history of autism in an individual's family, data regarding physical disabilities of the individual, and the individual's reaction in a social environment, which is classified as 'socially uncomfortable,' 'normal,' and 'extreme' behavior (Figs. 6 and 7).

The algorithms that were used for the Module 1 data are being used for the Module 2 data as well. Fabrication of data will lead to the yielding of results that might be too good to be true; hence, a waiting period was kept to hear back from the people who have received the questionnaire. More than 100 entries were received and were used to train the algorithm. This methodology can be used by an individual, or an autistic clinic/school, or by hospitals to diagnose or detect autism accurately in a cost-efficient manner.

For this data, k-means algorithm has been used, and the features that have been selected are 'Behavior in Social Environments (BISE)' and 'result.' In the above figure, two centroids have formed with two different types of clusters (Fig. 8).

After implementing k-means algorithm, the k-nearest neighbors (KNN) algorithm was implemented in order to classify the data into two distinct classes, yielding an accuracy of 92.78.

Logistic regression was also used for the given dataset. Due to the huge number of combinations of the dataset, the accuracy that has been achieved is lower than the accuracy of the dataset retrieved from the internet. Initially, the accuracy of the data

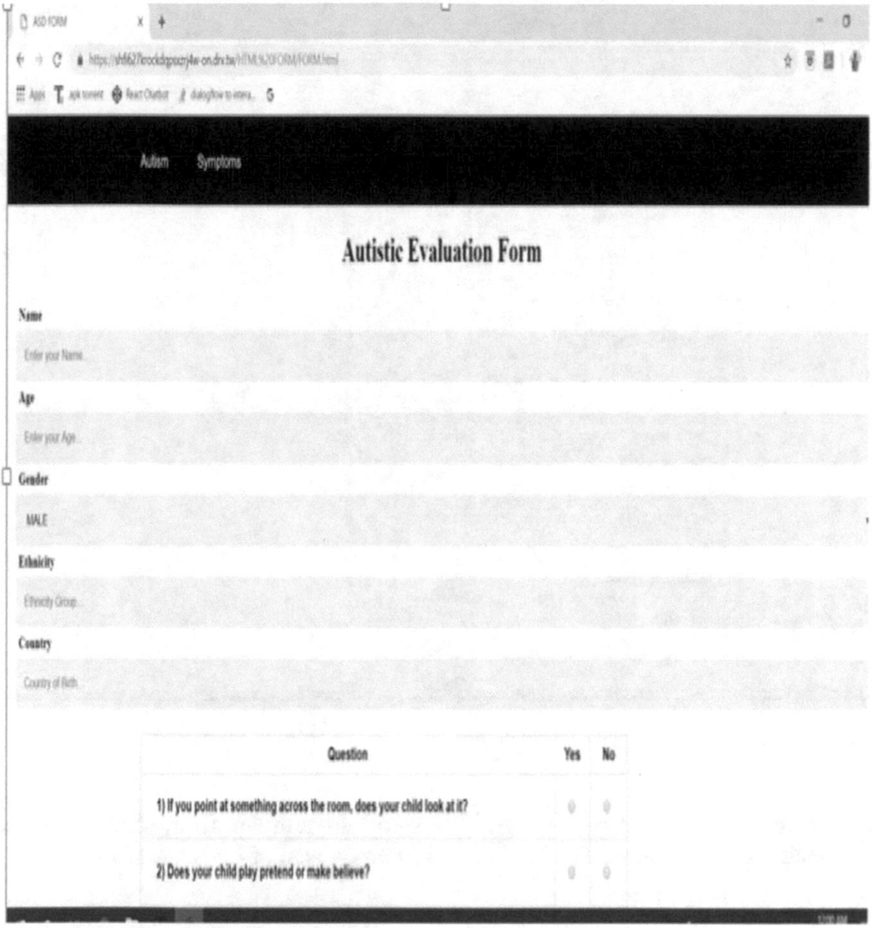

Fig. 6 Web page with autistic questionnaire

was below the half-way mark, but after the processing of data and clearing out the unwanted information, the algorithm yields an accuracy of 96.0. The SVM was also fed with the live collected data, which yielded an accuracy of 88%.

For the Module 2 data, three more algorithms were used to examine whether the algorithms could be trained with the newly collected data and perform accurate predictions. The accuracies of the algorithms exclusively used for the Module 2 data are listed in Table 2 and also compared with the accuracies of the same algorithms used in the research paper [6] (Table 1).

The accuracies of Module 2 data when compared to the research paper are:

A comparative study of the accuracies that were achieved had to be done and to carry that out, the K-means clustering algorithm, the support vector machine

Fig. 7 Questions to assess autistic individuals

algorithm and the naive Bayes algorithm were programmed from scratch and the accuracies were obtained. The Table 3 has a comparison of the accuracies.

4 Conclusion and Future Work

After running through all the different types of data, the algorithms that have been implemented have been trained efficiently and are successfully differentiating autistic individuals from non-autistic individuals. Also, these algorithms have been yielding a higher accuracy when compared to the accuracy obtained by the authors of the research paper that was referred, thus, improving the efficiency and accuracy of the algorithms. Logistic regression and K-nearest neighbor algorithms yielded the highest accuracy among the algorithms used for Module 2 data, thus narrowing down the algorithms that can be utilized. This becomes a cost-efficient, user-friendly tool to help doctors as well as a layman to determine whether someone is autistic or on the verge of being autistic or not. The reach among the public can be easily expanded with the help of the web page to collect the data as computers are available in almost every corner of the world. Smartphones also have not been eliminated

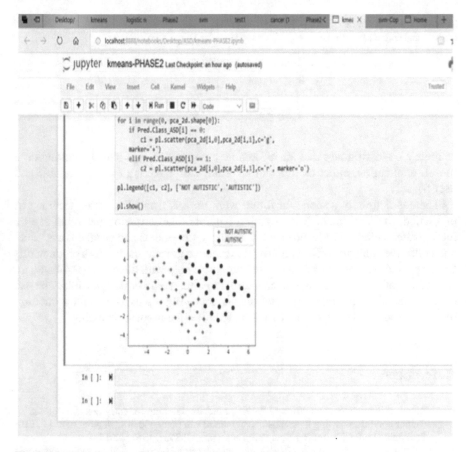

Fig. 8 Module 2 data after K-means clustering

Table 1 Accuracy of algorithms used for module 1 and module 2 data

Algorithm	Module 1 Data (%)	Module 2 Data (%)
K-means	57.19	92.7
SVM	–	88
Logistic regression	94.52	96

Table 2 Accuracy of algorithms used for module 2 data only

Algorithm	Module 2 Data (%)	Results in paper [6] (%)
K-nearest neighbors	96.00	–
Random forest	94.25	85.1
Naive Bayes	93.33	86.5

Table 3 Comparison of built-in and coded algorithms

Algorithms	Built-in algorithm (%)	Algorithm from scratch (%)
K-means classifier	92.70	92.70
Support vector machine	88	70
Naive Bayes	93.33	84

as there are smartphones that are affordable by everyone, and hence, working on a mobile application, either on iOS and Android, will make the module portable and user friendly.

There is definitely a scope for future work on this front. For better accuracy of prediction, the implementation of neural networks and attention networks can be incorporated. With this advancement, a huge amount of data can be taken in at once, and collective analysis can be carried out to produce greater results. As a product, this can be passed on to various medical institutions to help with the diagnosis of autism. Integration of hardware components in order to obtain data is also possible. The use of retina scanners and facial recognition cameras will help us to obtain information by processing the data with the help of techniques like image processing.

References

1. H. Abbas, F. Garberson, E. Glover, D.P. Wall, Machine learning for early detection of autism (and other conditions) using a parental questionnaire and home video screening. In 2017 IEEE International Conference on Big Data (Big Data), p. 35583561. IEEE (2017)
2. S.R. Dutta, S. Giri, S. Datta, M. Roy, A machine learning-based method for autism diagnosis assistance in children. In 2017 International Conference on Information Technology (ICIT), p. 3641. IEEE (2017)
3. W. Liu, X. Yu, B. Raj, L. Yi, X. Zou, M. Li, Ecient autism spectrum disorder prediction with eye movement: a machine learning frame—work. In 2015 International Conference on An active Computing and Intelligent Interaction (ACII), p. 649655. IEEE (2015)
4. M.J. Maenner, K. Van Naarden Braun Marshalyn Yeargin-Allsopp, D.L. Christensen, L.A. Schieve, Development of a machine learning algorithm for the surveillance of autism spectrum disorder. In—PLoS One (2016)
5. A. Pahwa, G. Aggarwal, A. Sharma, A machine learning approach for identification diagnosing features of neurodevelopmental disorders using speech and spoken sentences. In 2016 International Conference on Computing, Communication and Automation (ICCCA), p. 377382. IEEE (2016)
6. B. van den Bekerom, Using machine learning for detection of autism spectrum disorder. In—IEEE (2017)

Speech Emotion Recognition Using K-Nearest Neighbor Classifiers

M. Venkata Subbarao⊙, **Sudheer Kumar Terlapu, Nandigam Geethika, and Kudupudi Durga Harika**

Abstract The development of wide verity algorithms in artificial intelligence makes it easier in the identification of the emotion of a user. Speech emotion recognition (SER) systems may create a great impact on the working environment by timely identifying the mood of employees. Besides, it is also a key technology in next-generation security applications. This paper presents a verity of K-nearest neighbor (KNN) classifiers for automatic recognition of emotions in speech. For SER, a set of features such as mel-frequency cepstrum coefficients (MFCC) and its temporal derivatives are considered for training the proposed KNN classifiers. Performance analysis is carried on the Berlin dataset with different training rates and with different noisy conditions. Simulation results depict the superiority of the proposed classifier than the existing methods.

Keywords Emotion recognition · KNN classifiers · Distance function · Majority selection

1 Introduction

Emotion plays an important part in everyday social human relations. This is crucial to everyone's rational and smart decisions. It aids everyone to match and recognize the moods of others by passing on their feelings and giving response to others. The study has exposed the role that emotion shows in determining human social communication. Emotional presentations carry significant information about the psychological state of an individual. This makes a new research field in SER to know the desired emotions of a speaker. In the beginning, several researchers came up with a verity of approaches to recognize the speaker's emotions such as physiological signals [1], speech [2], and facial expressions [3]. The majority of investigators are showing interest in speech because of its easiness than other approaches. SER has many applications such as clinical studies, interface with robots, computer games,

M. Venkata Subbarao (✉) · S. K. Terlapu · N. Geethika · K. D. Harika
Department of Electronics and Communication Engineering, Shri Vishnu Engineering College for Women(A), Bhimavaram, AP, India

© The Author(s), under exclusive license to Springer Nature Singapore Pte Ltd. 2022   123
P. Shetty D. and S. Shetty (eds.), *Recent Advances in Artificial Intelligence and Data Engineering*, Advances in Intelligent Systems and Computing 1386,
https://doi.org/10.1007/978-981-16-3342-3_10

Table 1 Recent SER techniques [4–10]

Ref. no.	Features	Classifier	No. of emotions	Accuracy	Year
4	Pitch, format, phoneme	SVM, HMM	3	75–80	2017
5	LPCC, MFCC	GMM, HMM, SVM	5	82–93	2007
6	Mel spectrogram, harmonic percussive	DNN, KNN	9	90	2018
7	MFCC	SNN	6	83	2019
8	MFCCAcceleration, velocity	DTCNNLSTM	7	80	2017
9	Prosody,quality	SVMPSVM	4	80.7586.75	2017
10	Linguistic and non-linguistic	SVM,DT	6	84.39	2018

cardboard systems, banking, audio surveillance, commercial applications, and entertainment. For laboratory experimentation or online learning, improving the quality of teaching data about the emotional state of students plays a crucial role. SER may help an instructor for developing the new strategies to manage student emotions in the classroom. Some of the recent works on SER is tabulated in Table 1.

From the literature, it is observed that many authors considered a limited number of voice data in the dataset. So that the accuracy achieved is about 80%. There are many features in order to recognize emotion. But the existing techniques use a limited number of features to recognize emotion. For the existing technique, the classification accuracy is calculated only for fixed training rates. Consider these points as input; this paper analyzes the performance with various training rates.

The rest of the paper organization is as follows. Section 2 describes feature extraction and framework of the approach. Section 3 gives the idea about the proposed approach with necessary equations. Section 4 presents the simulation results. Finally, Sect. 5 concludes the work.

2 Framework

In recent approaches, AI techniques like machine learning and deep learning are used for SER. While considering the existing techniques, the performance is degrading due to external disturbances such as noise and environmental conditions. In order to increase the performance of a system, the evaluation is carried out under different noisy conditions with the help of MFCC features and proposed classifiers. The extraction of MFCC features from a voice signal is shown in Fig. 1

Figure 2 represents the framework of our SER systems. It consists of training and testing phases.

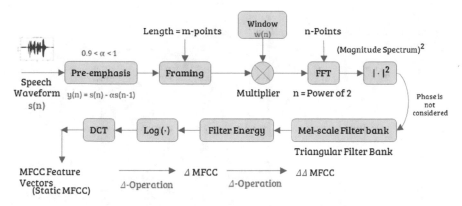

Fig. 1 Extraction of MFCC features

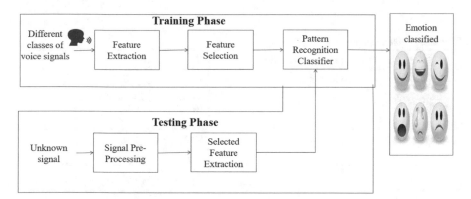

Fig. 2 Framework of SER system

3 KNN Classifiers

KNN classification is supervised classification approach. Figure 3 represents an example of KNN algorithm. It is a two-set problem, so it contains two reference classes, Class A and Class B. X1 and X2 are the two features of the classes, which are used in classification. The triangles indicate Class A, rectangles indicate class B, and the unknown class is represented by star. In KNN, firstly it measures the distance between all the reference classes to unknown class by using predefined distance function.

For the given example, initially K value is taken as 3 and it checks for the 3 nearest neighbors, the majority class among 3 is in Class B so the predicted class is Class B. For the next case, K value is changed to 6 so it checks for the 6 nearest neighbors and the majority in the group is Class A, so the predicted class is Class A. From this example, it is noted that K value and distance function affects the performance of the classifier. The best choice of K depends upon the data.

Fig. 3 Example of
recognition using KNN
algorithm

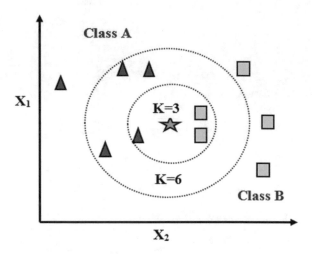

Table 2 KNN classifiers

KNN classifier	Distance function	K	d
Fine	Euclidean	1	$\sqrt{\sum_{i=1}^{n}(p_i - q_i)^2}$
Medium		10	$\sqrt{\sum_{i=1}^{n}(p_i - q_i)^2}$
Cosine	Cosine	10	$\frac{\vec{P}.\vec{Q}}{|\vec{P}||\vec{Q}|}$
Cubic	Cubic		$\left(\sum_{i=1}^{m}|p_i - q_i|^3\right)^{\frac{1}{3}}$
Weighted	Distance weighting		$w = \frac{1}{d^2}$

Based on the distance function and K value, the KNNs are subcategorized into different types shown in Table 2 [11]. Here, d is the formula for distance function.

4 Simulation Results

Table 3 represents the simulation parameters for the simulation environment. It consists of seven emotions which are anger, anxiety, boredom, disgust, happiness, sadness, and neutral. The performance is measured with additive white Gaussian noise as disturbance ranging from 0 to 40 dB. The size of dataset is 80000 × 82 which consists of 7 emotions with 39 statistical features with one label. The dataset for training is ranging from 90 to 50%, and the dataset for testing is ranging from 10 to 50%.

Table 3 Simulation parameters

Parameters	Description
Number of emotions	7
Emotion classes	1. Anxiety/fear 2. Anger 3. Boredom 4. Disgust 5. Happiness 6. Sadness 7. Neutral
Features	MFCC, MFCC Δ, and MFCC ΔΔ
Dataset	80,000*39
Rate of training	50–90%
Rate of testing	10–50%

Figure 4 represents the original dataset for the training phase at 0-40 dB SNR for MFCC at 90% training.

Figure 5 represents the performance of fine KNN. Figure 5a shows the confusion matrix, and Fig. 5b represents the accuracy of each emotion at 90% training. The overall accuracy obtained from the fine KNN is 90.1%.

Figures 6 and 7 represent the performance of medium and cosine KNN classifiers, respectively. The overall accuracy obtained from these two are 63.4% and 65.6%. From Fig. 6, it is observed that the medium KNN has very low accuracy in identifying the disgust emotion of a speaker, and the overall performance is poor as that of fine KNN.

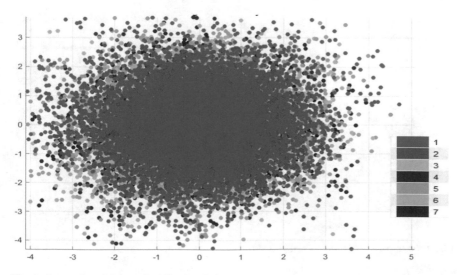

Fig. 4 Seven class dataset points for simulation

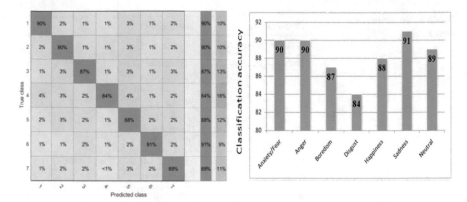

Fig. 5 Performance of fine KNN

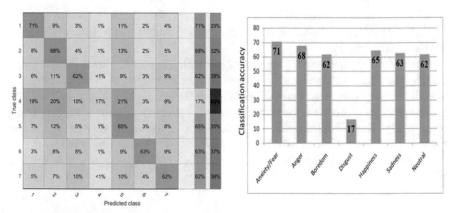

Fig. 6 Performance of medium KNN

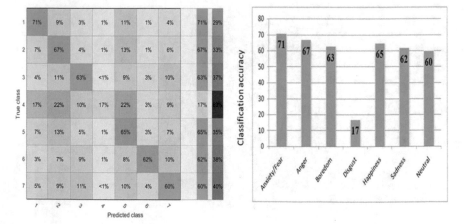

Fig. 7 Performance of cosine KNN

Table 4 Performance at different training rates

Classifier	Training rate (%)							
	50	55	60	65	70%	75	80	85
Fine KNN	75.9	77.7	79.3	81.2	83.1	84.7	86.5	88.0
Medium KNN	60.0	60.8	61.3	62.2	62.5	62.6	63.1	63.5
Cosine KNN	62.1	63.0	63.8	64.1	64.7	65.0	65.4	65.8
Cubic KNN	59.8	60.0	60.6	61.4	61.9	62.2	62.8	62.7
Weighted KNN	76.1	78.0	79.8	81.5	83.3	84.9	86.5	87.9

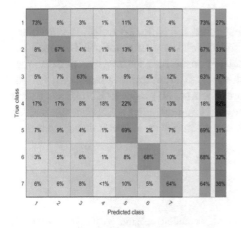 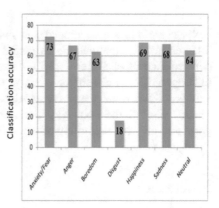

Fig. 8 Performance of cubic KNN

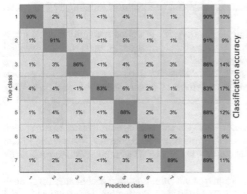 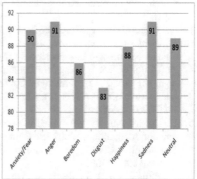

Fig. 9 Performance of weighted KNN

Figures 8 and 9 represent the performance of cubic and weighed KNN classifiers, respectively. The overall accuracy obtained from these two is 62.8 and 88.9%. Similarly, the performance of KNN classifiers with 50% to 85% training rates and 15–50% testing rates are shown in Table 4. From the simulation, it is shown that fine KNN and weighed KNN classifiers achieve better accuracy in recognition of emotion in a speech even at less training rates.

5   Conclusion

KNN classifiers are developed and investigated for SER. To train the classifiers, different MFCC features are extracted from each class of emotion. To verify the performance of classifiers, Berlin dataset is considered. Further, the performance of proposed KNN classifiers is analyzed under different training rates with various values of SNR. Finally, a performance comparison of all proposed KNN classifiers is made. Out of all the KNN classifiers, fine KNN and weighed KNN classifiers attain good accuracy in recognition of emotion in a speech even at less training rates.

References

1. H. Ali, M. Hariharan, S. Yaacob, A.H. Adom, Facial emotion recognition using empirical mode decomposition. Expert Syst. Appl. **42**(3), 1261–1277 (2015)
2. Z.T. Liu, M. Wu, W.H. Cao, J.W. Mao, J.P. Xu, G.Z. Tan, Speech emotion recognition based on feature selection and extreme learning machine decision tree. Neurocomputing **273**, 271–280 (2018)
3. M . Ragot, N. Martin, S . Em, N. Pallamin, J.M. Diverrez. Emotion recognition using physiological signals: laboratory vs. wearable sensors. In: International Conference on Applied Human Factors and Ergonomics. Springer, pp. 15–22
4. M.S. Likitha, S.R.R. Gupta, K. Hasitha, A.U. Raju, "Speech based human emotion recognition using MFCC," 2017 International Conference on Wireless Communications, Signal Processing and Networking (WiSPNET), Chennai, pp. 2257–2260 (2017), https://doi.org/10.1109/WiSPNET.2017.8300161.
5. H. Hu, M. Xu, W. Wu, "GMM supervector based SVM with spectral features for speech emotion recognition," 2007 IEEE International Conference on Acoustics, Speech and Signal Processing - ICASSP '07, Honolulu, HI, pp. IV-413-IV-416 (2007). https://doi.org/10.1109/ICASSP.2007.366937
6. K. Tarunika, R.B. Pradeeba, P. Aruna, "Applying machine learning techniques for speech emotion recognition," 2018 9th International Conference on Computing, Communication and Networking Technologies (ICCCNT), Bangalore, pp. 1–5 (2018). https://doi.org/10.1109/ICCCNT.2018.8494104
7. E. Mansouri-Benssassi, J. Ye, "Speech emotion recognition with early visual cross-modal enhancement using spiking neural networks," 2019 International Joint Conference on Neural Networks (IJCNN), Budapest, Hungary, 2019, pp. 1–8. https://doi.org/10.1109/IJCNN.2019.8852473
8. S. Basu, J. Chakraborty, M. Aftabuddin, "Emotion recognition from speech using convolutional neural network with recurrent neural network architecture," 2017 2nd International Conference

on Communication and Electronics Systems (ICCES), Coimbatore, pp. 333–336 (2017). https://doi.org/10.1109/CESYS.2017.8321292

9. Z. Han, J. Wang, "Speech emotion recognition based on Gaussian kernel nonlinear proximal support vector machine," 2017 Chinese Automation Congress (CAC), Jinan, pp. 2513–2516 (2017). https://doi.org/10.1109/CAC.2017.8243198

10. J. Mao, Y. He, Z. Liu, "Speech emotion recognition based on linear discriminant analysis and support vector machine decision tree," 2018 37th Chinese Control Conference (CCC), Wuhan, pp. 5529–5533 (2018). https://doi.org/10.23919/ChiCC.2018.8482931

11. M. Venkata Subbarao, P. Samundiswary, "K—nearest neighbors based automatic modulation classifier for next generation adaptive radio systems", Int. J. Secur. Appl. (IJSIA), NADIA, **13**(4), 41–50 (2019)

Object Detection and Voice Guidance for the Visually Impaired Using a Smart App

Ramya Srikanteswara, M. Chandrashekar Reddy, M. Himateja, and K. Mahesh Kumar

Abstract In the world, there are around 300 million people ("World Sight Day 2017:" Available: https://www.indiatoday.in/education-today/gk-current-aff airs/story/world-sight-day-2017-facts-and-figures-1063009–2017-10–12 (2017) ["World Sight Day 2017:" Available: https://www.indiatoday.in/education-today/ gk-current-affairs/story/world-sight-day-2017-facts-and-figures-1063009-2017-10-12 (2017)]), who have problems with vision, among those 40 million are completely blind and around 260 million have poor vision due to some cases being partial or complete visual disability. And among them around 95 percent of these stay in developed countries, where they find it very difficult to perform basic day-to-day activities like commuting. They are unable to read traffic warning signs and regulatory signs, also they cannot read the information signs to know their exact position and they often rely on other pedestrians to guide them to their destination. This proposed model aims to provide a method to solve this issue through an application that contains an image recognition system that detects nearby objects in surroundings. It makes the lives of visually impaired better by making them independent.

Keywords Object detection · Voice guidance · Visually impaired · Text to speech

1 Introduction

Human beings have the most useful senses, the eyes that help them look around the world. But there are many people around the world who are unfortunate and lack vision. Millions of people [1] around the world are dealing with the incapacities of understanding and navigating the environment due to partial or complete blindness. Even though they can challenge to perform their day-to-day work, they might possibly have difficulty in following or directing a route as well as social incommunicado. It is very difficult for them to find particular things in unexpected, unfamiliar, and

R. Srikanteswara (✉) · M. C. Reddy · M. Himateja · K. M. Kumar
Nitte Meenakshi Institute of Technology, Bengaluru, India
e-mail: ramya.srikanteswara@nmit.ac.in

© The Author(s), under exclusive license to Springer Nature Singapore Pte Ltd. 2022
P. Shetty D. and S. Shetty (eds.), *Recent Advances in Artificial Intelligence and Data Engineering*, Advances in Intelligent Systems and Computing 1386,
https://doi.org/10.1007/978-981-16-3342-3_11

dangerous environments. Blind and visually impaired people find the maximum difficulty than any other is in knowing whether a person is nearby or not. Computer vision plays an important role, by providing various algorithms for detecting objects. They can be categorized into two main methods, deep learning approach [2, 3] and the conventional approaches [4–6]. Computer vision technologies with unprecedented power of the deep convolutional neural network or simply ConvNets combined with computing power opened huge possibilities in computer vision. It is suitable to use the current technologies to give a ray of hope to the visually challenged people. The proposed model explores the possibility of helping the blind, by familiarizing them, with the world around them.

2 Background

From the past few decades, as image processing algorithms became more powerful, a number of technical projects around the world have been developed which help the blind. One of the methods [7] made use of the concept of local features extraction. Pictorial representation of the surrounding objects for these people makes use of recognizing an object in the video scene where state-of-the-art techniques like key points mining required data and structures matching are used for object identification. For simulating the results, SFIT algorithm was used. The method was successful in detecting the objects around. But the drawback of this method was that the outputs depend on the quality of the image.

Researchers have brought up wide-variety solutions like navigation systems [8] which are effectively inexpensive. This guides the visually impaired by sensing the hindrances in front of the person using sensors and microcontrollers. One more technique effectively recognizes the objects with wearable solutions [9] that uses microcontrollers. These are configurable and instantaneous objects that keep track of the surrounding to and give appropriate output. In this technique, a NOIR camera is being used. The TensorFlow API helps in creating different models which can identify and classify various objects. The Raspberry Pi3 helps in with the system. The system used may be quite expensive, and out of reach of common man.

When the visually impaired people want to lead a normal life, they need to be independent. Few models have been implemented, which help them in moving around. One of the procedures [10] makes use of IoT and Raspberry Pi to deliver sound direction over crosswalks. Recognition of image was made use of in this procedure. Identifying the light of the cross-walk images of significant situations was used as input data for training. These were captured and saved during the day and in the dark. The training of the model was done by Google Object Detection API. This had a limited dataset.

In this era, communication plays an important task. Data is available in most places, in the role of written text. It is a problem for the visually impaired people as they miss out this information. A technique [11] was introduced where the text was converted to speech. The main modules included voice module for searching, image module, and voice output processing module. Raspberry Pi and sensors were used here. This project mainly dealt with limited trained data, i.e., the bus number and bus name details. A text to read system [12] was described in one of the projects. This was helpful for the visually challenged people. This module was implemented using the OCR and scanner. This device was small scale, less expensive, and portable device. This made use of Raspberry Pi to convert text to speech. But one drawback was the use of scanner.

The systems developed are mostly travel aid for the visually impaired that guide them in travel from one place to another with the help of onboard maps or GPS. These systems are not reliable in certain situations like indoors or when the obstacles are not in the range of the sensors. There may be need of multiple sensors, microcontrollers, and devices in order to be configured. They may also require many other troubleshooting and expertise personnel, whereas a common man might not be suitable for the task. The most important function is that the system should provide the results in real time. Most of the systems are used for a specific purpose like obstacle avoidance which needs to be more generic that can be used for multipurposes like navigation, detection of the objects in and around the environment. Hence, there is a need for a system, that guides the blind people to find out objects around them in real time. The need for a speech output is the main requirement of the hour.

3   The Proposed System

The proposed system is capable of delivering on real time which must have the potential of delivering reliable object detection and also rough calculation of objects position pipeline, with the goal of informing the user with the synthesized audio about the surrounding objects, their spatial position with respect to the user along with the name of the objects.

The block diagram in Fig. 1 shows the pipeline in which the system works. The use of an input camera device is made for video capture, and these images are passed these video data frames to the framework which returns detection data which is communicated to the users.

The proposed system has the capability of being done, effected and also accomplished by using object detection, object relationship analysis to locate a blind person's environment, thus helping them along routes to destinations, providing them with reliable information about nearby points and objects of interest, and indeed with the synthesized audio guiding them.

The technologies used are as listed below.

A.   YOLO/Darknet Framework:

YOLO is a network model [13] which is used for detecting objects. It is a model that is very simple to understand. The method used here is straightforward. Prior to YOLO, many classifiers were used to achieve object detection. In just one evaluation,

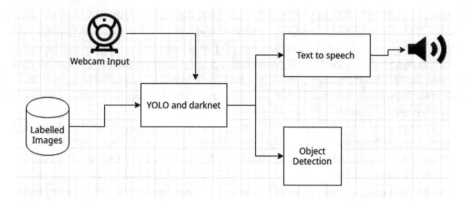

Fig. 1 Proposed system design

this framework predicts the object. YOLO is very fast and flexible. It achieves 51 ms on a [14] Titan X GPU. The YOLO network assimilates the various boxes extraction, extraction of features, and classification methods into a framework. Direct extraction [15] of the candidate boxes from the images is done for detecting objects.

B.   OpenCV:

OpenCV is a frequently used computer vision library which is open source with wide community support. Its original cause of growth was by the Intel corporation. OpenCV currently supports most of the programming languages and also many deep learning APIs or frameworks like Keras, TensorFlow, MXNet, Caffe, Gluon, torch according to a defined list of supported layers.

C.   Speech Engine:

Voice assistance is delivered using a cross-platform library [16], pyttsx3. This engine is used to synthesize text into audio users that will be able to hear through their speakers. This package works well in both Windows and Linux operating machines. This engine uses a naive speech driver which helps in good voice synthesis so it is the most flexible text-to-speech package in the open-source world of Python.

4   System Architecture and Implementation

4.1   Methodology

Once the application is started, the camera captures the frames. These frames are fed to the framework, which detect illustrations of objects in video input by inserting boundary containers around them. Then these annotated texts are converted to audio

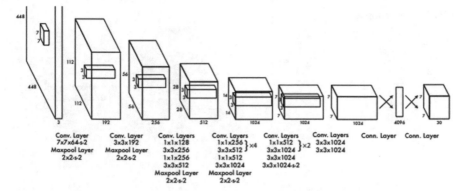

Fig. 2 Architecture of the framework

outputs. They give the state of each thing as the user can perceive. So, the detected objects will be all made to store in a data file, and this file later referred for the text-to-speech translation of the object names to voice, to communicate to the users.

4.2   Description of the Process

A.   **Training Data**:

The training of the model here is done with the Common Objects in Context (COCO) dataset which has 20 classes. The COCO dataset is an admirable dataset for object detection with 80 object categories, around 1 lakh training images.

B.   **Model**:

The prototypical design described in Fig. 2 is you only look once (YOLO) object detection framework that is framed of tremendously compound ConvNets which is convolutional neural network, where this architecture is named as dark net. The architecture is simple, and it is just a convolutional neural network. YOLO makes use of the total of the square errors among the predicted values and the value of the correct value which is the expected values [17], finally to calculate the losses which will help to decide the typical accuracy. The loss function will be having both the classification loss and the object's localization loss and the corresponding errors between the predicted bounding boxes, ground truth and also the confidence. It uses these multiple layer types which are ConvNets with a (3,3) kernel and max pooling layers with a (2,2) kernel [18].

C.   **Input Data**:

The input device used is a camera video stream as shown in Fig. 3. This camera feeds the images as frames every second (FPS) to the trained model, and it can be set up to

Fig. 3 Camera input

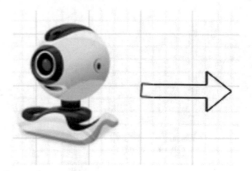

work with every alternate frame to fasten the procedure. This video capture is possible with OpenCV library's module called Video Capture. This class is for video sequence capturing from webcams, cameras, video format files, and sequences of images. It also provides multiple API calls for the video capturing from the above-mentioned input.

D. **Processing and calculations**:

In the classification and localization, the data normally that comes out of the framework is in the form (X,y), where X image data matrix and y is an array containing all the class labels that corresponds to image X, bx, by, bw, and bh [19] as shown in Fig. 4 above, where bx = Detections box x-coordinate, by = Detections box y-coordinate, bw = Detections box width, and bh = Detections box height.

The image is divided into boxes to do object localization tasks so the ConvNets [20–25] in place here. Then one more output layer predicts bounding box coordinates and do the required tweaking the loss function. The input image is passed on in the pipeline to the framework which then divides into grids in a single pass. The process of image object classification and determination of object location are applied on each of the grid. Then it predicts the bounding rectangle and their corresponding class probabilities for objects in the box.

If there is an object located in a grid, it will take the midpoint of the grid where there are objects. The corresponding detection data will then be allocated to the network which comprises of midpoint of these detected substances and the class label for the midcentral grid will be assigned. Even in some cases, if an object might be present in multiple grids, it will only be assigned to a single grid in which its midpoint is located. The x-coordinate of the detections box and y-coordinate of the detections box will be between 0 and 1 as the midpoint will always lie within the network, but width of detections box and height of detections box can be greater than one in few circumstances, when the magnitudes of the rectangle box have additional magnitudes compared to the grid.

Fig. 4 Finding the width of an object

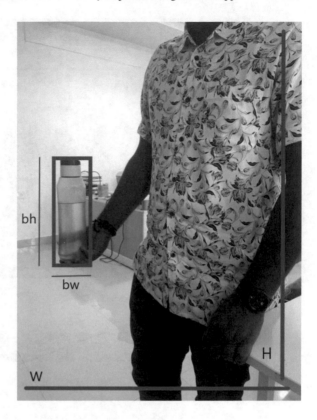

E.   **Voice Module**:

When the prediction of class is done, in each frame the objects found will be a set of object name. The coordinates of these things can be found in the picture and the location will be appended like "top," "mid," "bottom" and "left," "center," "right" to prediction of class of the object. One such example is given in Fig. 5.

The texts explanation is sent to Python Text to Speech application that uses the pyttsx3, a multiplatform library. It lets you synthesize audio from text that a user can hear through a speaker. Once the object is detected, the message "object found" is fed as text to the console and to the text-to-speech package.

The message is as shown in Fig. 6.

Figure 7 depicts the model used for text-to-speech conversion. The pyttsx3 Python module supports naive text-to-speech API packages.

Fig. 5 Finding the objects label and position

Fig. 6 Message displayed after object is detected

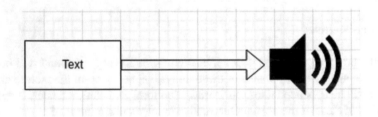

Fig. 7 Text-to-speech conversion

5 Results

The proposed model is proven to be more accurate than the existing models. This helps the visually impaired identify objects around them. Objects in the scene will be detected with their object class label, confidence on top of the bounding box, and this information along with the positions communicated to the users also printed in the console. As shown in Fig. 8, objects in the scene, a bag is detected with 87% confidence with blue color bounding box and a bottle in yellow box with 96% confidence.

Fig. 8 Objects in the scene a bag detected

The coordinates of the bounding box can also be obtained for every object detected in each frame passed. This will overlay the boxes on the objects detected and return the stream of frames as a video playback. This is scheduled to get a voice feedback on the first frame of each second with the object name along with its position of the camera view and also the voice feedback to the user.

Figure 9 shows predictions of objects with their relative positions, like the top/bottom, left/right. Figure 10 shows the output predicted and the statistics displayed in the console.

6 Test cases

The model was tested in several scenarios. All these are listed in Table 1, with input, expected result, and the output. There are four cases listed in the table. Out of this one case which fails is when the objects are not in the training dataset.

7 Conclusion

The proposed model will be helpful to the visually impaired people. A straightforward, flexible, configurable, easy to handle app with any requirement for technical expertise voice guidance system is the proposed system. This is capable of providing the valuable assistance and support for most blind and visually impaired people. The objects around them can now be heard. This enables them to know their position and helps them to decide where to go without help from anyone else. It not only makes

Fig. 9 Predictions of objects with their relative positions

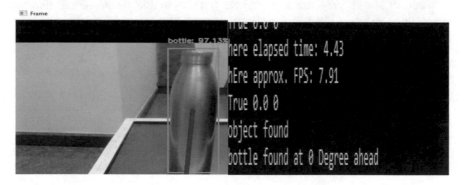

Fig. 10 Output predicted and the statistics displayed in the console

their travel easier but also saves a lot of time and energy and enhances their safety on roads.

It makes the lives of visually impaired better by making them independent. The outcomes of the proposed system show clear signs that it is flexible, reliable, effective, and individualistic capacity. It has shown 90% accuracy in specifying the objects that may be encountered by the blind. It is able to scan any environment which the input camera is pointed to by the user. This system is accessible to anyone through a simple interface.

Table 1 Test cases for the proposed system

Input	Expected result	Output (passed/failed)
Still objects shown to camera	Objects are detected and communicated to users	Passed
Moving objects	Objects are detected and communicated to users	Passed
Objects very far away from the camera or objects in fast motion	Objects are detected and communicated to users	Passed
New object which is not present in the training dataset	Object not detected	Failed

References

1. "World Sight Day 2017:" Available: https://www.indiatoday.in/education-today/gk-current-aff airs/story/world-sight-day-2017-facts-and-figures-1063009-2017-10-12(2017)
2. P. Dollar, R. Appel, S. Belongie et al., Fast feature pyramids for object detection. IEEE Trans. Pattern Anal. Mach. Intell. **36**(8), 1532–1545 (2014)
3. P. Dollar, Z. Tu, P. Perona, et al., Integral Channel Features. British Machine Vision Conference, BMVC, pp. 1–11 (2009)
4. Y, Freund, R.E. Schapire, A decision-theoretic generalization of on-line learning and an application to boosting. Comput. Learn. Theory, pp. 23–27 (1995)
5. D.G. Lowe, Object recognition from local scale-invariant features: International Conference on Computer Vision, pp. 1150–1157 (19999)
6. M.A. Hearst, S.T. Dumais, E. Osman et al., Support vector machines. IEEE Intell. Syst. Their Appl. **13**(4), 18–28 (1998)
7. H. Jabnoun, F. Benzarti, H. Amiri, Object detection and identification for blind people in video scene: Université de Tunis El Manar, Ecole Nationale d'Ingénieur de Tunis, In Proc. ResearchGate (Dec 2015)
8. A. Noorithaya, K. Kumar M, Dr. A. Sreedevi, Voice Assisted Navigation System for the Blind : R.V.C.E , Bangalore, India, In Proc. International Conference on Circuits, communication, control and computing (2014)
9. A. Nishajith, J. Nivedha, S.S. Nair, Prof. J.M. shaffi, Smart Cap—Wearable Visual Guidance System For Blind, In : Proc. International Conference on Inventive Research on Computing Applications (2018)
10. H. Park, H. Won, S. Ou, J. Lee (2019) Implementation of crosswalk lights recognition system for the blind's safety: Signal, Images et Technologies de l'Information (LR-SITI-ENIT), IEEE Eurasia Conference on IOT, Communication and Engineering (2019)
11. R.K. Rakshana, C. Chitra, A smart navguide system for visually impaired. Int. J. Innovative Technol. Explor. Eng. (IJITEE) ISSN: 2278–3075, **8**(6S3) (Apr 2019)
12. Dr. B. Muthusenthil, J. Joshuva, S. Kishore, K. Narendiran: Smart Assistance for Blind People using Raspberry Pi: International Journal of Advance Research, Ideas and Innovations in Technology, ISSN: 2454–132X, Volume 4, Issue 2, (2018)
13. J. Redmon, S. Divvala, R. Girshick, A. Farhadi: You Only Look Once:Unified, Real-Time Object Detection: IEEE Conference on Computer Vision and Pattern Recognition. (2016). https://doi.org/10.1109/CVPR.2016.91.
14. Joseph Chet Redmon: YOLO: Real-Time Object Detection: Available : https://pjreddie.com/darknet/yolo/
15. C. Liu, Y. Tao, J. Liang, K. Li, Y. Chen: Object Detection Based on YOLO Network: IEEE 4th Information Technology and Mechatronics Engineering Conference. https://doi.org/10.1109/itoec.2018.8740604 (2018)

16. NanoDano: Text-to-speech in Python with pyttsx3 2018, Avaiable: https://www.devdungeon.com/content/text-speech-python-pyttsx3
17. Z.Q. Zhao, P. Zheng, S.T. Xu, X. Wu:Object Detection with Deep Learning: A Review. In: Proc. arXiv:1807.05511v2 [cs.CV] (2019)
18. J. Redmon, S. Divvala, R. Girshick: You Only Look Once:Unified, Real-Time Object Detection. In : Proc. arXiv:1506.02640v5 (2016).
19. H. Yang, H. Wu, H. Chen: Detecting 11K Classes: Large Scale Object Detection without Fine-Grained Bounding Boxes, In: Proc. arXiv:1908.05217v1 [cs.CV] (2019) .
20. M.M. Kamal, A.I. Bayazid, M.S. Sadi, M.M. Islam, N. Hasan: Towards developing walking assistants for the visually impaired people , In: Proceedings IEEE Region 10 Humanitarian Technology Conference (R10-HTC), Dhaka, (2017), pp. 238–241.
21. S. Haque, M.S. Sadi, M.E.H Rafi, M.M Islam, M.K Hasan: Real-time crowd detection to prevent stampede: Proceedings of International Joint Conference on Computational Intelligence. Algorithms for Intelligent Systems, pp. 665–678. https://doi.org/10.1007/978-981-13-7564-4_56, (2020).
22. Y. Bengio, Learning deep architectures for AI. Found. Trends Mach. Learn. 2(1), 1–127 (2009)
23. J. Bai, S. Lian, Z. Liu, K. Wang, D. Liu, Virtual-blind-road following-based wearable navigation device for blind people: IEEE Trans. Consum. Electron. 64(1), 136–143 (2018)
24. W. Elmannai, K. Elleithy: Sensor-based assistive devices for visually impaired people: current status, challenges, and future directions :Sensors 17(3), 565–606 (2017).
25. V. Bansal, K. Balasubramanian, & Natarajan: Obstacle avoidance using stereo vision and depth maps for visual aid devices: SN Appl. Sci. 2, 1131 (2020). https://doi.org/10.1007/s42452-020-2815-z

Application to Aid Hearing and Speech Impaired People

Akshatha Patkar, Steve Martis, Anupriya, Rakshith, and Deepthi G. Pai

Abstract One of the most priceless gifts to a natural being is the capability of vision, hear, express and react correspondingly to the situations. Interaction between deaf–dumb and ordinary beings is an inspiring mission. The hearing-impaired and the mute society depends mainly on the hand gestures known as the sign language for communication. The sign language identification is one of the revolutions for serving the specially-abled society. The exploration of identifying sign gestures is successful but involves an exclusive charge to be commercialized. For the sign language identification system to be used widely, the data acquisition process varies largely depending on the cost of the system, the methods used, limitations, etc. The course of learning, recognizing the signs and interacting via the ISL can be simplified by the proposed system that converts speech to the sequence of sign language symbols. Speech processing embraces speech recognition, the learning of identifying the vocabularies being vocalized, irrespective of who the orator is. The proposed system practices template-based detection as the key tactic where the Voice to Sign (V2S) system initially requires to be skilled with a dialogue plan based on the predefined database of signs. It correspondingly translates speech to text via the dictation recognizer of the Unity 3D tool, processes the text and maps the phrases to animations that will assist to convey the desired message. Employing the proposed system trainers will be able to teach sign language effortlessly without explicit training.

Keywords Communication technology · Indian sign language (ISL) · Sign language · Recognition · Speech to text · Text to speech

1 Introduction

The Indian sign language (ISL) is one of the most important modes to exchange ideas for dumb and deaf individuals in India. Each country has its communicating sign language which is a finely structured code of gesture, with a unique meaning mapped

A. Patkar · S. Martis · Anupriya · Rakshith · D. G. Pai (✉)
Department of Computer Science and Engineering, Shri Madhwa Vadiraja Institute of Technology and Management, Bantakal, Udupi, India
e-mail: deepthi.cs@sode-edu.in

to each one. The ISL uses manual communication or body language or signs to convey thoughts and ideas. Communicating using the ISL, on the other hand, requires sophisticated training and practice. On the other hand, locating skilled and qualified teachers for their daily affairs is an extremely tough chore and also overpriced.

The latest statistics issued by the World Health Organization (WHO) demonstrates that for individuals aged above 50 years, the audible range gradually decreases over time all over the world. It is probable that above 900 million public throughout the globe will experience loss of hearing ability by 2050 [1].

Nearly 466 million individuals universally have difficulty to perceive sound and the count of youngsters' ranges in 34 million out of these. The 'Deaf' community comprises individuals of very diminutive or no hearing ability. Communication takes place through hand gestures. All around the world, different communities use variant sign languages in distinctive fragments of the globe. Verbal lingos are few in number [2].

There has been an exponential increase in the recognition and development of the ways to interact with the specially-abled. Those with physical frailties such as deafness, loss of sight, or deaf–blindness depend on their working senses to be in contact and have access to the world. The initial lingo the deaf children absorb is the sign language as their primary language. But only 386 schools for deaf and dumb are established in India as of 2017. They have an issue to express themselves or understand the printed text, numbers and also issues while making an image of abstract thought in mind and are unable to access the information which the normal people do very easily like television content, multimedia on the Web and individual public speeches. They desire to retrieve facts in the mode of hand gestures on the Internet instead of a written subject, but incredibly limited sites propose gist in sign language in the form of cinematic clips and can be retrieved by interpreting the script. Producing the material available on the Internet by means of a filmed methodology is expensive [3]. This demand for the need for a platform that acts as the intermediate between the hearing-impaired person and normal person even if he is not acquainted with the use of the internet or browser. Equipping dumb and deaf with qualified and trained teachers can be a challenging task mainly from a financial point of view.

The existing sign language applications are constrained to text inputs and cannot very clearly handle combined phrases or SOV (subject–object–verb) patterns. These issues are addressed with the assistance of natural language processing to break down the script into slighter comprehensible fragments. All the animations of Avatar will be created. The project's UI part includes a dictation recognizer. Text pre-processing part includes removing stop words, lemmatization, handling synonyms and transforming the given sentence into SOV form since ISL sentences have to be in SOV form mandatorily.

The proposed system converts speech by the conveyer to the sequence of sign language symbols for the dumb and deaf to communicate. It also converts speech to text, processes the text and maps the phrases to animations that will help convey the desired message using Avatars. There have been several considerable research activities in developing a system for the generation and automation of sign language.

This automated system takes either text or speech input to generate sign language as output.

About 4 million deaf and about 10 million find difficulty in hearing are present in India as reported by the All Indian Federation of the Deaf (AIFD). ISL is a way of interaction for a further 1 million deaf grownups and about 0.5 million deaf. The Sign-Language-Recognition (SLR) intends to equip algorithms and approaches to appropriately recognize an arrangement of shaped signs and deliver its gist in manuscript or dialogue. The two key classes of the SLR scheme: vision-based and sensor-based. Vision-based systems exploit pictures and films of symbols developed from cameras. Likewise, sensor-based systems need the custom of wearable strategies armed with sensors to abstract the hand gestures and wave of the signs [4]. The projected scheme makes use of the vision-based system to capture the gestures acted by deaf and dumb.

The SL can be staged via hand gesture either by one hand branded as isolated sign language or two hands recognized as continuous. Sign language. The sign language comprises a solo gesture consuming only a word whereas continuous ISL is an arrangement of signs that causes an expressive verdict. The proposed system operates on a continuous ISL sign recognition procedure using ISL. Continuous ISL signs are typically crafted of two-handed; besides, it is a blend of action as well as still gestures. Hence, the issues to identify signs in the actual setting increases [5]. The continuous SL recognition focuses on studying unsegmented signs of audio–visual recordings and is aimed at dispensation continuous motion films in actual organizations. Lately, deep convolutional neural networks include an incredible influence on associated errands on films. For example, Homo sapiens motion recognition, gesticulation recognition, and recurrent neural networks (RNNs) revealed the significant consequence of studying the time-based needs in symbol noticing [6]. In other words, neural networks are limited to acquiring bit-wise depictions. The hidden Markov models (HMMs) are intended for categorized learning. Though the bit-wise classification is not noise-free to exercise the deep neural networks. Besides, HMM may well be tricky to gather the intricate active dissimilarities given the partial illustration skills.

The idea can not only be used by teachers, but also by any person for that matter, who wants to communicate a given message through sign language without having any training for the same. This idea can be extended to multiple languages also, in which case, languages other than English can also be used. The proposed system can recognize hand gestures accurately with a single normal webcam and convert them into text and voice messages. The objective of this scheme is to distinguish the signs by utilizing the peak precision in addition to the minimum amount period and render the letters of ISL hooked on to the equivalent manuscript and speech in a vision-based fashion or commonly known as Avatars.

2 Literature Survey

Ruxandra Tapu et al. [1] projected a scheme that utilizes equally deep convolutional neural networks and computer vision algorithms carefully devised to sense as well as identify the characteristics of the dynamic orator in the film. As the numerous individuals stage in the identical act and occupies in a discussion, the problem that arises is to classify the individual who is presently voicing. An additional unclear condition points to the event in which the expression of the orator is undetectable. DEEPHEAR permits the audiences at ease to stick to the film information at the same time as the accompanying descriptions by locating the captions that cause to classify the involved orator. The DEEPHEAR skeleton equally utilizes deep convolutional neural networks known as CNN and computer vision procedures to accomplish the innumerable phase's essential, together with detection of a face, recognition, and tracking, film time-based breakdown, involved orator recognition, and identification, script discovery and description locating. The scheme is unqualified of being employed in circumstances of the absent appearance of a person in the cinematography. The outspread of the construction to deal with voice-to-script library for films in which the caption writes down is inaccessible in an early stage.

Richard E. Ladner [2] reviewed modules of message passing equipment for audible range enrichment machinery. Primary automated supports for hearing only improved the hearing capabilities to a little extend along with non-linear disturbance cancellation subjected with the help of some machinery tools. The assistance tools could be high-priced, smart-aids estimated in a couple of thousand dollars also cochlear transplant operation costs more than the reach of a common man. But in both the cases of automated support tools and cochlear transplants, there is a requirement of persistent safeguarding, chiefly replacement of battery has to be done regularly. For the reason of expense, a mere fraction of individuals may perhaps afford these automated hearing tools or manage to have a cochlear implant. Emerging with minimal budget automated tools is the key practical test. The two elementary modes using which the deaf community can interact: script and sign language. Each part of the world has its own distinct ways to sign language, on the contrary, there is no prevailing written form that is accepted by all.

The paper by Jestin Joy et al. [3] examined the Sign-Quiz that customs deep neural networks (DNN). The Sign-Quiz application will be capable of effortlessly being exercised by deaf people and also a normal people. Besides, it is economical, constructed to be used on a website, hand gestures sign education presentation for ISL, applying programmed identification of sign language procedure by means of DNN. Fingerspelling is represented via alphabets of a scripting scheme and digit schemes. It performs as an intermediate linking the SL and spoken lingo. The ISL is capable of signifying English letters A to Z with finger spelling. It characterizes texts that do not have any hand gesture corresponding to it or to manipulate an expression or allotted in the schooling of the SL. The operator has the privilege to hand-pick letters. For the respective letter, pictures and films are presented. But absorbing hand gestures are extremely tricky to be used by a learner lacking the assistant of a proficient SL expert. Sign-Quiz assessments examined volunteers who participated in it, and the results showed maximum volunteers were enthusiasts of tools-based presentations.

Suhail Muhammad Kamal et al. [4] present a methodological summary of the methods in scheming a Chinese Sign Language Recognition (CSLR) structure. The sign language recognition (SLR) schemes utilized are fingerspelling in circumstances in which different vocabulary, person's name, location and terms without recognized hand gestures are enacted by hand motions. The gesticulation identification examines an arrangement of pictures or motion made by hand activities via gloves embedded with sensors or just a webcam. The Chinese Sign Language Recognition (CSLR) employs neural network and hidden Markov models (HMM). The gloves embedded with sensors testified prominent precision but typically operated on isolated expression identification. Mining of only the applicable descriptions from statistics obtained via the sensor tools may not be generally exploited for the reason of its high price, although it is easy to collect the data from the sensors. Moreover, the dataset causes problems that are aimed at attaining an extraordinary routine of the SLR scheme. The list of datasets available is publicly available dataset and also does not contain much information that can allow to improve and compare various state-of-the-art techniques; incorporating facial expressions that convey extra meaning in a sign language sentence as the state-of-the-art German sign language dataset.

The primary language of deaf people is widely recognised by Runpeng Cui et al. [6], SL, and is typically recorded or transmitted in video form. The SL is also considered to be the gestural communication that is most grammatically organised. This makes SL recognition a perfect area of study for the development of techniques to resolve issues such as human motion analysis, human–computer interaction (HCI) and user interface design, and makes it receive great attention in multimedia and computer vision. This work establishes a continuous recognition system for sign language (SL) with deep neural networks, which transcribes videos of SL phrases directly to sequences of ordered gloss labels. Hidden Markov models (HMM) typically use previous methods dealing with continuous SL recognition with restricted capability to capture temporal information. SL recognition is all about learning long-term video streams with unsegmented gestures and is more suited for processing continuous gestural images in real-world systems. This training also does not require a costly annotation for each gesture on the temporal boundary. Recognizing SL suggests that gestural gestures and appearance attributes are simultaneously evaluated and incorporated, as well as disparate body parts, and therefore, potentially using a multimodal method. The emphasis is on the issue of continuous SL recognition on images, where it is important to learn both the spatiotemporal representations and their temporal matching for the labels.

3   Summary of Literature Review with Research Gaps

The sign language may be executed either by one hand known as independent sign language or by two hands known as continuous sign language using hand gestures. Sign language consists of a single gesture with a single word, whereas continuous ISL or continuous sign language is a set of gestures that produce a meaningful expression. We are working on continuous ISL gesture recognition techniques using ISL in the suggested framework. Continuous ISL movements consist mainly of two hands and sometimes of two hands [7].

Continuous SL recognition is all about learning long-term video streams with unsegmented gestures and is more suited for processing continuous gestural videos in real-world systems. Deep convolutionary neural networks have recently had a huge effect on similar video tasks, e.g. recognition of human behaviour, recognition of movements and sign spotting, and recurrent neural networks (RNNs) have shown substantial results in learning about temporal dependencies in sign spotting [8].

Neural networks are limited to studying frame-wise depictions in other works, and hidden Markov models (HMMs) are used for sequence learning. However, for the training of deep neural networks, the frame-wise labelling is noisy, and HMMs can be difficult to understand the complex dynamic variations, given their limited capacity for representation.

The idea can not only be used by teachers but also by any person for that matter who wants to communicate a given message through sign language without having any training for the same. This idea can be extended to multiple languages also; in which case, languages other than English can also be used.

With a single standard webcam, the proposed device will reliably interpret hand gestures and transform them into text and voice messages. The purpose of this project is to identify the gestures in a vision-based setup or generally known as Avatars with optimum accuracy and at least possible time and translate the alphabets of Indian Sign Language into corresponding text and speech.

4   Dataset Description

The dataset used is obtained from the link http://www.indiansignlanguage.org, which currently consists of all the existing words in the Indian sign language dictionary; 2785 words have been defined to date, and each of them is assigned to a particular category. There are various categories such as places, banking, arts and entertainment, shapes and educational terms. Each word is associated with its respective ISL action, and video for the same can be obtained by just clicking on that word. As we know, the ISL Dictionary is constantly being updated and new words which have their meaning in the ISL are being appended.

5   Methodology

The alphabets, numbers and hand gestures used in the Indian Sign Language varies from that of the other country's sign language. Each region inherits its own hand gestures for each alphabet and number to form words and sentences. The hand gestures for alphabets used in Indian Sign Language is given in Fig. 1.

The following steps are considered in the architectural design of the application:

a.   **Speech to text conversion**: Since the basic idea of the project is to convey spoken messages to ISL animation, the input in the form of speech is recorded and converted to text [10]. The input can also be in the form of direct text.

b.   **Text pre-processing**: The input obtained cannot be used directly for animations because spoken languages do not adhere to the grammar rules of the ISL. Hence, the input text has to be converted to its equivalent ISL form. Certain rules are to be followed to convey a message in ISL. Words are always used in their root form and their present tense. Undefined signs are split and shown as alphabets. The most significant rule is that every ISL sentence has to be in SOV (subject-object-verb) notation. To achieve this, the text is processed using NLP functionalities such as finding root words, lemmatization, removing stop words and identifying the parts of speech (POS). The final sentence will be in the ISL format.

c.   **Text to animation mapping**: The words in the processed input have to be mapped to their respective animations. In case the animation for that word is not present, its synonym is mapped. If the synonym does not exist too, it is split according to its alphabets.

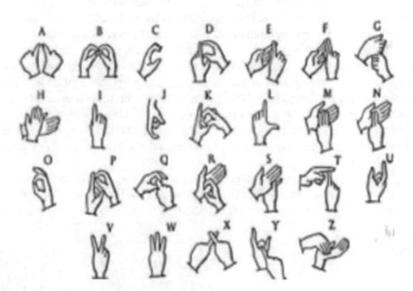

Fig. 1   The Indian sign language [9]

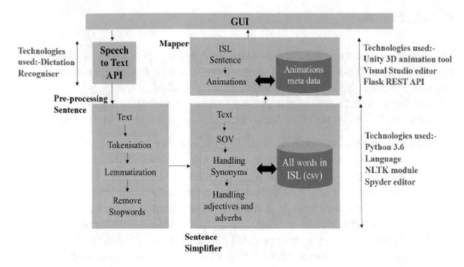

Fig. 2 The architecture design

d.   **Animation**: The output is conveyed using an Avatar. These animations are combined to communicate the ISL sentence. Hence, a given sentence is converted into its ISL form.

The application architecture diagram is a series of step-by-step as depicted in Fig. 2.

Figure 3 shows the process of speech to text conversion using Dictation Recognizer. Unity3D's dictation cognizer is used for converting speech to text. This process starts by enabling the microphone under capabilities followed by turning on online speech recognition and linking and typing under system settings. The user speaks through the microphone, which gets converted to text and displayed on the screen.

Figure 4 shows the process of pre-processing using Python. Python and NLTK are used for text pre-processing. This process starts with Tokenization where text is tokenized into words from a sentence, followed by Lemmatization which is the process of grouping together the different inflected forms of a word so they can be analysed as a single item. Finally, stop word removal where all the stop words in the tokenized sentence are removed.

Figure 5 shows the process of sentence simplification using Python. Output of the text pre-processing is now simplified and converted to ISL sentences. The sentence is brought to subject–object–verb (SOV) notation as per ISL grammar rules using

Fig. 3 Speech-to-Text API

Technologies used:-Dictation Recogniser

Speech to Text API

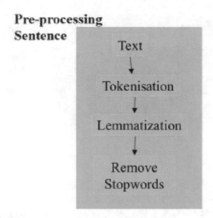

Fig. 4 Pre-processing sentence

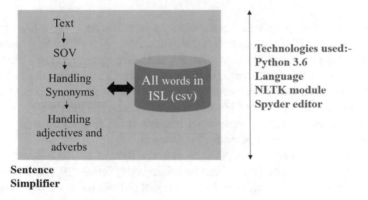

Fig. 5 Sentence simplify

the spacy module of Python, as all the words in English dictionary does not have a sign associated with it. In the absence of a word, its synonym sign is used. In absence of sign for a synonym, the word is split by letters. As per the ISL grammar rules, the adjectives and adverbs are handled.

Figure 6 shows the mapping of animations using Unity3D and Flask REST API. The output of the Sentence simplifier is called the ISL sentence. The mapping starts by obtaining the ISL sentence from the API call to the Python class which include the Fig. 6 methods. Animations associated with each word in the ISL Sentence is then triggered by the Animator object to display the animations. Flask REST API is used to coordinate between Python and C#. The animations are displayed in GUI which is built in Unity 3D.

Below is the sample output.

A story is dictated using the microphone and it appears as text on the screen as follows:

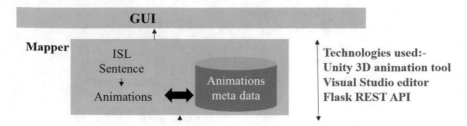

Fig. 6 Mapper

```
Enter split :   birbal tell to solve complaint
After split :   b i r b a l tell to solve complaint
string without verb['b', 'i', 'r', 'b', 'a', 'l', 'tell', 'to', 'complaint']
```

Fig. 7 Console output for unknown words

One day, a clever man sold his well to a farmer. The next day, when the farmer went to the well to fetch some water, the man refused him telling that he only sold the well and not its water. The farmer, with a sad heart, went to Akbar's court. Birbal was told to solve the complaint. The following day, the man and the farmer were called to the court. The clever man told—he sold his well, not the water in it. On learning this, Birbal said, 'My friend, in that condition, you either remove your water from the well or pay tax for your water because it is the farmer's well'. The man understood his mistake and asked for forgiveness.

After text processing, the output story is in the form of ISL as follows:

One day clever man his pond to farmer sell next day farmer to pond to some water go bring when man him that he only pond reject tell sell n o t it water farmer with sad heart to a k b a r court go b i r b a l tell to complaint solve follow day man farmer call to court clever man he his pond tell sell n o t water in it on this learn b i r b a l order my friend in that condition you e i t h e r your water from pond o r tax f o r your water b e c a u s e it farmer pond remove pay man his mistake understand f o r f o r g i v e n e s s ask.

Table 1 depicts the standard Indian Sign Language rules and regulations based on which the application is built.

This checks the first rule in the table. Fingerspelling is breaking each word to alphabets in case they do not have a sign. In the above case, 'Birbal' which is a proper noun does not have a specific sign in ISL. Hence, it is split to 'b i r b a l'.

Words that have no signs are checked for replacement with synonyms that exist in the ISL dictionary. If it exists, the synonym replaces the original word. In this case, 'fetch' is replaced by 'bring'. Else, the word will be split for fingerspelling.

This checks the Wh-word rule. It says that question words must always be used at the end of the sentence. In the above output, 'what' is appended to the end.

```
when farmer go to pond to fetch water
[('when', 'WRB'), ('farmer', 'NN'), ('go', 'VBP'), ('to', 'TO'), ('pond', 'VB'), ('to', 'TO'), ('fetch', 'VB'), ('water', 'NN')]
fetch replaced by bring
After synonyms :  when farmer go to pond to bring water
```

Fig. 8 Console output for synonyms

```
Enter interrogative handler :
[['what', 'you', 'do']]
[[('what', 'WP'), ('you', 'PRP'), ('do', 'VBP')]]
[('what', 'WP'), ('you', 'PRP'), ('do', 'VBP')]
Final sentence(if interrogative): you do what
Final Sentence: you do what
```

Fig. 9 Console output for wh-questions

```
Input string :  a clever man sold his pond to a farmer
After stop words removal: clever man sold his pond to farmer
```

Fig. 10 Console output for linking verbs and articles

```
Input : clever man sold his pond to farmer
Enter lemmatization with pos :
[['clever', 'man', 'sell', 'his', 'pond', 'to', 'farmer']]
After lemmatization : clever man sell his pond to farmer
```

Fig. 11 Console output for lemmatization

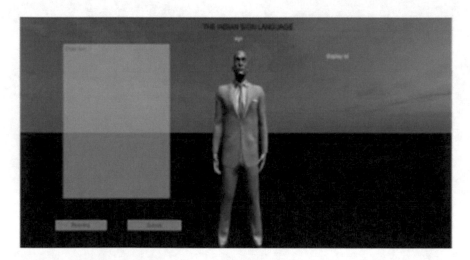

Fig. 12 UI of the application

ISL does not have signs or use linking verbs and articles. In the above case, these two cases are handled. Stop words are unnecessary words that are removed. The article 'a' is removed.

The existing live SLR approaches suffer from a few downsides of complexity in identifying composite hand gesticulations, minimal distinguishing precision for

Fig. 13 Working of the application

Fig. 14 Animations according to resultant output recording process

many active SLR, also the probable complications in the bulkier cinematic chain of information guiding. The SL put together a sequence of movements that take in swift gestures in addition to alike features. Thus, the dynamic SLR is problematic to transact in addition to the density and huge dissimilarities of terminology databases in hand gestures. Also, it possibly will misinterpret a variety of alternatives with one another. The dynamic SLR encounters difficulties in production besides the intricacy of gesture behaviours of finger movement of the wide-ranging frame environment. The mining of the utmost judicious characteristics from pictures and films is another added problem. Further, picking a suitable classifier is a desperate feature for influencing exact perception outcomes. Two of the numerous modes to resolve the setback with dynamic SLR are—the first one is the process grounded on hand

Fig. 15 Recording process

Table 1 The ISL rules and regulations

Rule	Input	Output
Finger spelling for unknown words	Birbal said	Birbal state (replaced by synonym)
Never use linking verbs (is, was etc.)	Birbal was told to solve the complaint	Birbal tell to complaint solve
Never use words with suffixes (s,'s etc.)	because it is farmer's pond	Because it farmer pond
Always sign in present tense	Clever man sold his well to a farmer	Clever man his pond to farmer sell (SOV format)
Articles are not used	The next day	Next day
Wh- questions are at the end	When are you leaving	YOU leave when
Never use gerunds	On learning this	On this learn

signs and movement curve of hand-gesticulations; the next one is the technique built on a cinematic arrangement of respective SL [3].

6   Results and Discussion

It aims to eliminate the explicit training which teachers have to undergo to teach in a Deaf and Dumb School. Using this project any person can teach the students in the deaf and dumb school. This project will ensure that deaf and dumb school students can get access to the majority of the learning materials and resources which were otherwise unavailable to them due to language barriers.

The end product is an open-source desktop application that is platform-independent. The UI is user-friendly which consists of an avatar, an input field, a button to record the voice in case of audio input, and a submit button. The ISL equivalent of the text along with the animation being played is displayed simultaneously after the input is processed. There is no considerable delay to display the animation through an avatar on submitting the text in the input field. The animations are clear with smooth transitions.

This application will simplify the job of teaching the mute community. It is quite hard to learn and master the Sign Language, as it is time-consuming. This application eliminates the requirement of knowing the ISL beforehand. This product can not only be used by teachers, but also by any person, who wants to convey a message through sign language without having any training for the same. The figures below illustrate an idea of the implementation of the application to aid hearing and speech impaired people.

For ages, very little importance has been given to the deaf and dumb. This can be noticed in the epics of this country. We cannot find many characters being deaf and dumb in any of the greatest epics of this country. Although they have the most perceiving sensory organ of the human body, which is the eye, using it to their fullest advantage has not helped them in achieving great feats [9]. The main reason is due to lack of resources and awareness among the people, that's why it's difficult to find the deaf and dumb achievers in this country. So, this project will help the people with such disabilities to learn on level ground along with the rest of the world.

There is no professional technical solution to teach the deaf and dumb until today. Few of the mobile applications which exist now merely convert each word into its respective sign without following ISL grammar. Also, there are hardly any applications that take in voice input also. Some of the existing applications are not open source. This application follows all the ISL grammar rules and regulations. Thus, this project can be of great use to the deaf and dumb and is a humble attempt to give something back to society.

7 Conclusion

The prevailing sign language applications are constrained to text inputs and cannot handle combined phrases or SOV (subject–object–verb) patterns. Many of them are in the form of mobile applications that do not consider audio input. These matters are addressed by means of the semantics of Natural Language Processing to break down the script into slighter comprehensible fragments, and through dictation recognizer which takes in speech as input through a microphone.

Linguistics is entertained as a crucial basis of interaction within human society. Signs can stem from the movement of a body part, usually initiated as the action of expressions made by face and hands. An extent of computer vision is Body Language

Identification, it takes into account understanding the signs via a diverse set of rules. In attendance is a shared delusion regarding the SL that it is global furthermore the society of unresponsive fitting to whichever chunk of the globe uses the identical lingo. However, the aforementioned is known that numerous gesture lingos use the resemblances of one another. The precise count of the prevailing sign languages is not known, nevertheless, the 2013 publication of Ethnologies' rolls them 137.

References

1. Ruxandra Tapu, Bogdan Mocanu and Titus Zaharia, "DEEP-HEAR: A Multimodal Subtitle Positioning System Dedicated to Deaf and Hearing-Impaired People" *IEEEAccess,* vol. 7, 2019.
2. Richard E. Ladner, "Communication Technologies for People with Sensory Disabilities", *IEEE,* vol. 100, 2012.
3. Jestin Joy, Kannan Balakrishnani and Sreeraj "SignQuiz: A Quiz Based Tool for Learning Fingerspelled Signs in Indian Sign Language Using ASLR", *IEEEAccess,* vol. 7, 2019.
4. Suhail Muhammad Kama, Yidong Chen, Shaozi Li and Xiaodong Shi and Jiangbin Zheng, "Technical Approaches to Chinese Sign Language Processing: A Review" *IEEEAccess,* vol. 7, 2019.
5. Formation Kumud Tripathi, Neha Baranwal and G. C. Nandi, "Continuous Indian Sign-Language Gesture Recognition and Sentence", Eleventh International Multi-Conference on Information Processing (IMCIP), *Elsevier,* 2015.
6. Runpeng Cui, Hu Liu, and Changshui Zhang, "A Deep Neural Framework for Continuous Sign Language Recognition by Iterative Training", *IEEE,* 2018.
7. Formation Kumud Tripathi, Neha Baranwal and G. C. Nandi, "Continuous Indian Sign Language Gesture Recognition and Sentence", Eleventh International MultiConference on Information Processing (IMCIP), Elsevier, 2015
8. Mu-Chun Su, Chia-Yl Chen, Shi-Yong Su, Chien-Hsing Chow, Hsiang-Feng Hsiu and Yu-Chine Wang, "Portable communication aid for deaf-blind people", Feb. 2001
9. B. Lakshmi, "Assistive SIGN LANGUAGE Converter for DEAF AND DUMB",International Conference on Internet of Things (iThings) and IEEE Green Computing and Communications (GreenCom) and IEEE Cyber, Physical and Social Computing (CPSCom) and IEEE Smart Data (SmartData), 2019.
10. Suharjito, Ricky Andersonb, Fanny Wiryanab, Meita Chandra Ariestab, Gede Kusumaa, "Sign Language Recognition Application Systems for Deaf-Mute People: A Review Based on Input-Process-Output", 2nd International Conference on Computer Science and Computational Intelligence 2017, Bali, Indonesia, *Elsevier,* October 2017.
11. Anish Kumar, Rakesh Raushan, Saurabh Aditya, Vishal Kumar Jaiswal, Mrs. Divyashree Y.V, "An Innovative Communication System for Deaf, Dumb and Blind People", International Journal for Research in Applied Science & Engineering Technology (*IJRASET*), Volume 5, Issue VI, June 2017.

Variants of Fuzzy C-Means on MRI Modality for Cancer Image Archives

C. K. Roopa, B. S. Harish, and R. Kasturi Rangan

Abstract The segmentation, identification and mining of contaminated tumor region from MRI images is a primary concern in medical image analysis. However, it is a monotonous and time-consuming process done by radiologists or clinical experts, and its precision is subjected to their expertise. To overcome these constraints, the use of supporting technology turns out to be very important. In this study, performance improvement and reduction of the complexity involved in the segmentation of medical images is been focused. We have investigated cancer cells using various fuzzy c-means methods and its different variants for Breast, Brain, Liver and Prostate dataset. The empirical outcomes of the various methods have been tested and corroborated on MRI for efficiency and quality analysis based on four well-known cluster validity indices. The results are very encouraging and robust in nature.

Keywords MRI · Cancer images · Fuzzy C-means · Cluster validity indices

1 Introduction

Computer aided diagnosis (CAD) has become essential in diagnosis of illness. It also plays major role in treatment planning. The other vital reason for the use of computers to assist is the development of medical imaging modalities like CT, MRI, PET and radiography. The main reason of using computer aided diagnosis is that clinical decision support system, where the output of CAD is taken as the second opinion and makes the final decision. CAD assists radiologist in interpreting the different types of abnormalities in medical images. It helps medical professionals in faster and automated diagnosis with higher rate of accuracy. The images obtained from the different medical imaging modalities are copied to CAD in a DICOM format.

C. K. Roopa · B. S. Harish (✉) · R. Kasturi Rangan
JSS Science and Technology University, JSS TI Campus, Mysore, India
e-mail: bsharish@jssstuniv.in

C. K. Roopa
e-mail: ckr@jssstuniv.in

© The Author(s), under exclusive license to Springer Nature Singapore Pte Ltd. 2022   161
P. Shetty D. and S. Shetty (eds.), *Recent Advances in Artificial Intelligence and Data Engineering*, Advances in Intelligent Systems and Computing 1386,
https://doi.org/10.1007/978-981-16-3342-3_13

This has to be prepared and analyzed using couple of measure like preprocessing, segmentation and evaluation.

Medical image is acquired using various modalities; one of its kinds is MRI. MR images give the structural details of Brain, Breast, Liver and Prostate which helps radiologist in monitoring and analysis for treatment.

The advantage of MRI is that it gives more details in the images when compared to other modalities. All though many research works are attempted on different application of MRI and MR image analysis, and the diagnosis is still a challenging and complex problem. In this paper, we attempted to explore the impact of variants of fuzzy C-means on MRI modality using four different cancer image archives [18] (i.e., Breast, Brain, Liver and Prostate). To mention some of the very well-known research works attempted are: Breast [3, 4, 6, 12, 13, 15, 21]; Brain [7, 9, 10, 25, 28]; Liver [2, 16, 17, 22, 23] and Prostate Cancer [5, 14, 20, 24, 26, 27].

2   Methodology

Accurate segmentation is a crucial step in most of the medical analytical procedures. It is also necessary to isolate the region of interest. If we can identify the boundaries of the ROI (image segment), we can also simplify decisions. Segmentation is a critical step that decides the final result of the proposal as a whole because the majority of the analysis depends solely on it. Since the image is segmented into coherent regions by segmentation, clustering procedures can be used for segmentation by extracting the global image characteristics in order to properly distinguish the region of interest from the context. Clustering can be done in numerous ways such as Density-based, Model-based and Systematic partitioning. In addition, due to the fact that most medical images are ambiguous and disordered data, Fuzzy approach becomes essential to manage uncertainty/vagueness in medical images during the segmentation process.

2.1   Fuzzy Approaches for MRI Segmentation

Fuzzy c-means (FCM) is a well-accepted clustering algorithm proposed by Bezdek [1], and it is based on traditional fuzzy set. FCM cluster the data element based on the membership value.

Though FCM showcased outstanding results on all the 4 dataset, unfortunately, it is sensitive to noise and ignored the neighborhood information. Further, FCM uses Euclidean metric that is sensitive to noise and in turn degrades the clustering results.

To handle the noise, researchers proposed a kernel version of FCM named as Kernel FCM (KFCM) [11]. The problem with KFCM-based methods is that they do not utilize the neighborhood information. To eliminate these limitations, various researchers proposed a variant of FCM named as Spatial Kernel FCM (SKFCM) [8,

19]. However, SKFCM still suffers the following limitations: work for single-feature inputs, high computation time and incorporate neighborhood information only to the objective function.

To alleviate these drawbacks, we found the way out to use two variants of FCM methods, namely Robust Spatial Kernel FCM (RSKFCM) and Generalized Spatial Kernel FCM (GSKFCM). The detailed derivation and explanation of RSKFCM and GSKFCM can be found in [10, 19], respectively.

2.1.1 Modified Intuitionistic Fuzzy C-Means (MIFCM)

The above RSKFCM and GSKFCM are based on the traditional fuzzy set where the decision for belongingness of the data point is founded using the membership value. Traditional fuzzy set assumes that the non-membership value is always the complement of the membership value. But, this assumption is not always true because of hesitation. This hesitation is caused by the uncertainty in defining the membership function. Intuitionistic fuzzy set (IFS) handles this uncertainty by considering the non-membership value and the hesitation degree. In the literature, based on the IFS, many clustering methods are proposed. The existing intuitionistic fuzzy c-means (IFCM) method employs the Sugenos and Yagers intuitionistic fuzzy complement generator function to compute the non-membership value. Unfortunately, for certain values, the aforementioned intuitionistic fuzzy complement generator fails to fulfill the fundamental requirement of intuitionism. To alleviate this limitation, the authors in [11] had proposed MIFCM by adopting a new intuitionistic fuzzy complement generator function. Further, employed the modified Hausdorff distance metric instead of the Euclidean metric. The same was used in this paper, and detailed tests were performed on all 4 datasets. MIFCM method comprises of two steps: Representation and Clustering. Next subsection explains these two steps in detail.

Representation
In IFS, each data point is associated with three parameters and it can be represented as:

$$\text{IFS}(X) = \{< x_i, \mu_X(x_i), \eta_X(x_i), \pi_X(x_i) > | x_i \in X\} \tag{1}$$

where $\mu_X(x_i)$ represents the membership value, $\eta_X(x_i)$ signifies the non-membership value, and $\pi_X(x_i)$ represents the value of hesitation degree associated with data point x_i.

As it is known that existing IFS complement generator fails to fulfill the fundamental requirement of the intuitionism. Thus, to alleviate this limitation, we employed IFS complement generator [11]. The non-membership value $\eta_X(x_i)$ is computed using the Bustince IFS complement generator:

$$\eta_X(x_i) = 1 - \phi(\mu_X(x_i)) \tag{2}$$

where ϕ refers to the Sugeno's or Yager's IFS complement generator. When we use Sugeno's function, Bustince IFS complement generator function becomes

$$\eta_X(x_i)_{\text{Bustince\_ Sugeno's}} = 1 - \left(\frac{1 - \mu_X(x_i)}{1 + \alpha\mu_X(x_i)} \right) \tag{3}$$

Similarly, if we use the Yager's function, Bustince IFS complement generator function becomes

$$\eta_X(x_i)_{\text{Bustince\_Yager's}} = 1 - (1 - (\mu_X(x_i))^\alpha)^{\frac{1}{\alpha}} \tag{4}$$

In this paper, we used both the Bustince IFS complement generator (i.e., Bustince–Sugeno's and Bustince–Yager's) function to compute the non-membership value. After applying the Bustince IFS complement generator function, IFS representation is given by:

$$\text{IFS}(X)_{\text{Bustince\_ Sugeno's}} = \left\{ \left\langle x_i, \mu_X(x_i), 1 - \left(\frac{1 - \mu_X(x_i)}{1 + \alpha\mu_X(x_i)} \right), \pi_X(x_i) \right\rangle | x_i \in X \right\} \tag{5}$$

$$IFS(X)_{\text{Bustince\_ Yager's}} = \left\{ \left\langle x_i, \mu_X(x_i), 1 - (1 - (\mu_X(x_i))^\alpha)^{\frac{1}{\alpha}}, \pi_X(x_i) \right\rangle | x_i \in X \right\} \tag{6}$$

The value of the hesitation degree is computed by using the below equations:

$$\pi_X(x_i)_{\text{Bustince\_ Sugeno's}} = 1 - \mu_A(x) - \eta_A(x)$$
$$= 1 - \mu_X(x_i) - \left(1 - \left(\frac{1 - \mu_X(x_i)}{1 + \alpha\mu_X(x_i)} \right) \right) \tag{7}$$

$$\pi_X(x_i)_{\text{Bustince\_ Yager's}} = 1 - \mu_A(x) - \eta_A(x)$$
$$= 1 - \mu_X(x_i) - \left(1 - (1 - (\mu_X(x_i))^\alpha)^{\frac{1}{\alpha}} \right) \tag{8}$$

In this method, we convert the crisp input data into intuitionistic fuzzy set (IFS) representation by employing the Bustince IFS complement generator. Let $X = \{x_1, x_2, x_3, \ldots, x_i, \ldots, x_n\}$ be the input data with samples. Firstly, we convert the input data into fuzzy form using the min–max normalization technique. The membership value for each data point is calculated using the below equation:

$$\mu_X(x_i) = \frac{x_i - \text{Min}(X)}{\text{Max}(X) - \text{Min}(X)} \tag{9}$$

Further, using these membership values, non-membership is computed using Eqs. (2) or (3) and the hesitation degree is computed using Eqs. (7) or (8). Finally, the input data is represented in the IFS form using all the three parameters (i.e., membership, non-membership and hesitation degree). Further, in the next step, these IFS converted input is clustered using the enhanced intuitionistic fuzzy c-means method.

Clustering
Improved intuitionistic fuzzy c-means (IIFCM) is the modified conventional IFCM algorithm which groups the input data by optimizing the subsequent objective function presented below:

$$J = \sum_{j=1}^{n} \sum_{i=1}^{c} \mu_{ij}^{m} \, d(X'_j, v_i) \tag{10}$$

with the constraints:

$$\sum_{i=1}^{c} \mu_{ij} = 1; \quad 1 \leq j \leq n \tag{11}$$

$$\mu_{ij} \geq 0; \quad 1 \leq i \leq c; \quad 1 \leq j \leq n \tag{12}$$

$$\sum_{j=1}^{n} \mu_{ij} > 0; \quad 1 \leq i \leq c \tag{13}$$

where $X'_j = \{\text{IFS}(x_j)\}$ denotes the IFS of data point x_j, $V = \{v_1, v_2, \ldots, v_i, \ldots, v_c\}$ denotes to c number of cluster centers and each v_i is related with the membership, non-membership and hesitation degree value, m denotes the fuzzification value, μ_{ij} signifies the membership value of jth data point to ith cluster, $d(X'_j, V_i)$ is the distance between jth data point to ith cluster center.

In IIFCM, the decision of belongingness of a data point to a cluster is made based on the membership value and it is computed using the distance function. Thus, the performance of the IIFCM depends on the selection of the distance metric. The conventional and existing variants of the IFCM method employ Euclidean as the distance metric which is very sensitive to noise. To alleviate this problem, the modified Hausdorff distance metric is used instead of the Euclidean distance metric. Modified Hausdorff distance metric is computed using the following equation:

$$d_H(X'_j, V_i) = \max\left\{ |\mu_{X'_j} - \mu_{V_i}|, |\eta_{X'_j} - \eta_{V_i}|, |\pi_{X'_j} - \pi_{V_i}| \right\} \tag{14}$$

Similar to fuzzy c-means, IIFCM optimizes the objective function iteratively by updating the membership and cluster centers. The membership value is updated subsequently as follows:

$$\mu_{ij} = \frac{1}{\sum_{k=1}^{c} \left(\frac{d\left(X'_j, V_i\right)}{d\left(X'_j, V_k\right)} \right)^{\frac{2}{m-1}}} \tag{15}$$

Further, based on the membership value, non-membership and hesitation degree are updated. In IIFCM, comparable to data point, cluster center V_i is also related with 3 IFS variables. Consequently, the center values are essentially to be updated for all the 3 variables. In the MIFCM, cluster centers are upgraded using (16)–(18) equations.

$$\mu_{V_i} = \frac{\sum_{j=1}^{n} \mu_{ij}^{m} \mu_{X_j}}{\sum_{j=1}^{n} \mu_{ij}^{m}}, \ 1 \leq i \leq c, \ 1 \leq k \leq n \tag{16}$$

$$\eta_{V_i} = \frac{\sum_{j=1}^{n} \eta_{ij}^{m} \eta_{X_j}}{\sum_{j=1}^{n} \eta_{ij}^{m}}, \ 1 \leq i \leq c, \ 1 \leq k \leq n \tag{17}$$

$$\pi_{V_i} = \frac{\sum_{j=1}^{n} \pi_{ij}^{m} \pi_{X_j}}{\sum_{j=1}^{n} \pi_{ij}^{m}}, \ 1 \leq i \leq c, \ 1 \leq k \leq n \tag{18}$$

IIFCM is an iterative process and it halts when the convergence condition is satisfied. Once the IIFCM converges, each data point is related with the membership values. Using the membership values, data points are grouped such that the data point is assigned to a cluster for which it has maximum membership value.

3   Empirical Evaluation

This section presents the empirical results of the proposed variants of FCM on all the four dataset. To find the optimal values of the parameters, we used cluster validity indices as a metric. We set the fuzzifier m value to 2 and stopping criteria G to 0.00001 empirically for all the four datasets. We varied the values of p and q from 0 to 3 with an increment of 1. If we set the values $p = 1$ and $q = 0$, RSKFCM becomes traditional FCM. On the other hand, if we set $p = 0$ and $q = 1$, then RSKFCM becomes Spatial FCM method. The experiments indicated that, if we set p and q to larger values, the results are deteriorating because of the loss of important information in the input data. Thus, we varied the value of p and q till 3.

We varied the window size from 3 to 9. When the window size is large, we had similar neighborhood data and the impact of neighborhood information was more. However, due to the increase in the computation time and bias results, the window size could not be too large. We varied the Kernel width σ value from 50 to 300. When we further increased the σ value, there was no improvement in the results. Thus, we set 300 as the maximum limit for σ.

In MIFCM, we set the fuzzifier m value to 2 and stopping criteria G to 0.00001. We varied α value from 0.1 to 30 with a uniform increment of 0.1. Since the results are not varied in between values, we presented the results only for those values which give different results. In experiments, we observed that there is no variation in the results when $\alpha > 30$. So, we set 30 as maximum limit for α value. In the proposed method, we adopted both the Sugeno's and Yager's function for the complement generator and presented results for the same. In the results, MIFCM_S represents Bustince with Sugeno's IFS complement generator and MIFCM_Y represents Bustince with Yager's IFS complement generator. To validate the proposed methods, we used Partition Coefficient (V_{pc}), Partition Entropy (V_{pe}), Fukuyama-Sugeno function (V_{fs}) and Xie-Beni function (V_{xb}) as a performance evaluation metrics [11]. Tables 1, 2, 3 and 4 present the empirically evaluated results on the four datasets.

Table 1 Cluster validity indices for Breast dataset

Breast dataset				
Methods	Metrics			
	V_{pc}	V_{pe}	$V_{xb}[1 \times 10^{-3}]$	$V_{fs}[-1 \times 10^{6}]$
FCM	0.816	0.286	83.93	185.77
RSKFCM	0.852	0.169	78.60	192.58
GSKFCM	0.897	0.932	71.29	205.69
MIFCM	0.916	0.108	63.35	235.85

Table 2 Cluster validity indices for Brain dataset

Brain dataset				
Method	Metrics			
	V_{pc}	V_{pe}	$V_{xb}[1 \times 10^{-3}]$	$V_{fs}[-1 \times 10^{6}]$
FCM	0.812	0.221	81.26	296.20
RSKFCM	0.826	0.125	76.45	310.67
GSKFCM	0.846	0.105	69.36	335.85
MIFCM	0.874	0.098	58.65	342.69

Table 3 Cluster validity indices for Liver dataset

Liver dataset				
Method	Metrics			
	V_{pc}	V_{pe}	$V_{xb}[1 \times 10^{-3}]$	$V_{fs}[-1 \times 10^{6}]$
FCM	0.624	0.170	79.45	170.85
RSKFCM	0.695	0.126	74.23	186.93
GSKFCM	0.762	0.108	70.60	206.58
MIFCM	0.819	0.099	66.35	211.60

Table 4 Cluster validity indices for Prostate dataset

Prostate dataset				
Method	Metrics			
	V_{pc}	V_{pe}	$V_{xb}[1 \times 10^{-3}]$	$V_{fs}[-1 \times 10^6]$
FCM	0.713	0.155	80.46	190.50
RSKFCM	0.765	0.119	76.53	198.65
GSKFCM	0.810	0.101	72.15	210.96
MIFCM	0.845	0.089	69.35	215.15

4 Summary

In this article, we conducted a comparative analysis on four datasets on variants of the FCM methods. From the results of the experimentation, we found that the RSKFCM and GSKFCM are less noise sensitive. Furthermore, MIFCM methods handled the uncertainty better compared to RSKFCM and GSKFCM and achieved better results. The study revealed that the FCM approaches outperform several other approaches that exist. We also presented the evaluation metrics used to assess the performance of the methods employed. Since the methods in this article are clustering-based, we used cluster validity indices as common performance evaluation metrics for all four datasets to validate the clustering results.

References

1. J.C. Bezdek, *Pattern Recognition with fuzzy Objective Function Algorithms* (Springer, 2013)
2. A. Das, S.K. Sabut, Kernelized fuzzy C-means clustering with adaptive thresholding for segmenting liver tumors. Procedia Comput. Sci. **92**, 389–395 (2016)
3. I. El-Naqa, Y. Yang, M.N. Wernick, N.P. Galatsanos, R.M. Nishikawa, A support vector machine approach for detection of microcalcifications. IEEE Trans. Med. Imaging **21**(12), 1552–1563 (2002)
4. S. Ellmann, E. Wenkel, M. Dietzel, C. Bielowski, S. Vesal, A. Maier, M. Hammon, R. Janka, P.A. Fasching, M.W. Beckmann et al., Implementation of machine learning into clinical breast mri: Potential for objective and accurate decision-making in suspicious breast masses. Plos one **15**(1), e0228446 (2020)
5. M.D. Greer, N. Lay, J.H. Shih, T. Barrett, L.K. Bittencourt, S. Borofsky, I. Kabakus, Y.M. Law, J. Marko, H. Shebel et al., Computer-aided diagnosis prior to conventional interpretation of prostate mpmri: an international multi-reader study. Eur. Radiol. **28**(10), 4407–4417 (2018)
6. A.E. Hassanien, Th. Kim, Breast cancer mri diagnosis approach using support vector machine and pulse coupled neural networks. J. Appl. Log. **10**(4), 277–284 (2012)
7. P. Hebli P, S. Gupta, Brain tumor detection using image processing: a survey 1 amruta (2017)
8. G. Hu, Z. Du, Adaptive kernel-based fuzzy C-means clustering with spatial constraints for image segmentation. Int. J. Pattern Recogn. Artif. Intell. **33**(01), 1954003 (2019)
9. M.S.S. Hunnur, A. Raut, S. Kulkarni, Implementation of image processing for detection of brain tumors, in *2017 International Conference on Computing Methodologies and Communication (ICCMC)* (IEEE, 2017), pp. 717–722

10. S.A. Kumar, B.S. Harish, Segmenting MRI brain images using novel robust spatial kernel fcm (rskfcm), in *Eighth International Conference on Image and Signal Processing* (2014), pp. 38–44

11. S.V.A. Kumar, B.S. Harish, V.N.M. Aradhya, A picture fuzzy clustering approach for brain tumor segmentation, in *2016 Second International Conference on Cognitive Computing and Information Processing (CCIP)* (2016), pp. 1–6

12. H.M. Moftah, A.T. Azar, E.T. Al-Shammari, N.I. Ghali, A.E. Hassanien, M. Shoman, Adaptive K-Means clustering algorithm for mr breast image segmenta- tion. Neural Comput. Appl. **24**(7–8), 1917–1928 (2014)

13. A.A. Nahid, Y. Kong, Involvement of machine learning for breast cancer image classification: a survey. Comput. Math. Methods Med. (2017)

14. Y. Peng, Y. Jiang, C. Yang, J.B. Brown, T. Antic, I. Sethi, C. Schmid-Tannwald, M.L. Giger, S.E. Eggener, A. Oto, Quantitative analysis of multiparametric prostate MR images: differentiation between prostate cancer and normal tissue and correlation with gleason scorea computer-aided diagnosis development study. Radiology **267**(3), 787–796 (2013)

15. S. Radhakrishna, S. Agarwal, P.M. Parikh, K. Kaur, S. Panwar, S. Sharma, A. Dey, K. Saxena, M. Chandra, S. Sud, Role of magnetic resonance imaging in breast cancer management. South Asian J. Cancer **7**(2), 69 (2018)

16. B.V. Ramana, M.S.P. Babu, N. Venkateswarlu et al., A critical study of selected classification algorithms for liver disease diagnosis. Int. J. Database Manage. Syst. **3**(2), 101–114 (2011)

17. M. Ramasamy, S. Selvaraj, M. Mayilvaganan, An empirical analysis of decision tree algo- rithms: Modeling hepatitis data, in *2015 IEEE International Conference on Engineering and Technology (ICETECH)* (IEEE, 2015), pp. 1–4

18. Repositories OAMI: http://www.aylward.org/notes/open-access-medical-image-repositories. Accessed 12 Jan 2020 (2020)

19. C.K. Roopa, B.S. Harish, S.A. Kumar, A novel method of clustering ECG arrhythmia data using robust spatial kernel fuzzy C-means. Procedia Comput. Sci. **143**, 133–140 (2018)

20. V. Shah, B. Turkbey, H. Mani, Y. Pang, T. Pohida, M.J. Merino, P.A. Pinto, P.L. Choyke, M. Bernardo, Decision support system for localizing prostate can- cer based on multiparametric magnetic resonance imaging. Med. Phys. **39**(7Part1), 4093–4103 (2012)

21. H. Shahid, J.F. Wiedenhoefer, C. Dornbluth, P. Otto, K.A. Kist, An overview of breast MRI. Appl. Radiol. **45**(19), 7–13 (2016)

22. S. Vijayarani, S. Dhayanand, Liver disease prediction using SVM and Nave Bayes algorithms. Int. J. Sci. Eng. Technol. Res. (IJSETR) **4**(4), 816–820 (2015)

23. H. Wang, Y. Liu, W. Huang, Random forest and bayesian prediction for hepatitis b virus reactivation, in *2017 13th International Conference on Nat- ural Computation, Fuzzy Systems and Knowledge Discovery (ICNC-FSKD)* (IEEE, 2017), pp. 2060–2064

24. J. Wang, C.J. Wu, M.L. Bao, J. Zhang, X.N. Wang, Y.D. Zhang, Machine learning-based analysis of mr radiomics can help to improve the diagnostic performance of PI-RADS v2 in clinically relevant prostate cancer. Eur. Radiol **27**(10), 4082–4090 (2017)

25. T. Xia, A. Kumar, D. Feng, J. Kim, Patch-level tumor classification in digital histopathology images with domain adapted deep learning, in *2018 40th Annual International Conference of the IEEE Engineering in Medicine and Biology Society (EMBC)* (IEEE, 2018), pp. 644–647

26. S. Yoo, I. Gujrathi, M.A. Haider, F. Khalvati, Prostate cancer detection using deep convolutional neural networks. Sci. Rep. **9** (2019)

27. Y. Yuan, W. Qin, M. Buyyounouski, B. Ibragimov, S. Hancock, B. Han, L. Xing, Prostate cancer classification with multiparametric MRI transfer learning model. Med. Phys. **46**(2), 756–765 (2019)

28. A. Zotin, K. Simonov, M. Kurako, Y. Hamad, S. Kirillova, Edge detection in mri brain tumor images based on fuzzy C-means clustering. Procedia Comput. Sci. **126**, 1261–1270 (2018)

A Review on Effectiveness of AI and ML Techniques for Classification of COVID-19 Medical Images

M. J. Dileep Kumar, G. Santhosh, Prabha Niranjajn, and G. R. Manasa

Abstract During the global urgency, the scientific world is searching for new innovation and technology to tackle the pandemic. It is the necessity to develop a tool that will monitor and reduce human intervention in treating COVID-19. Major techniques/tools are available to control the current situation by diagnosing the patients with the help of machine learning (ML) and artificial intelligence (AI). These techniques have helped researchers in identifying/controlling the epidemic in past decades. The use of ML and AI has helped medical fraternity to significantly improve screening, testing, predicting and treating and helped in developing vaccines for COVID-19 patients. This article presents a view of how techniques like AI and ML are used for detecting, classifying and clustering of novel coronavirus disease 2019 by using the various types of medical images.

Keywords Pandemic · COVID-19 · Machine learning · Artificial intelligence · Clustering

1 Introduction

Due to the ability of adapting to various forms of data which are not restricted by various assumptions like forms of decision-making functions, probability of distribution of variables, etc., the machine learning (ML) is emerging as a popular paradigm for many recent scientific researches. This versatile approach has enabled the use of ML in various fields ranging from forecasting financial variables [1] to analysing the

M. J. D. Kumar (✉) · P. Niranjajn
Department of ECE, NMAM Institute of Technology (VTU, Belagavi), Nitte, Karnataka 574110, India
e-mail: dileepmj@nitte.edu.in

G. Santhosh
Department of Mechanical Engineering, NMAM Institute of Technology (VTU, Belagavi), Nitte, Karnataka 574110, India

G. R. Manasa
Department of CSE, NMAM Institute of Technology (VTU, Belagavi), Nitte, Karnataka 574110, India

© The Author(s), under exclusive license to Springer Nature Singapore Pte Ltd. 2022
P. Shetty D. and S. Shetty (eds.), *Recent Advances in Artificial Intelligence and Data Engineering*, Advances in Intelligent Systems and Computing 1386,
https://doi.org/10.1007/978-981-16-3342-3_14

emotions of texts better known as sentiment analysis [2] and medical applications [3, 4]. Most recently, various medical professionals and medical councils across the globe are looking into a way of using this approach to detect new infectious samples by making use of X-ray images [5] and also to identify how the virus is spreading in the timeline by making use of confirmed cases [6].

With their capabilities to condensate high-level complex pattern to medium-to-low level complex pattern by making use of nonlinear interaction on the data set, the models developed exhibit a high degree of sensitivity to variations in their hyperparameters. These parameters are defined by the user and are specific for one model; generally, these parameters are used for training the process or as the sample data set.

As discussed by Claesen and De Moor [7], identifying the ideal combination of these parameters or predicting which of these trivial parameters to be used for a said task is not always required. With the emergence of new global pandemic COVID-19, it was observed that there is an urgent requirement for understanding and classifying the real-time information. While doing so, one must emphasize on origin of data, relevance of the information and about the characteristics of the virus [8]. The work of analysing, classifying and identifying the credibility of the data is to be done at a faster pace as observed, and the effective identification and early detection of how the virus spreads and how it affects can if not help in eradicating the virus will definitely help in reducing the speed of spreading of the virus. As identified by Leon et al. [9], it is important to have a large data set in order to assist decision-making on COVID-19's infections, and this will help in identifying various strategies to mitigate and/or suppress the spread of virus across various locations.

During the early stages of this pandemic outbreak, various scientific communities have published their findings [10, 11]. In those results, majority of the symptoms are classified as not so severe, but the clinical findings of these patients have varied in great proportions [12–14]. This causes a threat of normalizing the effect of virus as same across all group of people if we were to make the classification based on the simple features like age and gender alone. Apart from identifying the presence of virus, it is also required to detect/assess out of these identified people with the virus who is most likely to get severe illness and will face a greater risk including death itself. These are the important variables; if detected early, it can help in effective utilization of available resources (hospital beds, medical mask, respirator, capacity of the hospital, etc.) to save those who are severely affected, and as these are restricted or less in number, medical practitioners are made to take decision about a patient without having any knowledge to guide them. Due to all these constraints, an artificial intelligence (AI)-aided system can help them to make a decision based on available information limited not just to one place but across the globe.

Nowadays, AI forms a critical factor/component in healthcare systems, wherein it actively provides support in making several critical clinical decisions [15–17]. Effectiveness of the ML to identify the medical conditions like epilepsy [18, 19], nerve and muscle diseases [20, 21] and heart rhythms [22, 23] was all well documented. Also several major researches have happened to identify new data sets for developing new classifiers. Another effective predicting method is deep learning techniques which can be used in various clinical findings such as early detection of cancers, virus diseases and biomedical studies. Such techniques are proven to be efficient, and they can be used to early predict COVID-19 infection. With this as primary objective, this article presents several studies conducted by various medical professional bodies across the world and their results and accuracy in detecting the pandemic disease. This article is structured as follows: Sect. 2 discusses various case studies conducted by the researchers and their findings; Sect. 3 discusses the limitations and future developments.

2 Literature Review

COVID-19 has affected people in various ways. It has observed that majority of the people who are infected with the virus show the symptoms like fever, fatigue and dry cough [24], while others experience additional symptoms like aches and pains, nasal congestion, runny nose, sore throat and diarrhoea [25]. This pandemic has brought the healthcare system across the globe to its knee and also showed the inability of healthcare systems to manage the influx of patients in record number which resulted in the anxiety among the world population. The major cause for the rapid increase in the spread of COVID-19 is due to inability of early detection techniques or early diagnosis techniques [26]. Clinical approaches like real-time reverse transcription–polymerase chain reaction (rRT-PCR) [27] and other methods such as serologic tests [28] and viral throat swab testing [29] are necessary and widely utilized for the detection of COVID-19. But several research studies have suggested that chest radiographs (X-rays) [5] and chest computed tomography (CT) scans [30] can help in identifying the various types of lung-related anomalies including COVID-19. These medical imaging techniques can be used as an early detection tool to evaluate the degree of infection caused by COVID-19. Using these findings, one can monitor severely infected person and can predict the progression of disease [31]. But, the major constraint in such emergencies is the availability of time, which does not allow the medical examiner to run all these tests using the available traditional diagnosis [32]. Generally, this diagnosis technique requires an expert physician, but usually, these findings are prone to minor human error while taking the values or while reading or interpreting these values, but such errors even though are minor are not acceptable because such might cost a person's life. Due to the speed at which this COVID-19 has spread across the globe, all the hospitals are flooded with the patients; the conditions of these patients are either recovering from the infection or becoming severe (dying) [33]. In this case, CT scan and X-ray tests should be performed with maximum speed and efficiency to save as many lives as possible [31]. In such cases, a technology-enabled intelligent system which assists in diagnosing and classifying the patients would definitely help in reducing the burden caused on doctors and technicians. Also such intelligent systems would avoid unnecessary loading of hospitals which otherwise would have been occupied by those who does not require the critical care units [5]. The application of AI has been found in various fields, mainly in medical field wherein it is used as a tool to detect various diseases [34]. AI is used in order to get a better result or accurate detection of the disease and hence reduce strain on healthcare system [35]. By deploying AI in the field of healthcare system, it helps in reducing the time required to detect the disease if any when compared to traditional techniques [36]. As suggested in [37] in order to prevent future global health crises, we must develop an AI-enabled system which can be used to early recognize or detect the risk of epidemic diseases. Several studies have been carried out by researchers using the dataset of COVID-19 factoring in various case studies and targets [31]. Even though this software-enabled detection technique can be proven beneficial to detect and classify the COVID-19, choosing an suitable AI technique is what the major criterion which poses challenge for successful implementation of these techniques [38, 39]. The availability of diverse AI techniques is what making the task of selecting one among them to diagnose and classify the COVID-19 data a difficult procedure, as there is no profound AI method which is comparatively good from another. Along with this, most of these methods are less accurate but have least computational efficiency [40]. Another major hurdle in adopting these is associated with the method of evaluation and comparison due to availability of multiple evaluation criteria, and these criteria often have conflicts among them which further increase the complexity in selecting a suitable evaluation criterion [41]. The main goal of reviewing the articles on the topic of ML/AI techniques is to understand how the identification and classification of medical images targeted for COVID-19 can be implemented using AI/ML. Also while doing so, one can understand of what the need for research in this area and which topics are over-studied or under-evaluated.

In [31], authors reviewed the response of AI-empowered medical imaging techniques applied for COVID-19. The method of acquiring the image using the AI technique will help greatly in automating the procedure thereby reducing the contact of the medical personal with the patients and thus helping in preventing the spread of virus for these laboratory technicians. This study is mainly focused on all the process involved in diagnosing COVID-19 like medical imaging, segmentation, diagnostic report generation and finally suggesting for follow-ups to the clinic by making integrating AI with X-ray and CT images. In the study [42], it was shown that several deep learning methods can be used for diagnosing COVID-19 with the help of medical images obtained through various imaging techniques. In the decision-making confusion matrix, the authors relied on the class labels to identify the presence of infection. The identification of the disease is done on the rate of false-negative (FN) results, and this assumption affects the decision-making ability like whether to keep monitoring the patients or to discharge them. In this study, authors relied on the data set obtained by the available information from ten patients. But the results showed that out of ten negatively identified cases, two were identified as positive for COVID-19 by making use of rRT-PCR method of testing, thus resulting in 20% FN rate. Hence while designing an automated method or to create dataset for ML techniques, one must consider more number of patient information thereby reducing the FN rate of detection.

In the work [43], a scoring tool was designed targeted to understand the severity of COVID-19. The need for such tool is to provide necessary aide for the healthcare professionals in identifying and determining whether patients are suspected with or confirmed with COVID-19 infection and whether such patients are in need of respiratory interventions. In the study, the researchers followed the guidelines set by WHO in classifying the severity as mild/moderate and severe. It was identified that for a patient who was diagnosed at critical are the ones who require the ventilation, and the category of people who are diagnosed under severe category required the administration of oxygen, whereas the patients diagnosed as moderate level need not have to be administered with oxygen even though they show the symptoms of pneumonia. In this study, the researchers have utilized the information of 13,500 COVID-19 patients to form necessary data sets and to develop required decision-making matrices. The study showed the classification at 93.6%, also the underestimated value at 0.8% of patients and 5.7% of patients as overestimated. Another study involving the deep learning method for identification of COVID-19 using MobileNetV2 and Squeeze Net was carried out. Along with feature sets gathered by the deep learning method, the data was processed by applying social mimic optimization technique. This technique also made use of preprocessing technique to improve the quality of detection.

The preprocessing was done using fuzzy colour technique, thereby restructuring data classes, and structured images were stacked with the original images. After that, the required features were grouped together and then classified with the help of support vector machines (SVMs). This technique shows to be getting the rate of classification at 99.27%. A system including convolutional neural network along with patch-based technique was studied in [44]. In this, parameters are designed by making use of small numbers of training parameters to diagnose the COVID-19. This work emphasizes on the statistical study of biomarkers on the chest X-ray images, which along with identifying COVID-19 can potentially be helpful in detecting any other medical conditions also. The proposed method involves preprocessing of the images from cross-database, thus improving the accuracy of segmentation. From the result, it can be observed that preprocessing of database will result in increasing the performance of effective image segmentation. COVIDiagnosis-Net [45], an AI-enabled detection technique for COVID-19, was developed based on Bayes optimization technique. It makes use of deep learning technique to detect and classify the patients with the symptoms of pneumonia. This study showed the result with an accuracy of 98.3% in classifying the cases as normal, pneumonia and COVID-19. Using various lung images and considering multi-class and hierarchical approach, a classification method was designed in [46], to identify COVID-19. This technique used re-sampling and re-balancing for class distribution. This re-sampling method has improved macro-average F1 score to 0.89 from the earlier score of 0.65 with the use of multi-class method for the detection of COVID-19 in hierarchical classification scenario.

Classification models for AI techniques were developed using 3D imaging technique in [47], wherein one model considered entire lung region and another model considered mean score obtained by considering multiple regions inside the lung image with a fixed image resolution. The images were trained as full 3D: an original lung image with fixed resolution and hybrid 3D and a segmentation of original image with the fixed length. The training set considering full 3D showed the accuracy of

91.7%, while the hybrid model showed an accuracy of 92.4%. The other classification is detecting the other disease vs. COVID-19; the most accuracy was achieved when 3D model for classification was used; this model results in 90.8% of accuracy. While this model predicts the classification at the highest rate, there are several limitations to the proposed model, the major limitation is the dataset used for training, and all the data used for training are of those which have returned positive results from RT-PCR tests. But it was observed that the CT results were often found to be negative even when a person is found positive by RT-PCR test. So one cannot decide based on CT images only whether a person is infected with COVID-19 or not. In conclusion, using multi-national data set for training, one can develop an AI system capable of delivering a classification model with acceptable performance metrics for classifying CT images for COVID-19 disease. Also it is observed that CT image alone cannot be used to predict the COVID-19, and it can help in the assessment of the findings thereby helping in developing the variable to design a tool for AI approach. As proven by many studies, deep learning techniques have achieved far greater success in the field of radiology. In [48], a 3D deep learning framework was suggested to detect COVID-19. Using the proposed method, extracting both local and global features from the image was achieved. Various studies have applied and obtained the desirable results in detecting pneumonia in paediatric chest radiographs by making use of deep learning technique and further to differentiate viral and bacterial pneumonia in 2D paediatric chest radiographs. The study has been carried out by collecting a large number of CT samples from different locations, covering more than 1296 COVID-19 CT images. Along with those many positive samples, the data set also included 1735 CAP and 1325 non-pneumonia CT images as the control groups in the study, so that the model will be able to efficiently detect, differentiate and identify COVID-19 and other type of lung diseases. In this study, a deep learning model was designed and evaluated for detecting COVID-19 from chest CT images. On an independent testing data set, the model achieved high sensitivity (90% [95% CI: 83, 94%] and high specificity of 96% [95% CI: 93, 98%] in detecting COVID-19. The AUC values for COVID-19 and community acquired pneumonia (CAP) were 0.96 [95% CI: 0.94, 0.99] and 0.95 [95% CI: 0.93, 0.97], respectively.

3 Conclusion

The effect of COVID-19 has been felt at every corner of the world. All governments across the globe and various scientific communities are looking into a way to mitigate this new-found pandemic. Already various forms of medical examinations like X-rays and CT scans are conducted to detect the virus. The medical practitioners across the globe are limited with the knowledge about COVID-19; a computerized program can help in making the critical decisions. As observed in the study, an attempt is already being done to develop suitable model which can be used to monitor and predict the progression of the disease in patients.

References

1. B.M. Henrique, V.A. Sobreiro, H. Kimura, Literature review: machine learning techniques applied to financial market prediction. Expert Syst. Appl. **124**, 226–251 (2019)
2. J. Singh, G. Singh, R. Singh, Optimization of sentiment analysis using machine learning classifiers. HCIS **7**(1), 32 (2017)
3. M. Motwani, D. Dey, D.S. Berman, G. Germano, S. Achenbach, M.H. Al-Mallah, D. Andreini, M.J. Budoff, F. Cademartiri, T.Q. Callister et al., Machine learning for prediction of all-cause mortality in patients with suspected coronary artery disease: a 5-year multicentre prospective registry analysis. Eur. Heart J. **38**(7), 500–507 (2017)
4. J.A. Sidey-Gibbons, C.J. Sidey-Gibbons, Machine learning in medicine: a practical introduction. BMC Med. Res. Methodol. **19**(1), 64 (2019)
5. T. Ozturk, M. Talo, E.A. Yildirim, U.B. Baloglu, O. Yildirim, U.R. Acharya, Automated detection of covid-19 cases using deep neural networks with x-ray images. Comput. Biol. Med. 103792 (2020)
6. V.K.R. Chimmula, L. Zhang, Time series forecasting of covid-19 transmission in canada using lstm networks. Chaos Solit. Fract. 109864 (2020)
7. M. Claesen, B. De Moor, Hyperparameter search in machine learning. arXiv:1502.02127 (2015)
8. P. Song, T. Karako, Covid-19: real-time dissemination of scientific information to fight a public health emergency of international concern. Biosc. Trends (2020)
9. D.A. Leon, V.M. Shkolnikov, L. Smeeth, P. Magnus, M. Pechholdová, C.I. Jarvis, Covid-19: a need for real-time monitoring of weekly excess deaths. Lancet **395**(10234), e81 (2020)
10. K. Liu, Y.-Y. Fang, Y. Deng, W. Liu, M.-F. Wang, J.-P. Ma, W. Xiao, Y.-N. Wang, M.-H. Zhong, C.-H. Li et al., Clinical characteristics of novel coronavirus cases in tertiary hospitals in Hubei province. Chin. Med. J. (2020)
11. C. Huang, Y. Wang, X. Li, L. Ren, J. Zhao, Y. Hu, L. Zhang, G. Fan, J. Xu, X. Gu et al., Clinical features of patients infected with 2019 novel coronavirus in Wuhan, China. Lancet **395**(10223), 497–506 (2020)
12. J. Cai, J. Xu, D. Lin, L. Xu, Z. Qu, Y. Zhang, H. Zhang, R. Jia, X. Wang, Y. Ge et al., A case series of children with 2019 novel coronavirus infection: clinical and epidemiological features. Clin. Infect. Diseases (2020)
13. K. Kam, C. Yung, L. Cui et al., A well infant with coronavirus disease 2019 (covid-19) with high viral load. Clin Infect Dis. **10**
14. Y. Bai, L. Yao, T. Wei, F. Tian, D.-Y. Jin, L. Chen, M. Wang, Presumed asymptomatic carrier transmission of covid-19. JAMA **323**(14), 1406–1407 (2020)
15. F. Jiang, Y. Jiang, H. Zhi, Y. Dong, H. Li, S. Ma, Y. Wang, Q. Dong, H. Shen, Y. Wang, Artificial intelligence in healthcare: past, present and future. Stroke Vasc. Neurol. **2**(4), 230–243 (2017)
16. T. Davenport, R. Kalakota, The potential for artificial intelligence in healthcare. Future Healthcare J. **6**(2), 94 (2019)
17. S. Reddy, J. Fox, M.P. Purohit, Artificial intelligence-enabled healthcare delivery. J. R. Soc. Med. **112**(1), 22–28 (2019)
18. T.B. Alakus, I. Turkoglu, Detection of pre-epileptic seizure by using wavelet packet decomposition and artifical neural networks, in *2017 10th International Conference on Electrical and Electronics Engineering (ELECO)* (IEEE, 2017), pp. 511–515
19. N. Memarian, S. Kim, S. Dewar, J. Engel Jr., R.J. Staba, Multimodal data and machine learning for surgery outcome prediction in complicated cases of mesial temporal lobe epilepsy. Comput. Biol. Med. **64**, 67–78 (2015)
20. J. Yousefi, A. Hamilton-Wright, Characterizing emg data using machine-learning tools. Comput. Biol. Med. **51**, 1–13 (2014)
21. P. Karthick, D.M. Ghosh, S. Ramakrishnan, Surface electromyography based muscle fatigue detection using high-resolution time-frequency methods and machine learning algorithms. Comput. Methods Programs Biomed. **154**, 45–56 (2018)
22. M. Alfaras, M.C. Soriano, S. Ortín, A fast machine learning model for ecg-based heartbeat classification and arrhythmia detection. Front. Phys. **7**, 103 (2019)

23. C.A. Ledezma, X. Zhou, B. Rodriguez, P. Tan, V. Diaz-Zuccarini, A modeling and machine learning approach to ECG feature engineering for the detection of ischemia using pseudo-ECG. PloS One **14**(8), e0220294 (2019)
24. S. Kooraki, M. Hosseiny, L. Myers, A. Gholamrezanezhad, Coronavirus (covid-19) outbreak: what the department of radiology should know. J. Am. Coll. Radiol. (2020)
25. Y. Wang, Y. Wang, Y. Chen, Q. Qin, Unique epidemiological and clinical features of the emerging 2019 novel coronavirus pneumonia (covid-19) implicate special control measures. J. Med. Virol. **92**(6), 568–576 (2020)
26. T. Ai, Z. Yang, H. Hou, C. Zhan, C. Chen, W. Lv, Q. Tao, Z. Sun, L. Xia, Correlation of chest CT and RT-PCR testing in coronavirus disease 2019 (covid-19) in china: a report of 1014 cases. Radiology 200642 (2020)
27. Y. Fang, H. Zhang, J. Xie, M. Lin, L. Ying, P. Pang, W. Ji, Sensitivity of chest CT for covid-19: comparison to RT-PCR. Radiology 200432 (2020)
28. H. Zeng, C. Xu, J. Fan, Y. Tang, Q. Deng, W. Zhang, X. Long, Antibodies in infants born to mothers with covid-19 pneumonia. JAMA **323**(18), 1848–1849 (2020)
29. D.A. Schwartz, An analysis of 38 pregnant women with covid-19, their newborn infants, and maternal-fetal transmission of sars-cov-2: maternal coronavirus infections and pregnancy outcomes. Arch. Pathol. Lab. Med. **144**(7), 799–805 (2020)
30. M. Li, P. Lei, B. Zeng, Z. Li, P. Yu, B. Fan, C. Wang, Z. Li, J. Zhou, S. Hu et al., Coronavirus disease (covid-19): spectrum of CT findings and temporal progression of the disease. Acad. Radiol. (2020)
31. F. Shi, J. Wang, J. Shi, Z. Wu, Q. Wang, Z. Tang, K. He, Y. Shi, D. Shen, Review of artificial intelligence techniques in imaging data acquisition, segmentation and diagnosis for covid-19. IEEE Rev. Biomed. Eng. (2020)
32. H.S. Maghdid, A.T. Asaad, K.Z. Ghafoor, A.S. Sadiq, M.K. Khan, Diagnosing covid-19 pneumonia from X-ray and CT images using deep learning and transfer learning algorithms. arXiv: 2004.00038 (2020)
33. World Health Organization et al., Coronavirus disease 2019 (covid-19): situation report. 72 (2020)
34. D.D. Miller, E.W. Brown, Artificial intelligence in medical practice: the question to the answer? Am. J. Med. **131**(2), 129–133 (2018)
35. K.-H. Yu, A.L. Beam, I.S. Kohane, Artificial intelligence in healthcare. Nat. Biomed. Eng. **2**(10), 719–731 (2018)
36. A. Albahri, R.A. Hamid et al., Role of biological data mining and machine learning techniques in detecting and diagnosing the novel coronavirus (covid-19): A systematic review. J. Med. Syst. **44**(7) (2020)
37. Z. Yang, Z. Zeng, K. Wang, S.-S. Wong, W. Liang, M. Zanin, P. Liu, X. Cao, Z. Gao, Z. Mai et al., Modified SEIR and AI prediction of the epidemics trend of covid-19 in China under public health interventions. J. Thorac. Dis. **12**(3), 165 (2020)
38. M. Alsalem, A. Zaidan, B. Zaidan, M. Hashim, O. Albahri, A. Albahri, A. Hadi, K. Mohammed, Systematic review of an automated multiclass detection and classification system for acute leukaemia in terms of evaluation and benchmarking, open challenges, issues and methodological aspects. J. Med. Syst. **42**(11), 204 (2018)
39. M. Alsalem, A. Zaidan, B. Zaidan, O. Albahri, A. Alamoodi, A. Albahri, A. Mohsin, K. Mohammed, Multiclass benchmarking framework for automated acute leukaemia detection and classification based on BWM and group-VIKOR. J. Med. Syst. **43**(7), 212 (2019)
40. A. Zaidan, B. Zaidan, M. Alsalem, O. Albahri, A. Albahri, M. Qahtan, Multi-agent learning neural network and Bayesian model for real-time IoT skin detectors: a new evaluation and benchmarking methodology. Neural Comput. Appl. **32**(12), 8315–8366 (2020)
41. A. Zaidan, B. Zaidan, O. Albahri, M. Alsalem, A. Albahri, Q.M. Yas, M. Hashim, A review on smartphone skin cancer diagnosis apps in evaluation and benchmarking: coherent taxonomy, open issues and recommendation pathway solution. Health Technol. **8**(4), 223–238 (2018)
42. D. Li, D. Wang, J. Dong, N. Wang, H. Huang, H. Xu, C. Xia, False-negative results of real-time reverse-transcriptase polymerase chain reaction for severe acute respiratory syndrome

coronavirus 2: role of deep-learning-based CT diagnosis and insights from two cases. Korean J. Radiol. **21**(4), 505–508 (2020)

43. L.A. Wallis, Covid-19 severity scoring tool for low resourced settings. Afr. J. Emergency Med. (2020)
44. Y. Oh, S. Park, J.C. Ye, Deep learning covid-19 features on CXR using limited training data sets. IEEE Trans. Med. Imaging (2020)
45. F. Ucar, D. Korkmaz, Covidiagnosis-net: deep Bayes-squeezenet based diagnostic of the coronavirus disease 2019 (covid-19) from X-ray images. Med. Hypotheses 109761 (2020)
46. R.M. Pereira, D. Bertolini, L.O. Teixeira, C.N. Silla Jr, Y.M. Costa, Covid-19 identification in chest X-ray images on flat and hierarchical classification scenarios. Comput. Methods Programs Biomed. 105532 (2020)
47. S.A. Harmon, T.H. Sanford, S. Xu, E.B. Turkbey, H. Roth, Z. Xu, D. Yang, A. Myronenko, V. Anderson, A. Amalou et al., Artificial intelligence for the detection of covid-19 pneumonia on chest CT using multinational datasets. Nat. Commun. **11**(1), 1–7 (2020)
48. L. Li, L. Qin, Z. Xu, Y. Yin, X.Wang, B. Kong, J. Bai, Y. Lu, Z. Fang, Q. Song et al., Using artificial intelligence to detect covid-19 and community-acquired pneumonia based on pulmonary CT: evaluation of the diagnostic accuracy. Radiology **296**(2) (2020)

Medical Image Encryption Using SCAN Technique and Chaotic Tent Map System

Kiran, B. D. Parameshachari, and H. T. Panduranga

Abstract In today's digital world, security during medical image transmission is very important. The proposed method uses the concept of SCAN method and chaotic tent map for medical image encryption. It consists of bit plane decomposition, pixel rearranging stage and diffusion operation. Amount of information in the image is distributed into eight binary planes with a different proportional. Higher four bit planes are shuffled using spiral SCAN pattern, and lower bit planes are shuffled using diagonal SCAN pattern. Then perform pixel wise xor operation between shuffled image and the random image generated from the chaotic tent map. Experimental results show that proposed method achieves high level of security with lesser computation.

Keywords Medical image · SCAN · Tent map · Bit planes · Security

1 Introduction

Transmission of images through a communication channel requires high security which is attained by advanced encryption techniques. As the major communication is done through digital media, the importance for the security of digital images has been increased. In many applications such as military image database, medical imaging systems, confidential video conferencing, Pay-TV, etc., high end security is required for transmission as well as storing the digital images. Many consumer electronic devices such as mobile phones, tabs, laptops have an option of saving and transmitting the images with high end security on both ends via wireless networks by using service such as multimedia messaging.

Kiran (✉)
Department of ECE, Vidyavardhaka College of Engineering, Mysuru, India

B. D. Parameshachari
Department of Telecommunication Engineering, GSSS Institute of Engineering and Technology for Women, Mysuru, India

H. T. Panduranga
Department of ECE, Govt. Polytechnic, Turvekere, Tumkur, India

© The Author(s), under exclusive license to Springer Nature Singapore Pte Ltd. 2022   181
P. Shetty D. and S. Shetty (eds.), *Recent Advances in Artificial Intelligence and Data Engineering*, Advances in Intelligent Systems and Computing 1386,
https://doi.org/10.1007/978-981-16-3342-3_15

Akkasaligar et al. [1] explained DNA-based medical image encryption. Medical image pixels are encoded into DNA sequence then performed by permutation and diffusion operation. Zhang [2] proposed a fast image encryption scheme with the help of lifting scheme and chaos. Janani et al. [3] ROI-based secure image encryption has been proposed and which uses the concept of ROI encryption. Low security level due to only ROI part encryption. Akkasaligar et al. [4] DNA-based approach has been proposed for providing security to images. Zhang [5] proposed "an exchange and random access strategy" is employed to replace the traditional confusion operations. Ali and Ali [6] proposed conventional approach to encrypt the image along with Boolean operation. Anwar and Meghana [7] chaotic map-based image encryption scheme has been proposed. Ramasamy et al. [8] proposed block Scrambling and Modified Zigzag SCAN method for image encryption. Liu et al. [9] proposed a scheme for medical image encryption based on this new chaotic system. Khan and Masood [10] explained about multiple chaotic system-based image encryption scheme. For the confusion and diffusion operation, different kinds of chaotic system are employed. Nezhad et al. [11] proposed an hybrid method for encrypting an image based on DNA and tent mapping techniques. Rashmi and Supriya [12] describe encryption algorithm make use the technique of permutation and diffusion process. Dr. Parameshachari et al. [13] explained about controlling the amount of encryption using LSIC. In the exiting work, many algorithms use the concept of bit level permutation and ROI-based approach. So, due to this existing method suffers from low security level and to overcome proposed an high security for medical images.

Organization of the paper is as follows. Overview of SCAN basics and tent map are explained in Sect. 2. Working of proposed system as shown in Sect. 3. Performance parameters and their analysis for proposed work described in Sect. 4. Section 5 concludes the proposed work.

2 SCAN Method

SCAN method transforms 2D image to 1D form and those transformed results will be applied by SCAN language. There are many SCAN letters in this language. There are different SCAN orders representations for each SCAN letter. We can generate different secret images using different SCAN letters combinations. The SCAN string has to be generated once the SCAN letters combination is determined. The original image SCAN order is defined by these strings. Further, in the determined order, the original image is scanned by this method and also, the SCAN string is encrypted by commercial cryptosystems. The original image will be secured as the correct SCAN string cannot be obtained by illegal users. The image size is very large as there is no image compression technique used here. Therefore, it is very difficult to directly decrypt or encrypt the image.

The SCAN is a formal languages family which depends on 2D spatial accessing methodologies that can generate and represent a huge number of paths of scanning. Generalized SCAN, Extended SCAN and Simple SCAN are the formal languages

SCAN family that generate and represent specific scanning paths. The grammar rules which define the SCAN language are used to define simple patterns of SCAN, and the same is utilized to generate the complex SCAN patterns with the help of these simple SCAN patterns.

2.1 Chaotic Map

The chaos, one of the latest and basic modern science concepts, is the new concept which makes us understand better of our universe. The random and unorganized behavior of the real world events gives the meaning for the word chaos. For the information encryption, the new strategies are designed for chaotic dynamic systems which are widely used since 1990s. To be precise, the control parameters and the initial condition dependence with the growth of organic nature give a very good chance to use chaos for the cryptographic systems as a basis. Equation for chaotic is given by

$$X(n+1) = r * X(n) * (1 - X(n)) \tag{1}$$

where X is initial condition and its range is [0, 1], r is control parameter. If the system is chaotic system, then r value is in the range [3.5, 4].

2.2 Tent Map

The following equation gives the expression for Tent map:

$$X_{n+1} = \begin{cases} aX_n, 0 \le X_n < \dfrac{1}{2} \\ a(1 - X_n), \dfrac{1}{2} \le x < 1 \end{cases} \tag{2}$$

where the initial value is given by $X0$ and $X0 \ \varepsilon$ [0, 1]. A control parameter is between (0 and 1) and it is $a/ = 0.5$ when the system is chaotic. The tent map function is as shown in Fig. 2.

3 Proposed Method

Block diagram for proposed medical image encryption as shown in Fig. 3.

Original image having an m number of rows and n number columns undergo bit plane decomposition process. Where original image is decomposed into eight binary

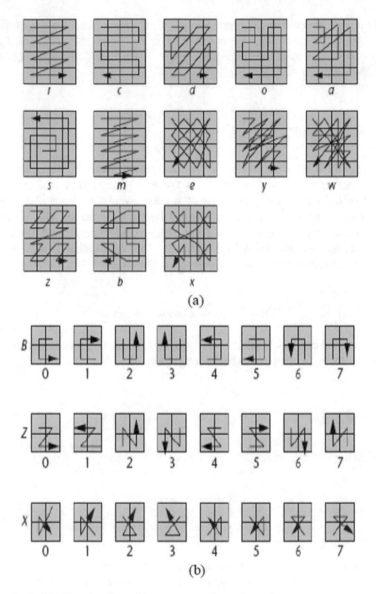

Fig. 1 **a** Basic SCAN pattern, **b** partition patterns and transformation

planes represented as B0, B1, B2, B3, B4, B5, B6, B7. B0 represents least significant bit plane and b7 represents most significant bit plane. Amount of information distributed in each plane as shown in Table 1. From Table 1, around 94% information presents in higher four bit planes and approximately 6% information presents in lower bit planes. Rearranging of the higher and lower bit planes as shown in Fig. 4. Then, pixels in the expanded higher four bit planes and lower four bit planes are shuffled

Fig. 2 Tent map signal with $a \in [0, 1]$

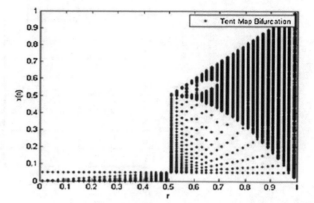

Fig. 3 Block diagram of proposed medical image encryption system

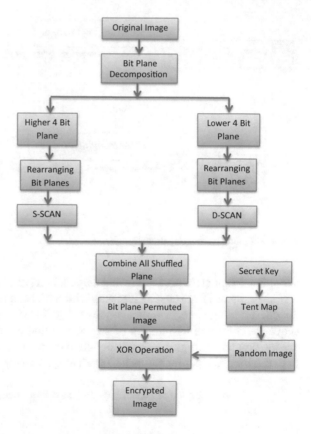

Table 1 Percentage of pixel information contributed by different bits	Bit position (i) in the pixel	Percentage $p(i)$ of the pixel information
	1	0.3922
	2	0.7843
	3	1.5686
	4	3.137
	5	6.275
	6	12.55
	7	25.10
	8	50.20

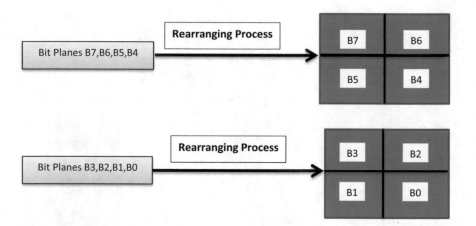

Fig. 4 Bit plane rearranging process

according to spiral SCAN pattern and diagonal bit pattern, respectively. Generate the chaotic sequence from chaotic tent map with initial condition. Later, these sequences are converted into integer value with the range 0–255 using following Eq. 3. Integer sequence converted into matrix of size $m * n$ that image matrix called as random image. Finally, encrypted image is obtained by performing pixel wise xor operation between bit plane permuted image and random image generated from tent map.

$$\text{Int\_ sequence} = \text{mod}((\text{tent\_ map sequence} * (10^9)), 256) \qquad (3)$$

4   Performance Analysis for Proposed Work

4.1   Histogram Analysis

Image histogram graphically represents the distribution of pixels based on their occurrence. For highly secured encryption technique, histogram of cipher image should be uniformly distributed.

4.2   Entropy

Entropy [14] gives the measure of information randomness in the image. The equation for entropy is given by

$$H(S) = \sum_{i=0}^{2^M-1} P(si) \log_2 \frac{1}{P(si)} \tag{4}$$

Ideal value of entropy is 8 for a random image. If it is less, the chance of predictability is more.

4.3   Mean Square Error

Mean squared error (MSE) [14] is defined as an average of the square of the difference between plain image and encrypted image. The MSE is given by the equation

$$\text{MSE} = \frac{1}{M \times N} \sum_{i=1}^{M} \sum_{j=1}^{N} [X(i, j) - Y(i, j)]^2 \tag{5}$$

4.4   Number of Pixel Change Rate (NPCR)

Plain image and encrypted image represented by C_1 and C_2, respectively. $C_1(i, j)$ and $C_2(i, j)$ are original image pixel and encrypted image pixel, respectively. The NPCR [15] is then defined as,

$$\text{NPCR} = \frac{\sum_{i,j} D(i, j)}{M \times N} \times 100\% \tag{6}$$

where D is bipolar array.

$$D(i, j) = \begin{cases} 1, & C_1(i, j) \neq C_2(i, j) \\ 0, & \text{otherwise} \end{cases}$$

4.5 Peak Signal to Noise Ratio (PSNR)

The peak signal to noise ratio is evaluated in decibels and is inversely proportional to MSE. It is given by the equation

$$\text{PSNR} = 10 \log_{10} \frac{255}{\text{MSE}} \tag{7}$$

4.6 Unified Average Changed Intensity (UACI)

It is used measure the intensity rate difference between the plain image and cipher image [15].

$$\text{UACI} = \frac{1}{N} \left[\sum_{i,j} \frac{|C_1(i, j) - C_2(i, j)|}{255} \right] \tag{8}$$

4.7 Experimental Results

Here, the proposed work has been evaluated and validated with its performance. We used MATLAB 2017b for the purpose of simulation. The simulation is carried out using one set of medical images as test images. We took these medical images from the database of Open-i image. The medical images size we used is 256×256. The medical images used for the simulation purpose are in the Table 2.

In our proposed scheme, Table 3 illustrates the histograms of the plain images, cipher images and decrypted encrypted image are evenly distributed, similar to white noise.

Table 4 illustrates the original image, encrypted image and decrypted image. In the encrypted images clearly shows that entire image is encrypted so that will not give any information about original image.

Table 2 Medical images used in the proposed system

| Medical Image 1 | Medical Image 2 | Medical Image 3 |

| Medical Image 4 | Medical Image 5 |

From Tables 5, 6 and 7 provides the performance parameter analysis of proposed work. Entropy values of encrypted images are very close to ideal value. MSE, PSNR, NPCR and UACI values are calculated between original image and encrypted image whose values also meet the security requirements, which shows that our encryption scheme meets the robustness requirements against differential attacks.

Table 8 provides the comparison of proposed work with existing method in terms of performance analysis like entropy, NPCR and UACI. From the parameter values, it is observed that proposed method achieves better values as compared to existing work.

5 Conclusion

In this paper, we proposed a medical image encryption based on SCAN pattern and tent map. Proposed method mainly consists of bit plane decomposition, SCAN-based shuffling and diffusion operation. SCAN method applied to higher and lower four bit planes separately to increase the randomness of the image. Then, xor operation is involved for the diffusion operation. Experimental results show that proposed method achieves higher security in terms of entropy, statistical analysis and differential attacks.

Table 3 Histogram analysis of proposed system

Medical images	input image histogram	Encrypted image histogram	Decrypted image histogram

Table 4 Original images, encrypted image and decrypted image of proposed system

Medical images	Encrypted image	Decrypted image

Table 5 Entropy analysis of proposed method

Medical images	Entropy of original image	Entropy of encrypted image
Medical image1	3.5963	7.9965
Medical image2	5.0030	7.9972
Medical image3	4.2654	7.9970
Medical image4	6.8129	7.9974
Medical image5	5.0923	7.9969

Table 6 NPCR and UACI of proposed method

Medical images	NPCR (%)	UACI (%)
Medical image1	99.6658	43.9856
Medical image2	99.6445	41.3774
Medical image3	99.6292	42.0747
Medical image4	99.6307	40.4066
Medical image5	99.6445	43.0771

Table 7 MSE and PSNR analysis of proposed method

Medical images	MSE	PSNR (db)
Medical image1	33.8168	22.8395
Medical image2	30.7966	23.2458
Medical image3	57.8667	20.5065
Medical image4	64.2554	30.0517
Medical image5	26.9480	24.8255

Table 8 Comparative analysis

Image	Proposed method			Existing method [16]		
	Entropy	NPCR	UACI	Entropy	NPCR	UACI
Medical image 2	7.9972	99.644	41.37	7.0425	99.56	33.46

References

1. Akkasaligar, T. Prema, S. Biradar, Medical image encryption with integrity using DNA and chaotic map, in *International Conference on Recent Trends in Image Processing and Pattern Recognition* (Springer, Singapore, 2018)
2. Y. Zhang, The fast image encryption algorithm based on lifting scheme and chaos. Inf. Sci. **520**, 177–194 (2020)
3. T. Janani, Y. Darak, M. Brindha, Secure similar image search and copyright protection over encrypted medical image databases. IRBM (2020)
4. P.T. Akkasaligar, S. Biradar, Selective medical image encryption using DNA cryptography. Inf. Secur. J.: Glob. Perspect. **29**(2), 91–101 (2020)

5. W. Zhang, Yu. Hai, Z.-L. Zhu, Color image encryption based on paired interpermuting planes. Optics Commun. **338**, 199–208 (2015)
6. T.S. Ali, R. Ali, A new chaos based color image encryption algorithm using permutation substitution and Boolean operation, in *Multimedia Tools and Applications* (Springer, Berlin/Heidelberg, Germany, 2020), pp. 1–21
7. S. Anwar, S. Meghana, A pixel permutation based image encryption technique using chaotic map. Multimedia Tools Appl. **78**(19), 27569–27590 (2019)
8. P. Ramasamy et al., An image encryption scheme based on block scrambling, modified zigzag transformation and key generation using enhanced logistic—Tent map. Entropy **21**(7), 656 (2019)
9. J. Liu et al., A new simple chaotic system and its application in medical image encryption. Multimedia Tools Appl. **77**(17), 22787–22808 (2018)
10. M. Khan, F. Masood, A novel chaotic image encryption technique based on multiple discrete dynamical maps. Multimedia Tools Appl. **78**(18), 26203–26222 (2019)
11. S.Y.D. Nezhad, N. Safdarian, S.A.H. Zadeh, New method for fingerprint images encryption using DNA sequence and chaotic tent map. Optik 165661 (2020)
12. P. Rashmi, Dr. M.C. Supriya, Encryption of color image to enhance security using permutation and diffusion techniques. Int. J. Adv. Sci. Technol. **28**(12), 375–384 (2019)
13. B.D. Parameshachari, R.P. Kiran, M.C. Supriya, Rajashekarappa, H.T. Panduranga, Controlled partial image encryption based on LSIC and chaotic map, in *ICCSP* (2019), pp. 60–63
14. Y. Wu, J.P. Noonan, S. Agaian, NPCR and UACI randomness tests for image encryption. Cyber J.: Multidisc. J. Sci. Technol. J. Sel. Areas Telecommun. 31–38 (2011)
15. J. Ahmad, F. Ahmed, Efficiency analysis and security evaluation of image encryption schemes. Int. J. Video Image Process. Netw. Secur. **12**, 18–31 (2012)
16. K.N. Madhusudhan, P. Sakthivel, A secure medical image transmission algorithm based on binary bits and Arnold map. J. Ambient Intell. Human Comput. (2020). https://doi.org/10.1007/s12652-020-02028-5

Utilization of Dark Data from Electronic Health Records for the Early Detection of Alzheimer's Disease

Sonam V. Maju and O. S. Gnana Prakasi

Abstract Information management and the presence of dark data is the most alarming and vulnerable topic that needs to be addressed amidst the rapid growth of technology. The unused, untapped, or unstructured data in our archives can be called as dark data. Under the HIPPA regulations, the clinical data needs to be stored and secured for years, so the amount of dark data in healthcare databases consumes a large amount of storage space. The paper shows the comparative performance of the random forest algorithm with and without dark data from the patient's health record for the early detection of Alzheimer's disease. The model executes in two ways, (1) considering only the Alzheimer's disease parameters and (2) including diabetes disease parameters which are considered as dark data for the Alzheimer's disease. The results of the research work clearly say that utilizing the health dark data has increased the accuracy by 16.3% and hence helps in making better decisions.

Keywords Dark data · Electronic health record · Machine learning · Random forest classifier

1 Introduction

We have discovered, designed, and invented supercomputers and expert systems to solve the complexity of every problem a human has to face in his/her daily life. The amount of data thus collected, processed, analyzed, and stored in our databases is beyond calculation and consumes more storage resources. Dark data is the data thus stored in our databases in a loose format, and we as users fail miserably to make use of it for other purposes. Dark data does not have any one true definition [1]; however, in reviewing the literature, common themes are found. Gartner originally

S. V. Maju (✉) · O. S. Gnana Prakasi
Computer Science and Engineering, School of Engineering and Technology, CHRIST (Deemed to be University), Bengaluru, India
e-mail: sonam.maju@res.christuniversity.in

O. S. Gnana Prakasi
e-mail: gnana.prakasi@christuniversity.in

© The Author(s), under exclusive license to Springer Nature Singapore Pte Ltd. 2022   195
P. Shetty D. and S. Shetty (eds.), *Recent Advances in Artificial Intelligence and Data Engineering*, Advances in Intelligent Systems and Computing 1386,
https://doi.org/10.1007/978-981-16-3342-3_16

coined the term 'Dark Data' and outlined it as, any data that is collected for a specific purpose but not used for other suitable purposes as well. Gartner describes dark data as every information asset that organizations/institutions collect, process, and store throughout the regular business activities, but usually fails to use for other purposes.

Majority of the dark data can be further labeled as unstructured content. Factually, 90% of all the information used in an organization on an everyday basis is in the unstructured format, and its escalating rate is much faster than structured data. So, dark data can be any data available and stored in our databases but not being used for multiple purposes.

Presently, almost every medical institution mostly depends on electronic health records (EHR) or the electronic medical record (EMR) to monitor a patient's present condition, to record diagnostic evidence, procedures performed on the patient, and past treatment results. EHR is one of the valuable resources for large-scale medical data analysis [2]. The development of technology and hospital information system is giving high importance and also been widely promoting EMR/EHR thus considered as a treasure for large-scale medical analysis of clinical/health data. This EMR can be classified into three kinds: structured data, semistructured data, and unstructured data. Structured data is the data stored that is easily available in databases in a particular framework which includes basic information like birth details, medications, vital signs, etc. Semistructured data will have a flow chart format, like resource description files, name, value, and time-stamp. Unstructured data is considered as one of the loosely structured data with no framework, and at the same time, it carries highly valuable information including handwritten clinical notes, surgical images, discharge reports, radiology, and pathology reports. By the conversion of these unstructured data to structured one, a huge amount of data can be made available for research analysis and learning from the last 20 years by the amalgamation of technology and health care. The results literally saved and are saving millions of lives as the diseases can be diagnosed early through data analysis on machine learning models and treatment can be done forehand.

Alzheimer's is one such disease that can be predicted in a very early stage considering different parameters. Along with these parameters, this research work makes it possible that early detection of diseases is possible by identifying the relationship between diseases and the presence of one disease can trigger the occurrence of another disease.

In this paper, dark data is extracted from the health records and is used to derive efficient and useful conclusions like identifying the relationship between Alzheimer's and its prediction at a very early stage using dark data.

2   Review of Literature

Dark data is an emerging term; the existence, identification, and utilization of dark data are one of the prominent research domains in the future.

Every electronically stored information (ESI) will come under focus to legal discovery if a threat of litigation arises, even in incomplete or trivial data [3]. The untagged/untapped or forgotten data called dark data can bring huge security risks if we did not manage it properly. The study [3] reveals the hidden risk of selling used electronics without improper deletion and management of data. The pre-owned users of the electronic devices can actually retrieve confidential information like the login passwords and misuse them, if it is not managed wisely.

IBM is doing an enormous amount of research on dark data and [4] IBM claims that 80% of the data around us is dark data and not all the collected IoT sensor data are being used for the analysis.

In the field of medical industry, the health informatics emerged back in 1960s, and the first system, MED-SYNDIKATE, on medical language processing which automatically retrieves clinically relevant information was released in 2000s [5]. Four different supervised multi-label classifiers were experimented [5], which are decision trees, multinomial naive Bayes, conditional random fields (CRFs), and support vector machines (SVM). From the observations, the performance of conditional random fields (CRFs) and support vector machines (SVM) was significantly better than other classifiers like naive Bayes and decision trees. Also, comparing CRFs and SVMs, CRFs performed better because it can identify entities in longer spans.

Identifying the relationship between diseases is an important parameter in the early detection of diseases. Patients with seizures, epilepsy, and Parkinson's are at increased risk of developing Alzheimer's disease. Friedman et al. [6] suggested the positive association between two neurological disorders, epilepsy and Alzheimer's, which are seen commonly in elderly people. Asadollahi et al. [7] discussed some additional parameters such as medications, down syndrome, dementia, and cognitive impairment as the potential risk factors that leads seizures in Alzheimer's disease.

Ott et al. [8, 9] related the association of diabetics and dementia based on the renowned large population study conducted in Rotterdam and the observations from the study shows that diabetic patients treated with insulin are having high risk of Alzheimer's disease with 4.3 fold. In the same time frame, another longitudinal cohort study was conducted with T2DM patients [10] and shows a bigger risk of all dementias and Alzheimer's (2.27 fold for men and 1.37 fold for women), supporting the Rotterdam study by proving it again that Alzheimer's and diabetics are related. The studies [11, 12] concluded that T2DM can increase the risk of Alzheimer's amid 1.3 fold and 1.8 fold. Schilling [13] analyzes and integrates multiple approaches used in different genre of studies, making it possible to unravel the research opportunities on Alzheimer's disease.

3 Proposed Model

Any data which is collected and stored but fails to use for improving the performance of other purposes is called dark data. Electronic health records are the systematic collection of patient details in electronic format which contains the personal information, medical reports, scanning images, consultation reports, etc., of a patient.

Diabetes parameters stored in EHR are considered as dark data in this research as it is not generally used in Alzheimer's prediction. As defined, this research focuses on the early prediction of Alzheimer's disease by considering some diabetics parameters in addition to the parameters of Alzheimer's disease.

Generally, for early prediction of Alzheimer's, features like Mini-Mental State Examination (MMSE), normalized whole-brain volume (nWBV), estimated total intracranial volume (eTIV), and Atlas scaling factor (ASF) are used from the EHR records.

However, the EHR data of any health record includes many other features; for example, the diabetes-related parameters may be recorded but not utilized for calculating the results, and thereby becoming dark data. In this research, along with the Alzheimer's features, the relation of Alzheimer's with other diseases is identified, and those parameters are also included in training the prediction model to increase the accuracy of the model and for better predictions.

The correlation of diabetes and Alzheimer's is discussed [14, 15], and the parameters of diabetes like glycated hemoglobin (Glyhb), also known as A1C test and stabilized glucose (stab.glu) from the EHR, are also included along with the Alzheimer parameters.

The above-mentioned parameters are extracted from electronic health records, preprocessed, and divided into training and testing data, and then, the data will undergo random forest machine learning classifier, and Alzheimer's is predicted more accurately. Later, area under the curve is used along with accuracy and recall as the performance metrics. The workflow of the proposed model is represented in Fig. 1.

3.1 Data Collection

As the research work requires identification of dark data and usage of the same for better decision making, manipulation of the dataset is required, and it is mandatory to make a dataset of EHR which contains all data. Thus, a new dataset is created in order to identify the relation between two or more diseases, and so, the merging of independent disease datasets is important.

The initial dataset used in the research work is longitudinal MRI data in non-demented and demented older adults by the Open Access Series of Imaging Studies (OASIS) [16]. This dataset includes a longitudinal collection of patients who are aged from 60 and above. All the patients are right-handed, and both men and women are taken into consideration. From the total number of patients, 72 patients were categorized as non-demented, 64 patients were diagnosed as demented, and remaining 51 individuals had mild to moderate Alzheimer's disease. The rest of the 14 patients were categorized as non-demented initially, and they do have the chances of showing signs of dementia in the later period.

The variables of the dataset are Patient.ID, MRI.IDs, Group (non-demented/demented), Number of Visits and MR.delay. The dataset also includes

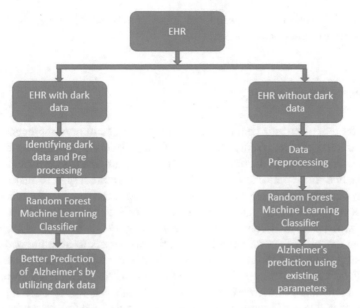

Fig. 1 Workflow of proposed model

Gender, Hand, Age, Years of Education (EDUC), Socioeconomic status (SES) ranging from 1 to 5 (highest status and lowest status). Other variables that are considered important in the model are Mini-Mental State Examination (MMSE from 0 to 30) and Clinical Dementia Rating (CDR, 0 = no dementia, 0.5 = very mild dementia, 1 = mild dementia, 2 = moderate Dementia). The derived anatomic volumes or the estimated total intracranial volume (eTIV), normalized whole-brain volume (nWBV), which is the percent of all voxels labeled as the gray or white matter in the atlas-masked image taken from the automated tissue segmentation process and finally ASF—Atlas scaling factor—which is a computed scaling factor that changes the native-space brain and skull to the atlas target.

The original data came from the biostatistics program at Vanderbilt [17], and the dataset used here is a manipulated dataset with diabetes parameter. Patients without a hemoglobin A1c test is excluded. If the patient's hemoglobin A1c is 7 or greater, then they are labeled with diabetic. Glycated hemoglobin (Glyhb) or the A1C test is a blood test that specifies the average blood sugar level of a patient for the past few months. Standard levels are in the range of 4–5.6%, and a result between 5.7 and 6.4% is measured as pre-diabetic, and an A1C level of 6.5% or higher indicates that the patient is diabetic. A1C test is a commonly used blood test to diagnose type-1 and type-2 diabetes. A1C test is used to measure the percentage of hemoglobin. As the A1C level increases, the blood sugar control is very poor, and thereby the risk of diabetes complications is high. The new variables of the EHR datasets other than OASIS are cholesterol (chol.), stab.glu, high-density lipoprotein (hdl) also known as good cholesterol, Glyhb, location, height, weight, waist, hip, bp.1s, bp.2s, bp.2d, time.ppn, results and condition.

3.2 Data Preprocessing

The database of electronic health record is a composition of non-homogeneous data sources which is obtained from the datasets of clinical records, and the data contained is redundant, diverse, and incomplete, which directly affects the performance of the model. Hence, the electronic clinical data must be preprocessed to make sure the accuracy, completeness, and consistency of the EHR dataset [6]. As our research work is a comparison, the data preprocessing steps are the same for both datasets. In the first dataset, SES column is having missing values in eight rows, and this issue can be dealt with two methodologies; the first one is to drop the rows having the missing values, and the second is to substitute the missing values from the eight rows with different matching values; the second methodology is called 'imputation.' While implementation, before imputation, all the missing values were replaced as NaN values and was triggering errors, and thus, we decided to impute the detected missing values by inferring them from the known parts of the available data, and we are using median for the imputation. As the dataset contains only less than 500 data, imputation is the right option to accelerate the performance of our proposed model. In the second dataset, as the diabetes features are also added, other than 8 rows of SES missing values, there are missing values in other features of EHR dataset. In the selected features, only SES and stab.glu had missing values. So, both the datasets will undergo same preprocessing steps.

3.3 Machine Learning Classifiers

Machine learning approaches transform clinical IE tasks into classification problems. In this, random forest machine learning classifier is used for the early detection, and this classifier can perform both regression and classification. Since the main objective is to focus the utility of dark data in EHR, we compare the performance of random forest classifier in Alzheimer disease dataset with and without dark data.

4   Results

In this, the prediction of the Alzheimer's disease with and without dark data is evaluated using random forest classifier. Both the datasets are first read and will undergo all the preprocessing steps, and then, the data will be split into training and testing sets, finally the model is trained and tested with random forest classifier. As

Table 1 Performance metrics

Performance metrics	Alzheimer dataset	EHR dataset
Accuracy of the validation set	0.7785	0.9927
Accuracy with the best parameters	0.8421	0.9893
Recall with the best parameters	0.7	0.9787
AUC score	0.85	0.9893

the performance metric, the accuracy and recall of best parameters, accuracy of the validation set and the AUC score is calculated. The observations of implementation are shown in Table 1, and the accuracy of the model increased by 17% and AUC score by 16.3%

The importance of each feature used in training and testing the model is also calculated. The best four features of Alzheimer datasets are MMSE, nWBV, ASF, and eTIV respectively. After identifying and utilizing the dark data of EHR from this application, which is diabetics parameters, the best performing features of second dataset for the early prediction Alzheimer's are, glyhb, MMSE, nWBV, and ASF. The ROC curves of both the datasets are shown in Figs. 2 and 3.

The importance of each feature used in training and testing the model is also calculated. The best four features of Alzheimer datasets are MMSE (0.385), nWBV (0.159), ASF (0.125), and eTIV (0.088), respectively. After identifying and utilizing the dark data of EHR from this application, which is diabetics parameters, the best performing features of the second dataset for the early prediction of Alzheimer's are Glyhb (0.9), MMSE (0.02), nWBV (0.01) and ASF(0.01). The ROC curve of both the datasets is shown in Figs. 1 and 2.

Fig. 2 ROC of Alzheimer's dataset

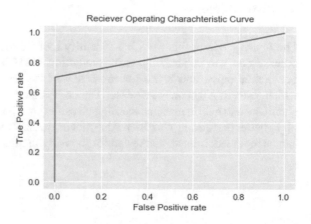

Fig. 3 ROC of EHR dataset

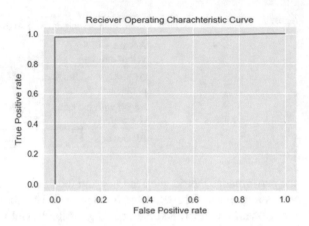

5 Conclusions

In this paper, the utilization of dark data helps to improve the prediction of Alzheimer's in a very early stage. Thus, the additional parameters of diabetes from the electronic health records increase the AUC score to 0.98 from 0.85 which clearly shows the better performance of the proposed model. Along with the AUC score, other performance metrics were accuracy and recall. Accuracy of the validation set got increased from 0.77 to 0.99, the accuracy of best parameters increased from 0.84 to 0.98, and the recall score increased from 0.7 to 0.9. Overall, the performance of the model increased when the diabetes feature got added to the Alzheimer's features for the prediction of early Alzheimer's.

6 Future Scope

The future scope of this work is to identify more associativity of Alzheimer's and other diseases to enhance the early prediction of Alzheimer's in a broad manner. Seizure, epilepsy, and Parkinson are a few of the factors that can be considered, which leads to Alzheimer's disease. The comparison of different machine learning classifiers and hybrid machine learning classifiers can be used on the same datasets to find a better performing classifier. Also, not only Alzheimer's but similar architecture can be modeled, trained, and tested for the early prediction of other diseases.

References

1. V. Seehusen, E. Maldonado, Using a roadmap in the back alleys of dark data. J. Technol. Res. **9** (2020)
2. W. Su, Z. Cai, Y. Li, F. Liu, S. Fang, G. Wang, Data processing and text mining technologies on electronic medical records: a review. J. Health Care Eng. (2018)
3. W. Dimitrov, C. Сярова, L. Petkova, Types of dark data and hidden cyber-security risks. Technical Report (2018). https://doi.org/10.13140/RG.2.2.31695.43681
4. https://developer.ibm.com/technologies/analytics/articles/ba-data-becomes-knowledge-3/
5. M. Tao, Clinical information extraction from unstructured free-texts. PhD thesis (Department of Information Science, University at Albany, 2018)
6. D. Friedman, L.S. Honig, N. Scarmeas, Seizures and epilepsy in Alzheimer's disease. CNS NeuroSci. Ther. **18**(4) (2012). https://doi.org/10.1111/j.1755-5949.2011.00251
7. M. Asadollahi, M. Atazadeh, M. Noroozian, Other Demen, seizure in Alzheimer's disease: an underestimated phenomenon. Am J Alzheimers Dis. **34**(2), 81–88 (2019). https://doi.org/10.1177/1533317518813551
8. A. Ott, M.M.B. Breteler, F. Van Harskamp, D.E. Grobbee, A. Hofman, Incidence of Alzheimer's disease and vascular dementia in the Rotterdam study. 970–973 (1996). https://doi.org/10.1136/bmj.310.6985.970
9. A. Ott, R. Stolk, F. Van Harskamp, H. Pols, A. Hofman, M. Breteler, Diabetes mellitus and the risk of dementia: The Rotterdam study. Neurology **10**, 1937–1942 (1999)
10. C.L. Leibson, W.A. Rocca, V.A. Hanson, R. Cha, E. Kokmen, P.C. O'Brien, P.J. Palumbo, The risk of dementia among persons with diabetes mellitus: a population-based cohort study. Ann N Y Acad Sci **826**, 422–427 (1997)
11. R. Peila, B.L. Rodriguez, L.J. Launer, Type 2 diabetes, APOE gene, and the risk for dementia and related pathologies—The Honolulu-Asia aging study. Diabetes **51**, 1256–1262 (2002)
12. W.L. Xu, C.X. Qiu, A. Wahlin, B. Winblad, L. Fratiglioni, Diabetes mellitus and risk of dementia in the Kungsholmen project—A 6-year follow-up study. Neurology **63**, 1181–1186 (2004)
13. M.A. Schilling, Unraveling Alzheimer's: making sense of the relationship between diabetes and Alzheimer's disease. J Alzheimer's Dis. **51**(4), 961–977 (2016). https://doi.org/10.3233/JAD-150980.PMID:26967215;PMCID:PMC4927856
14. N.R. Rajeshkanna, S. Valli, P. Thuvaragah, Relation between diabetes mellitus type 2 and cognitive impairment: a predictor of Alzheimer's disease. Int. J. Med. Res. Health Sci. **3**(4), 903 (2014)
15. A.M. Ortiz Zuñiga, R. Simó, O. Rodriguez-Gómez, C. Hernández, A. Rodrigo, L. Jamilis, L. Campo, M. Alegret, M. Boada, A. Ciudin, Clinical applicability of the specific risk score of dementia in Type 2 diabetes in the identification of patients with early cognitive impairment: results of the MOPEAD study in Spain. J Clin Med. **9**(9), 2726 (2020). https://doi.org/10.3390/jcm9092726.PMID:32847012;PMCID:PMC7565958
16. https://www.kaggle.com/jboysen/mri-and-alzheimers
17. https://data.world/informatics-edu/diabetes-prediction

Brain Tumor Segmentation Using Capsule Neural Network

Jyothi Shetty, Shravya Shetty, Vijaya Shetty, and Chirag Rai

Abstract In medical science, one of the deadly diseases demonstrated by specialists is a brain tumor. Early identification of brain tumors is responsible for improving treatment possibilities, and chances of survival of the patients can be increased. Deep learning techniques can evaluate the huge amounts of MRI image data. In this paper, capsule neural network (CapsNet)-based model is used for the image segmentation to extract tumor regions. The proposed method was assessed using BraTS18 data consisting of two sets, HGG and LGG. The model is using dice coefficient, sensitivity and specificity metrics and also is compared with the U-Net model.

Keywords Brain tumor · Deep learning · Segmentation · Capsule neural network

1 Introduction

In the last decades, brain tumor segmentation has been the most trending topic in the research field. In today's world, brain tumor has become one of the reasons in rise of mortality among the people. The neoplasm called gliomas can affect the functionality of the brain depending on its location and size of growth [1]. Medical imaging technique termed as magnetic resonance imaging (MRI) is used to examine the brain tumor in the nervous system. Brain tumors cannot be identified at the beginning stage, and some tumors cannot be recognized ever after the appearance of the symptoms. Deep learning method helps in the early detection of tumor. In this work, BraTS 2018 [2] datasets consist of four different MRI modalities as shown in Fig. 1. These modalities help to give information about the irregular shaped tumors. The four modalities of an MRI image differ in their structure based on their shape, tumor border or color [3]. It is possible to find several methods in brain tumor

J. Shetty (✉) · S. Shetty · C. Rai
NMAM Institute of Technology, Nitte, India
e-mail: jyothi_shetty@nitte.edu.in

V. Shetty
NMIT, Bangalore, India
e-mail: vijayashetty.s@nmit.ac.in

P. Shetty D. and S. Shetty (eds.), *Recent Advances in Artificial Intelligence and Data Engineering*, Advances in Intelligent Systems and Computing 1386,
https://doi.org/10.1007/978-981-16-3342-3_17

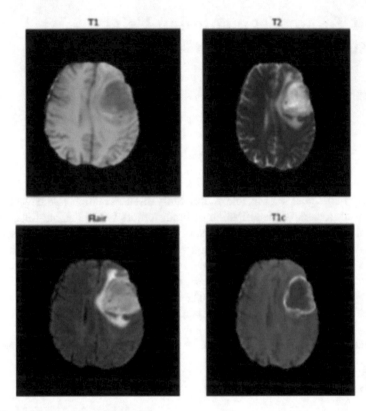

Fig. 1 Four different MRI modalities [3]

segmentation that are gaining importance in medical science and are complicated because of its unstable tumor shape [4].

The brain tumor is detected by reading the MRI image of brain and applying segmentation method to identify the tumor region [5]. Various machine learning-based segmentation methods were implemented based on their similarity or dissimilarity measurements. In this paper, a segmentation model is proposed using a new architecture called capsule neural network (CapsNet) with dynamic routing has been introduced. Capsule neural network model stores the information about the input by replacing max pooling layers with convolutional strides and dynamic routing [6].

This model is evaluated using dice coefficient, sensitivity and specificity metrics for HGG and LGG cases for these different models. This model is also compared with U-Net model, and results show that CapsNet model performs better than the U-Net model [7]. The U-Net and CapsNet designs are assessed using leaderboard BraTS 2018 datasets consisting of LGG and HGG patients. The predicted result is differentiated with manual described as ground truth; CapsNet model acquired good results.

2 Literature Review

Segmentation is the method where objects having same kind of properties are put in a group together and dissimilarity ones are separated. Segmentation of brain tumor has a great consequence in medical image segmentation. The researchers have proposed several papers on brain tumor segmentation. Several machine learning algorithms are used to analyze and understand the concept of brain tumor segmentation. K-means clustering gave best result when comparing to other algorithms [8].

Subbanna and Arbel [9] used Markov random fields (MRFs) for segmenting the brain tumor. Wu et al. [10] introduced superpixel features. To segment brain tumor images, superpixel features were employed using conditional random fields framework, but the results were not up to the mark as it did not perform well with LGG cases. Randomized forest was introduced for classifying features that achieved dice score of 83% [11]. Yang et al. [12] merged classification technique randomized trees with another algorithm named superpixel-based segmentation techniques. Single FLAIR sequence-based MRI is used for the performance of the model. Pereira et al. [13] implemented CNN algorithm for segmentation. The fully connected layers were used as classifiers in their model.

3 Dataset

The experiments were conducted on multimodal brain tumor segmentation challenge (BraTS) 2018 datasets to assess the performance of the model. It consists of two sets of data containing 210 high-grade glioma (HGG) and 75 low-grade glioma (LGG). It includes four MRI modalities such as T1-weighted (T1), T2-weighted (T2), FLAIR and T1-weighted imaging with gadolinium enhancing contrast (T1c) images with corresponding ground truth images. These four modalities of MRI are included in both HGG and LGG data. The three modalities of MRI are T1, T2 and FLAIR images of each patient cataloged into the MRI modality of T1c image. These four modalities differ from each other based on their features and also contain instructions to predict the area and tumor type. Voxels of each image are classified into five categories: enhancing core, necrosis, non-enhancing core, edema and background noise. BraTS18 data consists of 3D image, each with size of 240 × 240 × 155.

4 Preprocessing

Preprocessing of data is used to transform unprocessed data to clean data for better performance. It helps to train the network very smoothly and also to enhance the overall execution of the model to obtain good results. The code has been executed using Colab notebooks which support free GPU. The preprocessing techniques used are bias correction, 2D slicing, cropping and data augmentation. As smooth and less frequency noise signal is attached to the image, N4ITK method is used to remove this effect. 2D slicing BraTS 2018 datasets consist of 3D images. These 3D images are sliced as 2D NumPy array. Each image generated 155 NumPy arrays. All 155 slices do not show tumor region, and hence, only the mid-portion of slices needs to be considered. We took from 30th slice to 120th slice for creating the final data. All are resized and cropped to center with final dimension of (N1,192,192,4). Data augmentation technique is used to solve data imbalancing problem that results in unequal distribution of classes. Augmentation helps in improving balancing classes in the training datasets. Augmentation approach like rotation, flipping and enhancement of image is being used by the experts as a preprocessing method. In this work, flipping method is used. Finally, the data was randomly split into training, validation and test data with 60%:20%:20% of ratio, respectively.

5   Evaluation Metrics

Dice similarity co-efficient:

It measures the comparability between the ground truth label and the predicted label.

$$DSC = 2 * \frac{TP}{FP + 2 * TP + FN}$$

Sensitivity:

Sensitivity calculates the amount of true positive rate or recall rate (risk class).

$$Sensitivity = \frac{TP}{TP + FN}$$

Specificity:

Specificity computes the amount of true negatives (normal class).

$$Specificity = \frac{TN}{TN + FP}$$

6   Results

In capsule neural network architecture, the model is assessed using training and validation sets for HGG and LGG cases. Firstly using HGG data, the model is trained on 3780 samples, validated on 1260 samples. The results obtained for CapsNet model are shown below.

Figure 2a, b shows the result of dice coefficient score and dice coefficient loss for training and validation of HGG set, respectively.

In Fig. 3a, b, results of sensitivity and specificity for validation and training HGG set are plotted.

Next, using LGG set of data that consists of 75 patients scan images. We have trained CapsNet model on 4050 samples, validated on 1350 samples.

The result of dice coefficient score and loss for training and validation set is plotted, respectively, in Fig. 4a, b.

In Fig. 5a, b, sensitivity and specificity results are plotted. Figures 6 and 7 show the prediction of HGG and LGG data from BraTS18 datasets using CapsNet model.

Table 1 shows the dice similarity coefficient, sensitivity and specificity for HGG and LGG test sets using CapsNet model. We have compared these results with the dice similarity coefficient (DSC), sensitivity and specificity results for HGG and LGG test sets using U-Net model. Results of U-Net model are as shown in Table 2. This study shows that the CapsNet model has given best result on testing data.

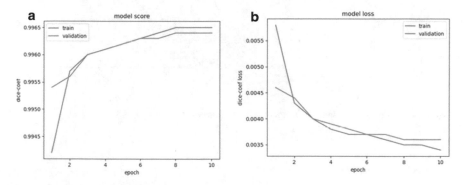

Fig. 2 **a** Model score for HGG set. **b** Model loss for HGG set

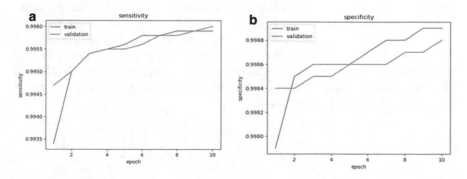

Fig. 3 **a**. Sensitivity for HGG set. **b** Specificity for HGG set

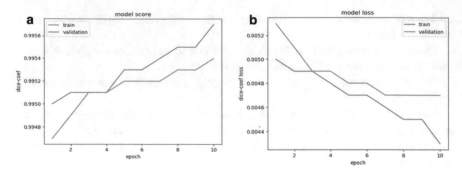

Fig. 4 **a**. Model score for LGG set. **b** Model loss for LGG set

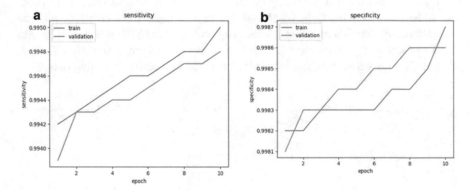

Fig. 5 **a**. Sensitivity for LGG set. **b** Specificity for LGG set

7 Conclusion

In biomedical engineering, brain glioma segmentation is the task of segmenting tumors from normal brain tissues. Fully convolutional network (U-Net) algorithm and capsule neural network (CapsNet) algorithm are constructed to extract the tumor regions. Validation is done using BraTS2018 leaderboard dataset holding two sets: 70 high-grade gliomas (HGG) and 75 low-grade gliomas (LGG) patient scans each set split into training, validation and testing data. U-Net and CapsNet networks are trained and assessed using dice coefficient, sensitivity and specificity of HGG and LGG data. Comparing the U-Net results with the CapsNet results, capsule neural network model gave a promising result.

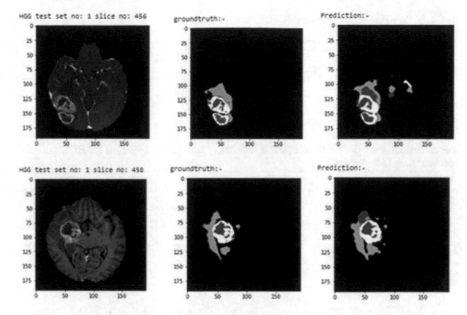

Fig. 6 Prediction of a HGG subject using CapsNet model

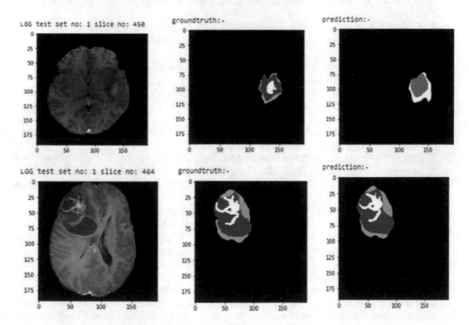

Fig. 7 Prediction of a LGG subject using CapsNet model

Table 1 Dice similarity coefficient (DSC), sensitivity and specificity results for HGG and LGG test sets using CapsNet model

Dataset	Model name	Metrics	HGG data	LGG data
BraTs 2018 Leaderboard	CapsNet model	DSC	0.9841	0.9953
		Sensitivity	0.9826	0.9947
		Specificity	0.9944	0.9953

Table 2 Dice similarity coefficient (DSC), sensitivity and specificity results for HGG and LGG test sets using U-Net model

Dataset	Model name	Metrics	HGG data	LGG data
BraTS 2018 Leaderboard	U-Net model	DSC	0.9796	0.9792
		Sensitivity	0.9780	0.9797
		Specificity	0.9928	0.9927

References

1. M. Lyksborg, O. Puonti, M. Agn, R. Larsen, An ensemble of 2d convolutional neural networks for tumor segmentation, in *Scandinavian Conference on Image Analysis* (Springer, 2015), pp. 201–211
2. B.H. Menze, A. Jakab, S. Bauer, J. Kalpathy-Cramer, K. Farahani, J. Kirby et al., The multi-modal brain tumor image segmentation benchmark (BRATS). IEEE Trans. Med. Imaging **34**(10), 1993–2024 (2015)
3. B. Menze et al., The multimodal brain tumor image segmentation benchmark (BRATS). IEEE Trans. Med. Imaging **34**(10), 1993–2024 (2015)
4. J. Long, E. Shelhamer, T. Darrell, Fully convolutional networksfor semantic segmentation in *2015 IEEE Conference on Computer Vision and Pattern Recognition (CVPR)* (IEEE, 2015), pp. 3431–3440
5. F. Milletari, N. Navab, S.-A. Ahmadi, V-Net: fully convolutional neural networks for volumetric medical image segmentation (2016), pp. 1–11
6. S. Sabour, N. Frosst, G.E. Hinton, Dynamic routing between capsules, in *Proceedings of the 31st Conference on Neural Information Processing Systems (NIPS)* (2017) pp. 3859–3869
7. O. Ronneberger, P. Fischer, T. Brox, U-Net: convolutional networks for biomedical image segmentation. arXiv, cs.CV (2015)
8. K.M. Nimeesha, R.M. Gowda, Brain tumour segmentation using K-means and fuzzy c-means clustering algorithm. Int. J. Comput. Sci. Inf. Technol. Res. Excell. **3**, 60–65 (2013)
9. N.K. Subbanna, T. Arbel, Probabilistic Gabor and Markov random fields, segmentation of brain tumours in MRI volumes, in *Proceedings of the MICCAI-BRATS* (2012)
10. W. Wu, A.Y.C. Chen, L. Zhao, J.J. Corso, Brain tumor detection and segmentation in a CRF (conditional random fields) framework with pixel-pairwise affinity and superpixel-level features. Int. J. Comput. Assist. Radiol. Surg. **9**(2), 241–253 (2013)
11. L. Szilágyi, L. Lefkovits, B. Benyó, Automatic brain tumor segmentation in multispectral MRI volumes using a fuzzy c-means cascade algorithm, in *2015 12th International Conference on Fuzzy Systems and Knowledge Discovery (FSKD)* (2015), pp. 285–291
12. M. Soltaninejad, G. Yang, T. Lambrou, N. Allinson, T.L. Jones, T.R. Barrick, F.A. Howe, X. Ye, Automated brain tumour detection and segmentation using superpixel-based extremely randomized trees in FLAIR MRI. Int. J. Comput. Assist.
13. S. Pereira, A. Pinto, V. Alves, C.A. Silva, Brain tumor segmentation using convolutional neural networks in MRI images. IEEE Trans. Med. Imaging **35**, 1240–1251 (2016)

Forgery Detection and Image
Recommendation Systems

Using Machine Learning for Image Recommendation in News Articles

Rohit Jere⒟, Anant Pandey⒟, Hasib Shaikh⒟, Sulochana Nadgeri⒟, and Pragati Chandankhede⒟

Abstract The recent developments in the machine learning domain have changed the journalism workflow significantly. A growing interest is being observed in adapting to technologies that can scan, process, filter, and even suggest photos for news articles. This research paper focuses on image selection by harnessing keyword extraction. The terms present in the article are first converted to a machine-readable format and then ranked based on three characteristics, namely first occurrence, term frequency, and entity category. Four variations of machine learning approach and simple statistical calculations are applied for the image selection process followed by that. Then, the highest-ranking terms are used for search query generation that involves a heuristic method predicting threshold probability values for the terms.

Keywords Machine learning · Open Calais entity detection · Statistical calculation

1 Introduction

Artificial Intelligence is streamlining the workflow for media outlets globally for curating as well as writing news articles. Microsoft has been using AI to scan for content and then process and filter it and even suggest photos for human editors to pair it with. Microsoft had been using human editors to curate top stories from a variety of sources to display on Microsoft News, MSN, and Microsoft Edge [1].

However, selecting suitable images for these articles is not an area that is gaining traction in this domain. Images considerably impact the reader's mind and studies show that they help in recalling the article better. Selecting a large number of images for the articles can be tedious and automating this process can be immensely beneficial for the newsroom. This research leverages keyword extraction mechanisms, computationally efficient statistical calculations, and machine learning to produce search queries that return suitable images for news articles. Keyword extraction is an automated process that collects a set of terms, illustrating an overview of the

R. Jere (✉) · A. Pandey · H. Shaikh · S. Nadgeri · P. Chandankhede
Excelsior Education Society's, K. C. College of Engineering and Management Studies and Research, Kopri, Thane (East), Maharashtra, India

© The Author(s), under exclusive license to Springer Nature Singapore Pte Ltd. 2022   215
P. Shetty D. and S. Shetty (eds.), *Recent Advances in Artificial Intelligence and Data Engineering*, Advances in Intelligent Systems and Computing 1386,
https://doi.org/10.1007/978-981-16-3342-3_18

document. The term is defined as how the keyword identifies the core information of a particular document [2]. The images for this research are collected from Getty Images and the news articles from BBC News Website.

The plain article text is first converted into a machine-readable format through tokenization, n-gram selection, stopword removal, stemming, and feature assignment. These terms are then ranked based on first occurrence, term frequency, and entity categorization. The image's metadata is preprocessed to find better keywords for searching the right images. Metadata is "data that provides information about other data" [3]. Also, in this research paper, "Keywords" are described as terms that best illustrate a news article.

The training data is generated with a matcher component that assigns Boolean values to the terms upon comparing article terms with image tags. These terms are transformed into a vector that forms the input layer of the neural network. A neural network is a network or circuit of neurons, or in a modern sense, an artificial neural network, composed of artificial neurons or nodes [4]. A training data generator component is used to convert all the vector values that help in identifying fully activated and inactivated neurons.

Pre-trained feedforward neural networks were used for term classification in this research and they used different types of term feature combinations to distinguish keywords from non-keywords. Both machine learning and computationally efficient statistical approach were used to predict the probability values of the terms acquired through ranking mechanisms to assess their suitability as a keyword.

Furthermore, a heuristic method is applied to enhance the image query generation. A heuristic function, also called simply a heuristic, is a function that ranks alternatives in search algorithms at each branching step based on available information to decide which branch to follow. For example, it may approximate the exact solution [5]. In the end, the results are evaluated to identify the best performing approach.

2   Methodology

2.1   Data Collection

Data is sourced from the records of the popular visual media company Getty Images and the BBC News website. A web harvesting application is designed based on this data that stores articles gathered from BBC News website locally. Also, only the articles that consisted of images from Getty were used as training data. These are those images that convey the whole idea of the article and the application checks if its Getty ID is exposed. The application then requests image metadata from the Getty API also. Thomson Reuters Open Calais is also used to identify people, events, organizations, and other entities in the corpora. In the end, a dataset of 500,000 terms was accumulated through this process. Figure 1 describes the methodology followed through a flow chart.

Fig. 1 Flow diagram for image selection

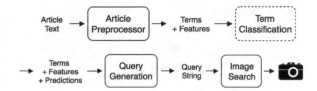

2.2 Structuring the Data

2.2.1 Article Conversion Process

The article is converted into a machine-readable format through the following process.

Tokenization The strings of plain text in the article are first transformed into a sequence of words with the help of a tokenizer.

N-gram Selection N-grams are sequences of up to four consecutive words that are taken into consideration by the pre-processor. N-grams with punctuations are not considered though.

Stopword Removal Stopwords are the words that do not confer meaning individually such as pronouns. All such words are removed during preprocessing.

Stemming Stemming is applied for the purpose of inflection associated with a particular word or a group of words. Porter's stemming algorithm was used for performing stemming.

Feature Assignment Finally, every term is classified with a set of features. It distinguishes the terms that can be instrumental in finding an image that conveys the main idea of the article.

2.2.2 Term Labels

After the article is converted to the desired format, the extracted terms are ranked based on the following.

Term Frequency The number of times a particular term repeats in the article is recorded to measure its importance. Repeating words are combined to one list entry by the preprocessor. A term frequency of value 1 is assigned to each term and every time the term repeats in the text, the count is incremented.

First Occurrence The relative position at which a term is introduced for the first time in the article is also tracked. The term occurring earlier is identified as more important in comparison. This is inspired from the concept of inverted pyramid which involves placing the most important information at the top in order to captivate reader's attention. It is implemented by searching for the term first and then counting

Table 1 Detections

Entity name	Terms	Entity type	Entity category
Reddit	Reddit	Organization	OrganizationProtagonist
	Tech-Firm		
	Company		
Alexis Ohanian	Alexis Ohanian	Person	HumanProtagonist
	Co-founder him		

the number of characters that occur before it. The value obtained is divided by the total number of characters in the text that transforms it into a relative measure. Consequently, 0 savors the first character in the article and 1 the last.

Entity Category Automated semantic tagging was used to improve the term classification to consolidate terms that refer to the same entity. Entities can include individuals, organizations, products, and people, among others. This was achieved by using the web service—Thomson Reuters Open Calais, which assigns intelligent metadata tags to the entities. These entities are attached to the entity category, which helps in describing a term better.

This further helps in creating a better search query as it segregates names, locations, events, and other entities, and helps finding better search terms for finding suitable images. Seven entity categories are developed for this purpose that includes: Event, HumanProtagonist, OrganizationProtagonist, Position, Location, Product, and Other.

The input text and Table 1 are a representation of Open Calais entity detection. Input Text: *Alexis Ohanian—co-founder of Reddit, has resigned from the tech-firm's board and urged the company to replace him with a black candidate.*

2.3 Training the Model

Figure 2 depicts the training procedure using the flow diagram. An ideal training data for the system pertinent to this research should be the search queries that human editors use to find a particular image for their article. However, due to limitations of time and resources, it is assumed that combinations of the terms used in the image description can be utilized to fetch images similar to that image. Although this is a weak approach, it can help generate search queries that might even be identical to what human editors search. Figure 3 is image used for image preprocessing and Table 2 is a representation of the image metadata.

The image metadata is preprocessed to make the terms in the article consistent with the image description. This process is fairly identical to the Open Calais entity detection explained in Sect. 2.2.2 under term labels. However, a new list is introduced here named "image tags" comparable to the article terms. Image metadata

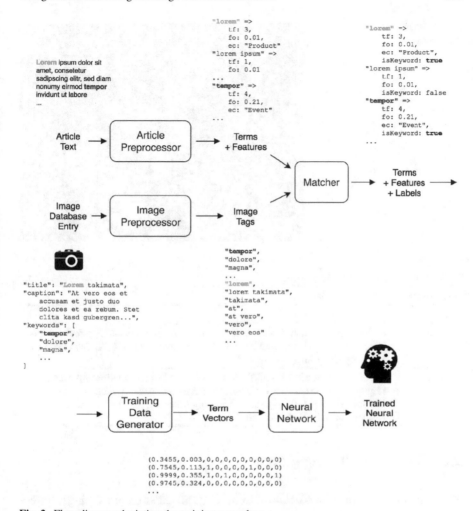

Fig. 2 Flow diagram depicting the training procedure

preprocessing is composed of three fields: title, caption, and image keywords. The data is converted to machine-readable format using the process described in Table 2. However, only stemming is applied to the image keywords since most of them represent fixed terms describing topics, and features such as color and orientation. Both these lists are merged to eliminate repeating terms in the output.

Fig. 3 Image used for image
preprocessing

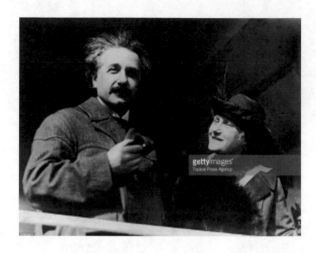

Table 2 Image metadata preprocessing

Representation	Image metadata
Title	Einstein and Wife
Caption	circa 1921: German–Swiss–American mathematical physicist Professor Albert Einstein (1879–1955) with his wife Elsa in Egypt. (Photo by Topical Press Agency/Getty Images)
Keywords	1990–1929, Adult, Africa, Black And White, Couple- Relationship, Data, Egypt, German Culture, Germany, Mathematical Symbol, Mathematics, People, Physicist, Science and Technology, Swiss Culture, Wife, Women, Albert Einstein, Elsa Einstein

2.4 Generating Training Data

The resultant list of image tags obtained from image metadata preprocessing is further processed by the matcher component. The matcher component assigns the Boolean feature is keyword to every term upon comparing the image tags with the article terms. The Boolean value is true if the article term matches with the image tag and false otherwise.

Each of these terms is transformed into a vector that forms the input layer of the neural network. A training data generator component is used to convert all the values to numbers ranging from 0 to 1. 0 indicates inactivated neuron and 1 represents fully activated neuron. Figure 4 represents an example vector for one term.

Term Frequency The sigmoid function is used to normalize the values in the input layer of the neural network. Sigmoid was chosen since its value increments swiftly for natural numbers greater than 1.

$$\text{sig}(x) = \frac{1 - e^{-x}}{1 + e^{-x}} \tag{1}$$

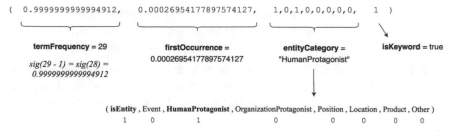

Fig. 4 One term as vector in the training data

Since the lowest possible value for term frequency is 1, the value is decreased by 1 before applying the sigmoid function. This implies that a term frequency of 1 will result in an inactivated neuron ($\text{sig}(1 - 1) = \text{sig}(0) = 0$).

First Occurrence Values for first occurrence range from 0 to 1 by design and do not require normalization. They are applied to the term vector without alterations.

Entity Category This feature is represented as an eight-dimensional vector, wherein, the first value, is Entity, indicates whether the terms is an entity or not and the other seven values represent the entity categories. If the term is not identified as an entity all the values are assigned 0, else, they are assigned 1. An entity is better suited as a search term as compared to a term that is not an entity. If a term qualifies as a keyword, it acquires the value 1, else it is 0.

2.5 Training Neural Networks

Pre-trained feedforward neural networks were used for term classification in this research because they provide a simple and effective mechanism for performing classifications. Four networks were trained that used different types of term feature combinations to distinguish keywords from non-keywords which include:

- term frequency + first occurrence
- term frequency + entity category
- first occurrence + entity category
- term frequency + first occurrence + entity category.

2.6 Image Selection Process

The terms identified by the preprocessor that qualify as possible keywords are further classified regarding their suitability as search queries. Probability values were predicted for these keywords. 1 indicates that the system the considered term is a 100% search term and 0 indicates otherwise.

Five different methods are applied for predicting the probability values which include the four variations of machine learning approach described in Sect. 2.5 and another method based on statistical approach that is explained below.

Statistical Approach The calculation applied here is based on term frequency and first occurrence. Therefore, the term that earliest occurring keyword that is repeated the most in the articled can be classified as the ideal keyword. This can be described as follows:

$$p(\text{tf}, \text{fo}) = \frac{\text{tf}}{\max(\text{TF})} * (1 - \text{fo}) \tag{2}$$

Here, tf denotes the frequency of a term, fo its first occurrence value, max(TF) the highest term frequency value of all terms in an article, and p the probability of this specific term being a keyword. $\frac{\text{tf}}{\max(\text{TF})}$ equals 1 when the term under consideration is the most frequent one in the article, in all other cases, it is less than 1. The difference $(1 - \text{fo})$ converges toward 1 if fo converges toward 0. The ideal keyword in this model would therefore have a probability value of $\frac{\max(\text{TF})}{\max(\text{TF})} * (1 - 0) = 1*1 = 1$.

2.7 Query Generator Component

After keyword classification, the list generated is passed on to the query generator component which converts the ranked terms into a query string. To eliminate the problem of polysemy, two highest-ranking terms are selected by default and linked together to form a query. It is observed that queries consisting of more than three words become too specific, so the shorter term is disregarded if the resultant query is longer than three words. It was also conspicuous that abbreviations and short forms were higher ranked and resulted to be more suitable for a query. However, the short forms are followed by spelled-out terms in the ranking list soon. Thus, a simple heuristic is applied to replace short forms with the spelled-out terms, if they are ranked reasonably high. This method is illustrated below in Fig. 5.

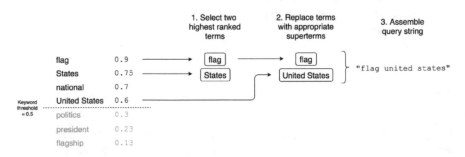

Fig. 5 Term replacement at the time of search query generation

This heuristic method defines a threshold probability value first and the terms that fall above this value are considered suitable search queries. The threshold was set to 0.5 for the statistical approach and 0.1 for the machine learning approaches, respectively, after conducting manual benchmarks. Two highest-ranked terms were selected followed by that. The query generated also looks for super terms after selecting the search terms which indicates the terms that have the same meaning as the selected terms. Then, the search query is generated by connecting the resulting terms and transforming them into lowercase letters. Finally, this query is sent to Getty API, which returns a broad range of images sorted by their popularity. The most popular image is selected for the article.

3 Results

Figure 6 depicts the results of image selection process. It is observed from Fig. 6 that the machine learning approach renders appreciable results when term frequency and first occurrence are both used as features. Term frequency + First Occurrence returned 40% correct images and Term frequency + First Occurrence + Calais returned 42% correct images. Therefore, it is recognized as the best performing approach in this research. A comparison between statistical and machine learning approaches is shown in Fig. 7. From Fig. 7, it is observed that the statistical approach returns the highest percentage of incorrect images (59%).

Figure 8 shows the power of features in sculpting the performance of neural networks. Also, adding the First Occurrence feature to the networks improved the results from 30% correct images in the Term frequency + Calais network to an average of 39.66% in all other networks that use first occurrence. This shows how impactful the first occurrence of a word is in an article. Apart from that, term

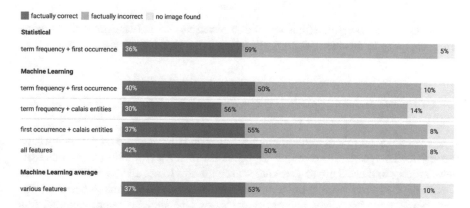

Fig. 6 Results of image selection process

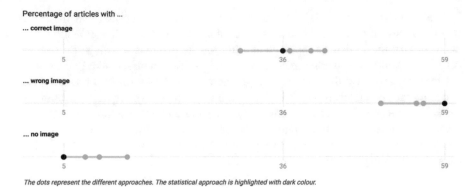

Percentage of articles with ...

... correct image

... wrong image

... no image

The dots represent the different approaches. The statistical approach is highlighted with dark colour.

Fig. 7 Statistical approach versus machine learning approach

■ factually correct ■ not factually correct (wrong or no image)

term frequency

with tf	37.3%	62.7%
without tf	37%	63%

first occurrence

with fo	39.7%	60.3%
without fo	30%	70%

term frequency + first occurrence

with both tf+fo	41%	59%
without both tf+fo	33.5%	66.5%

entity category

with calais	36.3%	63.7%
without calais	40%	60%

Fig. 8 Influence of features on the performance of neural network

frequency has a powerful effect when combined with first occurrence and adding entity category to this network further increases the successes rate.

4 Conclusion

Images play a vital role in influencing the reader and dictates how they may interpret the article. Therefore, this research proposes a model for extracting keywords from news articles and use them to generate search queries to acquire suitable images for the article. First, terms are extracted from the articles and then ranked based on first occurrence, term frequency, and entity category. One statistical approach and four variations of machine learning approach were applied to distinguish the keywords from the non-keywords in these terms. Image query was then generated based on the highest-ranked terms. Finally, it was found that the machine learning approach performed best with Term frequency + First Occurrence + Entity Category network, returning an accuracy of 42%. This system can be useful for suggesting images to the human editors, which can improve the overall productivity of the newsroom. The future work of this research involves comparing the performance of these approaches on different article topics and varying the type and design of the networks.

References

1. T. Warren, Microsoft lays off journalists to replace them with AI (2020). Retrieved 19 Aug 2020 from https://www.theverge.com/2020/5/30/21275524/microsoft-news-msn-layoffs-artificial-int elligence-ai-replacements
2. H.M. Hasan, F. Sanyal, D. Chaki, Md. Ali, An empirical study of important keyword extraction techniques from documents (2017). https://doi.org/10.1109/ICISIM.2017.8122154
3. Merriam Webster. Archived from the original on 27 Feb 2015. Retrieved 17 Oct 2019
4. J.J. Hopfield, Neural networks and physical systems with emergent collective computational abilities. Proc. Natl. Acad. Sci. U.S.A. **79**(8), 2554–2558 (1982). Bibcode: 1982 PNAS 79.2554H. PMC: 346238, PMID: 6953413. https://doi.org/10.1073/pnas.79.8.2554
5. J. Pearl, *Heuristics: Intelligent Search Strategies for Computer Problem Solving* (Addison-Wesley Pub. Co., Inc., Reading, MA, US, 1984), p. 3. OSTI 5127296

An Approach to Noisy Synthetic Color Image Segmentation Using Unsupervised Competitive Self-Organizing Map

P. Ganesan, B. S. Sathish, L. M. I. Leo Joseph, B. Girirajan, P. Anuradha, and R. Murugesan

Abstract The complex data is transformed as the simple but meaningful smaller data groups in the segmentation process. It is the utmost phase of the data exploration process. It is the method of allocating a tag to each pixel to make them as groups (clusters) and the pixels using the same tag have the common characteristics such as color, texture, or intensity. It is challenging to decide on the optimal segmentation method. For noisy images, segmentation becomes more difficult one. This is due to both the image and noisy pixels are considered as the same category. In this work, an artificial neural network based unsupervised self-organizing maps utilized to analyze and cluster the noisy synthetic images. The projected technique employed three levels (competition, cooperation, and adaptation) of competitive learning to segment the data into meaningful regions. The investigational end result undoubtedly revealed the proficiency of the suggested methodology to cluster the noisy images.

Keywords Synthetic image · Self-organizing map · Competitive learning · Clustering · Noise

P. Ganesan (✉)
Department of Electronics and Communication Engineering, Vidya Jyothi Institute of
Technology, Aziz Nagar, C.B. Post, Hyderabad, India
e-mail: gganeshnathan@gmail.com

B. S. Sathish
Department of Electronics and Communication Engineering, Ramachandra College of
Engineering, Eluru, Andhra Pradesh, India

L. M. I. Leo Joseph · P. Anuradha
Department of Electronics and Communication Engineering, SR University, Warangal, Telangana,
India
e-mail: leojoseph.lmi@sru.edu.in

B. Girirajan
Centre for Social Innovation, SR University, Warangal, Telangana, India
e-mail: dir.csi@sru.edu.in

R. Murugesan
Department of Electronics and Communication Engineering, Narsimha Reddy Engineering
College, Secunderabad, Telangana, India

© The Author(s), under exclusive license to Springer Nature Singapore Pte Ltd. 2022   227
P. Shetty D. and S. Shetty (eds.), *Recent Advances in Artificial Intelligence and Data
Engineering*, Advances in Intelligent Systems and Computing 1386,
https://doi.org/10.1007/978-981-16-3342-3_19

1   Introduction

Segmentation, the clustering technique, is the utmost phase of the data exploration process. The complex image is simply transformed into number of small but meaningful images [1]. It is the method of allocating a tag to each pixel to make them as groups (clusters), and the pixels using the same tag have the common characteristics such as color, texture, or intensity [2]. Segmentation finds many applications in image processing. In clustering, the whole data is huddled as an amount of small data groups (clusters) in a significant comportment [3]. All the image pixels in the same cluster should share the common and unique characteristics among them. These characteristics may be color, texture, or intensity [4]. The accomplishment of the higher level image processing techniques solely depends on the outcome of the segmentation process. It is challenging to decide on the optimal segmentation method [5]. Especially for the noisy images, segmentation becomes more difficult one. This is due to both the image, and noisy pixels are considered as the same category. This is the major reason that the segmentation process is failed to achieve its goal, i.e., leads to either under or over segmentation. For image segmentation, number of methods are developed, but most of them are application oriented. The noise in image processing is considered as the superfluous, unnecessary information which corrupts the original information. This useless information is the result of the instability of image characteristics such as color or intensity [1]. The proposed method analyzed the impact of three additive noises (salt and pepper, Gaussian, and Poisson) and one multiplicative noise (Speckle). The salt and pepper (impulse) noise is mainly due to the arbitrarily dispersed black (pepper) and white (salt) pixels over the image. In this work, the impact of salt and pepper on synthetic images is analyzed before segmentation. The noise simply destroys the original content of the image. The optimal selection of the filter, to remove the unnecessary components (pixels) vehemently added to the image, mainly depends on its behavior, the nature of the assignment carried out, and the kind of the data. The proposed work utilized the median filter, powerful nonlinear filter, to remove the unnecessary impulse noisy pixels. The principle behind this filter is that all the pixel values are arranged into ascending order, and then, the median value is computed.

2   Proposed Methodology for Noisy Synthetic Image Clustering

The unsupervised self-organizing maps (SOM) changes its neuron weights using the competitive learning [6, 7]. The structural design of SOM is represented in Fig. 1. SOM consists of one input layer and feature map [8, 9].

The segmentation of color synthetic image clustering based on SOM instigate with the weight vectors initialization. Then, SOM arbitrarily selected a test vector [9]. All the weight vectors are compared with test vector to opt the best (similar) one [10].

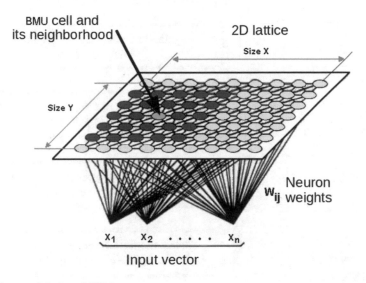

Fig. 1 Structural design of SOM

We know that every weight vector is surrounded by similar neighboring weights. The opted weight and its neighbors are recognized as finest alike units [11]. From this step, the amount of neighbors and the learning rate decreases over time. The entire procedure is reiterated a huge amount of times with respect to the complexity of the data. This unsupervised ANN using three levels (competition, cooperation, and adaptation) of competitive learning to revise its weights [12]. In competition level, the SOM estimate the space between every neuron (Output layer) and the input data using Euclidean distance [13]. The neuron with the minimum distance is considered as the winner of this process. In the second level (cooperation), the opted weight and its neighbors are updated. The neighbor's selection is exclusively based on the kernel function and its factors such as time and the space between the winner and other neurons [14]. In the adaptation process, the winner neuron and its neighborhood simply updated using (1).

$$W_k = W_k + \eta(t)h_{ik}(t)\left(X^{(n)} - W_k\right) \qquad (1)$$

The learning rate (η) changes the weights after every iteration. The learning rate gradually decreases and converges to zero after time 't'.

$$\eta(t) = \eta_0 \exp\left(-\frac{t}{T_1}\right) \qquad (2)$$

The neighborhood kernel function, ($h_{ik}(t)$) and decay rule ($\sigma(t)$) is computed using (3) and (4), respectively

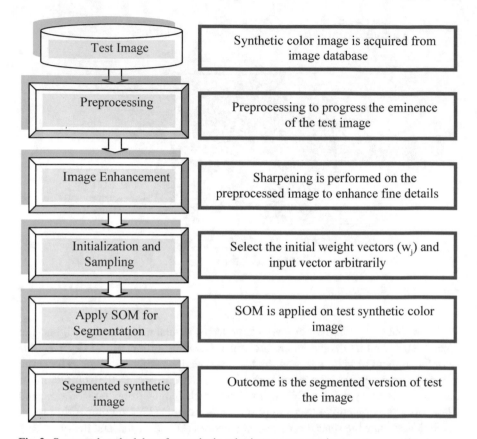

Fig. 2 Suggested methodology for synthetic color image segmentation

$$h_{ik}(t) = \exp\left(-\frac{d_{ik}^2}{2\sigma^2(t)}\right) \qquad (3)$$

$$\sigma(t) = \sigma_0 \exp\left(-\frac{t}{T_1}\right) \qquad (4)$$

d_{ik} = the distance between W_i and W_k.

The suggested methodology for synthetic color image segmentation illustrated in Fig. 2.

3 Experimental Outcomes and Discussions

The test image is acquired from the database, a collection of fifty synthetic images. The test image and its corresponding noisy and filtered version illustrates in Fig. 3.

Fig. 3 Test image 1 and its corresponding noisy and filtered version

The end result of this process is depicted in Fig. 4. The input image (Fig. 4a) is in RGB space and noise free. It is resized to smaller size to reduce the execution time. The test image is clustered into 3, 4, and 5 clusters according to its color attribute as illustrated in Fig. 4b, c, and d, respectively.

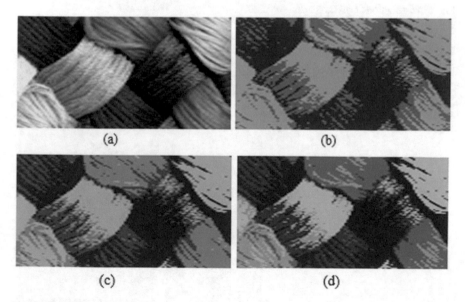

Fig. 4 Test image 1 and its clustered versions

The SOM neighbor distance and its hits (clusters) are demonstrated in Fig. 5. The total number of pixels in the test image 1 is 10,000. They are fragmented into three clusters as illustrated in Fig. 5. The cluster one comprised of 3509 pixels. The cluster two and three included 2642 and 3859 pixels, respectively.

The weight positions of the proposed methodology for this cluster process are clearly represented in Fig. 6.

The image 1 is divided into four clusters by 0.2631 s. This process completed within 12 iteration and PSNR is 14.1282 dB. The complete image clustered into four segments as shown in Fig. 7. Out of 10,000 pixels, the cluster one has 3035 pixels. The cluster two, three, and four comprised of 2169, 2391 and 2405 pixels, respectively.

The weight positions of the proposed methodology for four clusters are clearly represented in Fig. 8.

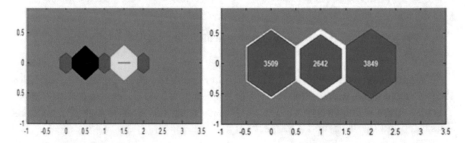

Fig. 5 Distribution of pixels for three clusters

Fig. 6 Weight positions for three clusters (image 1)

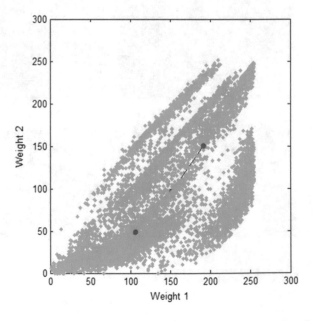

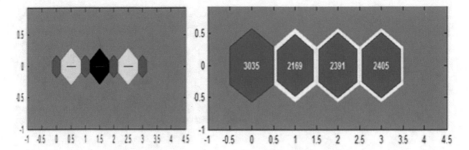

Fig. 7 Distribution of pixels for four clusters

Fig. 8 Weight positions for
four clusters (image 1)

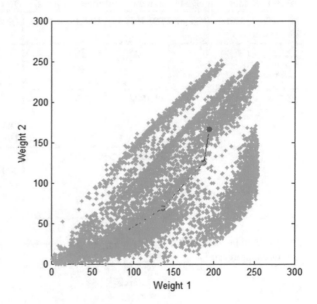

Figure 9 demonstrated that the image 1 is divided into five clusters by 0.2769 s. This process completed within 12 iteration and produced PSNR is 14.0587 dB. The complete image clustered into five segments as shown in Fig. 9. Out of 10,000 pixels, 1892 belonged to the cluster one. The cluster two, three, four, and five comprised of 1377, 1823, 2163, and 2745 pixels, respectively (Table 1).

The weight positions of the proposed methodology for four clusters are clearly represented in Fig. 10.

The test image 2 and its corresponding noisy and filtered version are illustrated in Fig. 11.

The test image 2 (Fig. 12a) is resized to smaller size to reduce the execution time. The test image 2 clustered into 3, 4, and 5 clusters according to its color attribute as illustrated in Fig. 12b, c, and d, respectively.

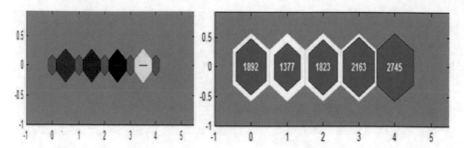

Fig. 9 Distribution of pixels for four clusters

Table 1 Synthetic image clustering for different clusters (image 1)

No. of iterations	No. of clusters	PSNR (dB)	Execution time (s)	Centroid		
12	3	14.3025	0.2186	192.1001	150.5383	92.3365
				152.5753	96.5127	61.1608
				106.7617	48.8623	36.7069
12	4	14.1282	0.2633	90.9490	40.1693	34.1045
				136.8896	69.0841	42.9025
				188.3347	124.6539	70.5705
				194.836	166.1575	106.5883
12	5	14.0587	0.2769	183.7212	182.3987	132.1818
				200.3471	149.9198	82.3908
				189.4729	99.1911	44.0536
				128.0811	62.3581	41.8389
				84.9170	39.9321	35.4926

Fig. 10 Weight positions for four clusters (image 1)

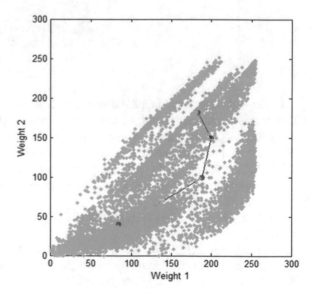

Fig. 11 Test image 2 and its corresponding noisy and filtered version

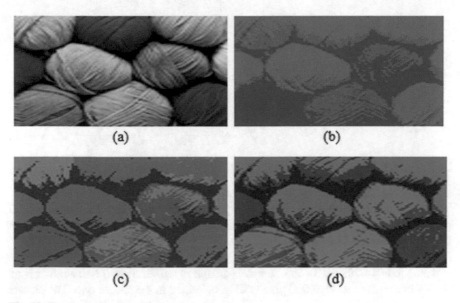

Fig. 12 Test image 1 and its clustered versions

The SOM neighbor distance and its hits (clusters) are demonstrated in Fig. 13. The total number of pixels (10,000) in the test image is segmented into three clusters as illustrated in Fig. 13.

The cluster one is comprised of 3715 pixels. The cluster two and three included 2373 and 3912 pixels, respectively. The weights position for this process is illustrated in Fig. 14.

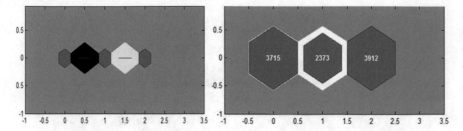

Fig. 13 Distribution of pixels for three clusters

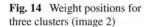

Fig. 14 Weight positions for three clusters (image 2)

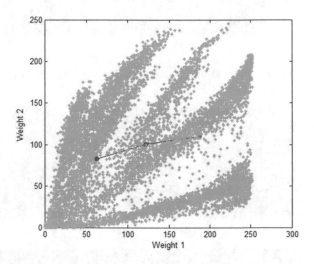

The image 2 is segmented into four clusters by 0.2786 s, and PSNR is 14.7792 dB. The complete image clustered into four segments as shown in Fig. 15. Out of 10,000 pixels, the cluster one has 2686 pixels. The cluster two, three, and four comprised of 2320, 1593, and 3401 pixels, respectively.

Figure 16 demonstrated that the image 2 is segmented into five clusters by 0.2681 s. and PSNR is 14.7927 dB. The complete image is clustered into five segments as shown in Fig. 5. Out of 10,000 pixels, 2316 belonged to the cluster one. The cluster

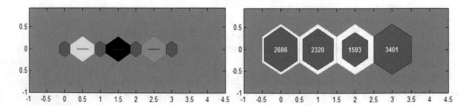

Fig. 15 Distribution of pixels for four clusters

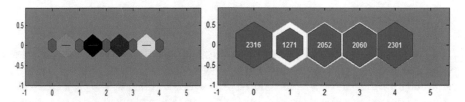

Fig. 16 Distribution of pixels for four clusters

two, three, four, and five comprised of 1271, 2052, 2060, and 2301 pixels, respectively (Table 2).

The illustrative representation of the impact of the proposed method for synthetic color image 1 and 2 displayed in Figs. 17 and 18, respectively.

Table 2 Synthetic image clustering for different clusters (image 2)

No. of iteration	No. of clusters	PSNR (dB)	Execution time (s)	Centroid		
12	3	14.9976	0.2611	186.8795	110.2171	94.5572
				122.0374	100.3241	96.9722
				62.4593	82.9755	93.1347
12	4	14.77972	0.2786	63.5368	60.1775	66.9544
				71.1992	101.8814	111.801
				141.2413	127.3624	122.236
				196.4517	110.0399	93.8887
12	5	14.7927	0.2681	66.0415	133.4127	153.2588
				60.8333	88.3334	100.0013
				97.9488	45.3848	45.0791
				165.0670	74.9134	62.9312
				202.6202	120.4674	95.5374

Fig. 17 Pictorial representation (PSNR) of the suggested approach

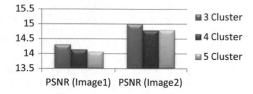

Fig. 18 Pictorial representation (execution time) of the suggested approach

4 Conclusion

The proposed methodology presented an artificial neural network-based unsupervised self-organizing maps to cluster the noisy synthetic images. The test image is acquired from the database, a collection of fifty synthetic images. The projected technique employed three levels (competition, cooperation, and adaptation) of competitive learning to segment the data into meaningful regions. The investigational end result tabulated and undoubtedly revealed the proficiency of the suggested methodology.

References

1. P. Ganesan, V. Rajini, Segmentation and denoising of noisy satellite images based on modified Fuzzy C means clustering and discrete wavelet transform for information retrieval. Int. J. Eng. Technol. **5**(5), 3858–3869 (2013)
2. B.S. Sathish, P. Ganesan, S. Khamar Basha, Color image segmentation based on genetic algorithm and histogram threshold. Int. J. Appl. Eng. Res. **10**(6), 5205–5209 (2015)
3. V. Kalist, P. Ganesan, B.S. Sathish, J.M.M. Jenitha, Possiblistic-fuzzy C-means clustering approach for the segmentation of satellite images in HSL color space. Procedia Comput. Sci. **57**, 49–56 (2015)
4. P. Ganesan, B.S. Sathish, G. Sajiv, A comparative approach of identification and segmentation of forest fire region in high resolution satellite images. in *2016 World Conference on Futuristic Trends in Research and Innovation for Social Welfare (Startup Conclave)* (2016), pp. 1–6
5. P. Ganesan, K. Palanivel, B.S. Sathish, V. Kalist, K.B. Shaik, Performance of fuzzy based clustering algorithms for the segmentation of satellite images-a comparative study. in *IEEE Seventh National Conference on Computing, Communication and Information Systems (NCCCIS)* (2015), pp. 23–27
6. T. Kohonen, S. Kaski, K. Lagus, J. Salojärvi, J. Honkela, V. Paatero, Self-organization of a massive document collection. IEEE Trans. Neural Netw. **11**, 574–585 (2000)
7. J. Sun, Y. Yang, Y. Wang, L. Wang, X. Song, X. Zhao, Survival risk prediction of esophageal cancer based on self-organizing maps clustering and support vector machine ensembles. IEEE Access **8**, 131449–131460 (2020)
8. S. Wang, X. Zhang, Analysis of self-organizing maps (SOM) methods for cell clustering with high-dimensional OAM collected data. in *IEEE 5th International Conference on Cloud Computing and Big Data Analytics* (Chengdu, China, 2020), pp. 229–233
9. P. Yang, D. Wang, Z. Wei, X. Du, T. Li, An outlier detection approach based on improved self-organizing feature map clustering algorithm. IEEE Access **7**, 115914–115925 (2019)
10. V. Chaudhary, R.S. Bhatia, Anil K. Ahlawat, The self-organizing map learning algorithm with inactive and relative winning frequency of active neurons. HKIE Trans. **21**(1), 62–67 (2014)
11. M. Hagenbuchner, A. Sperduti, A.C. Tsoi, A self-organizing map for adaptive processing of structured data. IEEE Trans. Neural Netw. **14**, 491–505 (2003)
12. J. Vesanto, E. Alhoniemi, Clustering of the self-organizing map. IEEE Trans. Neural Netw. **11**, 586–600 (2000)
13. Kohonen, T.: Self-Organizing Maps. Springer Series in Information Sciences, vol. 30, 3rd edn. (Berlin, Germany, 2001)
14. X. Xiao, E.R. Dow, R. Eberhart, Z.B. Miled, R.J. Oppelt, A hybrid self-organizing maps and particle swarm optimization approach. Concurrency Comput. Pract. Experience **16**, 895–915 (2004)

Building Dataset and Deep Learning-Based Inception Model for the Character Classification of Tigalari Script

Sachin S. Bhat, Alaka Ananth, Rajashree Nambiar, and Nagaraj Bhat

Abstract Image classification and optical character recognition are important research areas in computer vision. With advancement in machine learning and deep learning techniques, these fields are attracting lot of researchers to develop models with near human perfection. Many character recognition models are available for modern languages. But, it is still a challenging task to analyze the handwritten text in Indian scripts. It is further complex for the scripts with large alpha syllabary and complex nature. This paper proposes a technique for the recognition and classification of ancient Tigalari characters from the handwritten text. Tigalari is widely used in coastal Karnataka and Kerala for documenting Sanskrit, Tulu, and Malayalam languages. Method involves the creation of database, design of deep convolution neural network (DCNN)-based architecture to classify the text, training the model with the data and recognizing text using test set. Being an inception model for this script, proposed method classifies 46 basic Tigalari characters with an impressing accuracy of 98.55%.

Keywords Tigalari · Tulu · Kannada · Character recognition · Convolutional neural network · VGG16

1 Introduction

Image classification is a process of assorting the images into classes in accordant with their resemblance. This is an extensive application of image recovery that simplifies the search for those images with specific visual content across an image dataset. Classification techniques assume that certain features represent the image in the problem domain, and that individual features belong to one of the many discrete and sole classes.

S. S. Bhat (✉) · R. Nambiar · N. Bhat
Shri Madhwa Vadiraja Institute of Technology and Management, Bantakal, Karnataka 574115, India

A. Ananth
NMAM Institute of Technology, Nitte, Karkala, Karnataka 574110, India

Classification model uses two methods: testing and training. Prominent attributes of the image features are extracted in the training level and a unique representation of individual classification family called training class. In the next level, these feature-space partitions are used to classify image features. From the last few years to solve the classification problem, the research community is enthusiastic on machine learning (ML) and deep learning (DL) problem. Since 2015, deep learning algorithms are becoming widely popular and getting an increasing attention.

ML and DL enable the machines to mimic the human intelligence. It is all about making machines to do things at which humans are better at present. Alan Turing in his famous work starts with a question, 'Can machines think?' Which question is still unanswered. If the behavior of a machine is indistinguishable from that of a human being, then we must accept that the machine can indeed think. This is probably an important step, in the ongoing journey to further improve the abilities of computing machines. The main objective of these techniques is to make the machines more smart and effective in dealing the issues important to humans. Some of the major ML algorithms used for text classification so far are support vector machines (SVM), Ridge classifier, Naive Bayes, K-nearest neighbor, etc. Apart from these, convolutional neural network-based deep learning approaches are swiftly overtaking the traditional ML models in the recent past.

2 Tigalari Script

Having evolved from Tamil-Brahmi, Tigalari finds its origin in Grantaezhuttu. Tigalari was widely used to write Tulu, Kannada, and Malayalam languages in olden days. It was also used by the Shivalli (Tulu speaking), Havyaka (Kannada/Malayalam speaking), and Kota (Kannada speaking) communities to write Sanskrit religious texts. Also known as Arya-Ezhuttu in Kerala, this is a familiar script seen since 1200 CE scripted on almost all old manuscripts found along Karavali (coastal), Malenadu (Western Ghats) of Karnataka and northern Kerala [1]. The orthography of Tigalari is influenced by other dominant languages/scripts like Nandinagari, Kannada, Tamil Grantha, and Malayalam. Cladwell [2] in his magnum opus called this as one of the most evolved Indian scripts. In 1841, the first Karnataka printing press was started by German missionaries in Mangaluru. They started to print the literature in the Kannada script instead of Tigalari. Tigalari was hardly used in printed format which led to the disuse of the script turning to it extermination in a gradual manner. But, a vast amount of documents and manuscripts can be found in this script, which is of great importance because of diverse subjects addressed by them. Large number of these manuscripts are being preserved, catalogued, digitized, and studied by several institutions today.

There is a lack of material or research available for accurately dating the introduction of this script to these regions and the reasons for doing so when there were several other scripts that were actively used here at the time to write Sanskrit. It is commonly stated by those who have studied this script that due to geographic isolation of this region created by the Western Ghats to the east and the Arabian Sea to its west, Tigalari further evolved independently from the Chola Grantha Script. Besides this, if we take into account several prominent character constructions, it retains from the Chola Grantha script and the fact that it was a fully formed script when it was used in Sarvamoola-Grantha of Madhwacharya, and it seems possible that Tigalari might have existed at least hundred years before twelfth century.

The regions where Tigalari was used have been centers of learning for centuries and still continues to have high levels of literacy. It naturally follows that the manuscripts written in these regions, majority of which are in Tigalari script, to be of great literary and scientific value. These manuscripts cover a wide range of subjects such as: medicine, various sciences, Vedas, Sutras, Upanishads, mathematical formulae, daily accounts, astronomy, aesthetics, and philosophy to name a few.

Govt. of Karnataka has introduced Tulu and Tigalari in schools of Mangaluru and Udupi districts through Tulu Sahithya Academy. Text books, Web lessons, and manuals are being provided to learn this script. Scholars like Dr. K. P. Rao and Vaishnavi Murthy are working hard to introduce Unicode complaint typeface for Tigalari.

Tigalari has 16 swaras (vowels) including an Anusvara (am) and Visaraga (ah) which do not have an existence of their own and are pronounced with consonants. Letters vocalic rr and ll are rare and only used in Sanskrit texts. We have excluded these four swaras of 'am', 'aha', 'rr', and 'rr' from our dataset. Independent swara has their corresponding dependant vowel signs except for 'a' which combine with the pure vyanjanas to produce the consonant sound. These vowel signs can be termed as semi-vowels. Tigalari also has 34 vyanjanas (consonants) from 'ka' to 'la' like similar to the scripts of Kannada and Malayalam. Consonants with semi-vowels together constitute 'Gunitakshara'. In the orthography of Tigalari, these Gunitakshara are phonetically ordered similar to other Indic scripts. 'Virama' symbol is used to present the consonant in its pure form without inherent vowel 'A'. There are three basic ligatures in Tigalari. First one is a conjunct character which is a combination of a consonant/semi-vowel and another consonant/semi-vowel. This is called as a vattakshara. Here, an independent consonant is vertically stacked below a base character. A large number of vattaksharas can be produced in Tigalari by the combination of two or three consonants. Second is a conjunct character where vowels U and Ru used as a vattu with consonant conjuncts are placed below the vattakshara. Third case occurs when a special character called Rephais followed by a conjunct or consonant and ends with a virama. Table 1 shows the orthography of Tigalari script. Characters are digitally represented in Srihari font of Pada keypad [3].

3 Preparation of Dataset

As no work has been reported yet on Tigalari, major task in OCR will be the creation of character patterns. Handwritten characters are collected from 42 native Tuluvas

Table 1 Tigalari character classes

Vowels (*Swara*)								
a	ā	i	ī	u	ū	r̥	rr	!
e	ai	o	au	am	ah			

Consonants(*Vyanjana*)								
ka	kha	ga	gha	ṅa	ca	cha	ja	jha
ṭa	ṭha	ḍa	ḍha	ṇa	ta	tha	da	dha
pa	pha	ba	bha	ma				
ya	ra	la	va	śa	ṣa	sa	ha	ḷa

between the ages of 18–45. Among the writers, 19 are Male and 23 are Female. They are either graduates or in the final year of graduation. All writers have Tulu as their mother tongue and minimum knowledge of Tigalari script. They were asked to write down the Varnamala (character set) on a form as shown in Fig. 1 for several number of times in different styles. Later, these papers are scanned using Canon E560 scanner and saved in .jpg format. Characters are extracted from the sheets using machine learning segmentation tool. These characters are manually grouped into different classes. 94% of the characters are obtained in this fashion. No data augmentation is used for this dataset.

For remaining 6% of the characters, ancient Tigalari manuscripts are scanned and saved in the earlier mentioned way. Noise removal, preprocessing, and character reconstruction are done on the manuscripts. Phase congruency-based binarization is used to binarize the images of manuscripts, whereas geodesic morphological operators are applied to reconstruct the shape of characters in the script [4, 5]. Similar ML-based segmentation tool is used to extract the text. Characters belonging to the earlier mentioned 45 classes are picked and are added to the database. Nearly 150 palm leaf inscriptions are scanned in this way, and around 5200 characters are constructed.

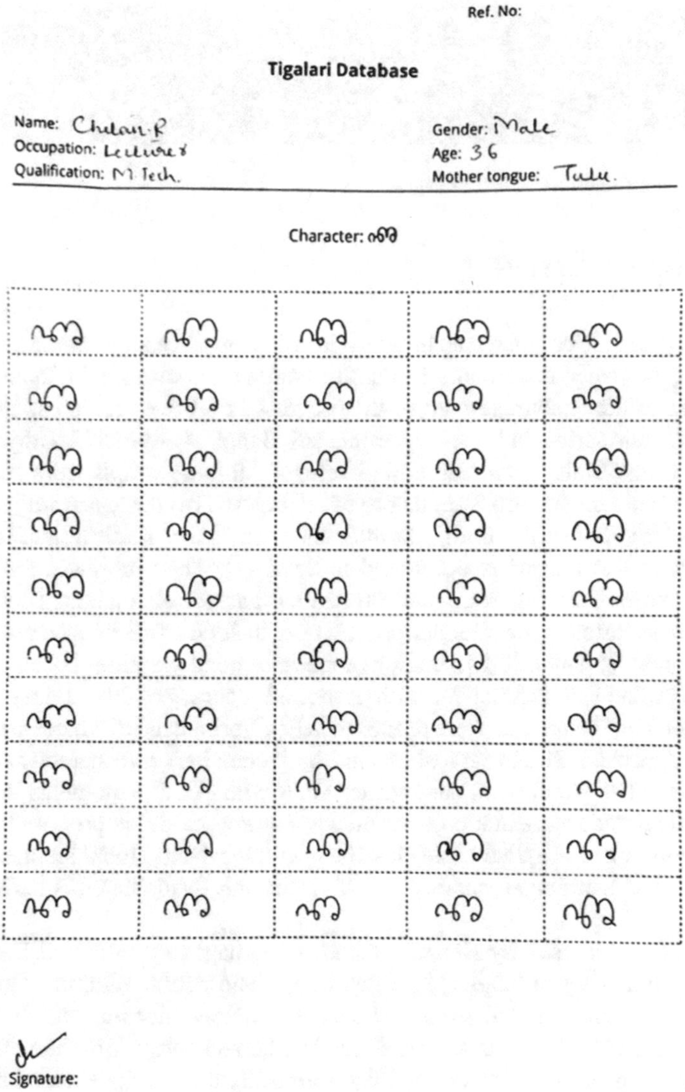

Fig. 1 Sample of Varnamala character set

92 characters of 46 classes are used to develop the database. Training, test, and validation sets are chosen in a proportion of 3:1:1. All the images are resized to 32 × 32 format. Few random samples taken from six different classes of the dataset are shown in Fig. 2.

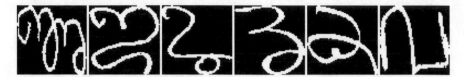

Fig. 2 Randomly chosen character images

4 Literature Review

With the spread of computer technology in working places and homes, automatic processing of paper documents is rapidly gaining importance in India. A short description of the encouragement of the progress in OCR of Indian scripts can be seen in [6]. Main scripts in India are Devanagari, Bangla, Gurumukhi, Odiya, Gujrati, Urdu, Kannada, Malayalam, Tamil, and Telugu. All these scripts were acquired by ancient Brahmi except Urdu. Devanagari which evolved from Nagari family of scripts is used to write Sanskrit, Hindi, Marathi, and many other north Indian languages. First work in Devanagari is accounted to Sethi [7]. They proposed an inception model to extract features from hand printed Devanagari characters. Though many works were reported in the later period, it is only in 2006 a full-fledged classifier has been designed by Sharma [8]. Author proposed a quad classifier-based method to recognize offline Hindi characters with a reported accuracy of 80%. Fuzzy model for handwritten Hindi numerals were proposed in [9]. Normalized distances are obtained using box approach. Structural features like horizontal bars and character strokes are employed in [10]. Compound handwritten OCR with Zernike moments is proposed in [11]. Integrated segmentation with hidden Markov model is proposed for Hindi word recognition in [12]. Multiple classifiers with local and global features are used in [13] to detect Devanagari characters. This is the only available OCR in Devanagari using CNN.

Bangla script is used by Bengali, Axomia languages in India and Bangladesh. Urdu is an India-Aryan language written in Nastaliq script. Kahmiri, Gurumukhi, and Shahmukhi are derived from Sharada. It is noteworthy that the first copy of Mahabbharata manuscript discovered was in Sharada script. Southern Brahmi or Tamil-Brahmi gave rise to five Dravidian scripts. Two from the Kadamba script of Halegannada namely the modern Kannada and Telugu. Tamil, Malayalam and Tulu scripts developed through Grantha-lipi.

Bangla is the only Indian language where remarkable amount of work is reported related to OCR. It is second widely spoken language in India. Bangla is an official language of Bangladesh, West Bengal, Tripura, Assam, Jharkhand and Sierra Leone a West African country. In general, Bengali has 11 vowels and 39 consonants. Recently, there are few applications of deep learning techniques for Bangla numeral classification [14, 15]. Halima Begum et al. [16] use artificial neural network to work on their customized dataset with 79.4% accuracy rate. A CNN approach [17] delivered 85.36% test accuracy. A combination of multilayer perceptron and SVM algorithms have shown 80.9% recognition accuracy on their dataset [18].

Beside these two scripts, very little work is reported in other Indian languages. Unavailability of database is the main reason to be attributed. Sheshadri et al. [19] proposed a multilayer approach to classify Kannada characters. Few others have used traditional image processing approaches with machine learning like SVM [20, 21], divide and conquer [22], wavelet [23], SDTW [24], Tesseract tool [25], and principle component analysis [26] for Kannada or Telugu. Rahiman et al. used structural pattern-based classifier for Malayalam characters [27]. Apart from this statistical [28], gradient [29] and transform domain [30] features are also employed for this task. Even traditional ML and neural network-based algorithms also proved with better results [31–34]. CNN-based model was used for basic Tulu character classification which was an inception model in this language [35].

5 Methodology

Proposed work is inspired by the most popular and efficient DCNN model of transfer learning, called VGGNet. It is considered as one of the best vision models even now that outperforms many of the other networks. In 2015, Vision-Geometry-Group introduced it on ImageNet database for ILSVRC. ImageNet database consists more than 14 million annotated images ranging up to thousand classes. It is the first model to use the smallest receptive field. VGG has a combination of 16 convolutional layers (CL) and fully connected layers (FCL). Many versions of VGG like VGG19 are introduced later. In every CL, it uses a filter size of 3X3. Compared to this, earlier deep learning models used larger receptive fields like 7×7 or 11×11. Small-size filters help broaden network scope with greater discrimination.

The proposed network takes input images that are transferred over a series of layers. The layer represents one processing step which can be a CL, FCL, or a pooling layer. Equation (1) illustrates how CNN runs through layers in forward pass.

$$i_1 \to l_1 \to i_2 \to \cdots \to i_{n-1} \to l_{n-1} \to i_L \to l_n \to z \qquad (1)$$

The input i_1 goes through processing in various layers. i_2 is an output of first layer which is fetched as an input to second layer and so on. Parameters involved in processing of each layer are denoted by tensor $l_1, l_2 \ldots ln$. This operation continues until the completion of all layers in the CNN, which outputs in. Yet one additional layer is introduced for backpropagation, a process that learns good values of parameters in the CNN. Suppose the issue is a grouping of images with classes C. A commonly used strategy is to output in as a C dimensional vector, whose ith entry encodes the prediction. If you know all the parameters in the model, then it is ready to be used for prediction, which involves running the network forward. Pseudocode of training the proposed model is given in Algorithm 1.

Algorithm 1 : Pseudo code for training the proposed model

1. Input: Model, T, t // Model = Keras sequential model, T = training dataset, t = test dataset
2. Output: Trained model
3. Start: // adding convolutional and pooling layers
    Model.add(xConvy), //x = number of filter, y=kernel, Conv=convolution
    Model.add(xAF), //x = convolution output, AF = activation function
    Model.add(xPy), //x = pooling size, y = stride, P = pooling layer
    // adding convolution and pooling layers until optimum test results are obtained(iterative investigation)
    Model.add(xFc) //x = number of neurons, FC = fully connected layer
    Model.add(softmax) // final layer for the classification of 46 classes
    Model.compile(Optimizer, accuracy measure)
    Model.fir(T, epochs, validation split)
4. Stop:

Like VGGNet, there are numerous small sequential convolutional and FC layers in our model too. Instead of 224 × 224, we used 32 × 32 grayscale images in input layer. After repeated investigation, we have observed that even after extracting a few convolutionary layers, the test results remain the same. A rigorous iterative investigation has been carried out to identify these layers are being removed. Customized FC layers have two dense, a flatten, and a dropout layer. A number of channels in the dense layers are fixed to 1024 with a dropout ratio of 0.5. Output is a softmax layer with 46 classes. Different hyperparameters are tuned and performances are measured. Activation function, optimizers, and learning rates are chosen to obtain the optimal solution. The architecture of this model is shown in Fig. 3.

6 Results and Evaluation

This section deals with the obtained result through the proposed model and their comparison with the available literature. The Tigalari database consists of 92,000 images, and the experimental models were developed in Anaconda programming environment using deep learning API Keras. This was implemented with the support of Google Colab GPU platform. The training, test, and validating sets are taken in the proportion of 3:1:1. Network was trained with a batch size of 500 samples for 15 epochs. On validation set, model showed an impressing accuracy of 98.55% on the validation dataset. It can be concluded from Fig. 4 that the model is neither underfitting nor overfitting. Top five misclassified points with the probability of misclassification are shown in Table 2.

Following three standard measurement systems were employed in the performance evaluation of the model.

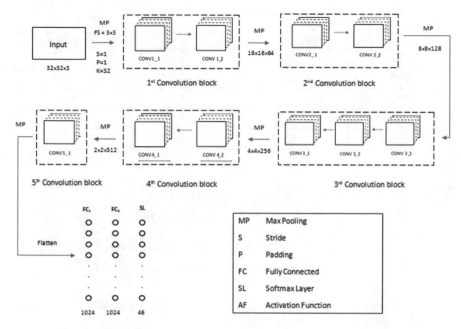

Fig. 3 Proposed architecture

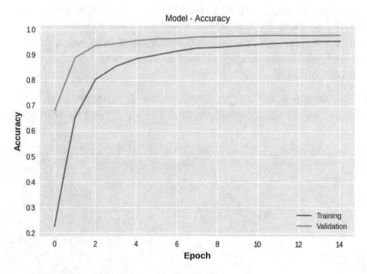

Fig. 4 Train and validation accuracy

Table 2 Top five misclassified classes

Actual class	Predicted class	Loss	Probability
യ	ഥ	2.64	0.07
അ	ആ	3.07	0.05
ഞ	ന	3.14	0.04
ശ	ഗ	3.30	0.04
ഊ	ഉ	4.70	0.02

Table 3 Confusion matrix of classifier

Confusion matrix	Predicted	
Actual	True positives (TPs)	False negatives (FNs)
	False positives (FPs)	True negatives (TNs)

(i) Confusion matrix: is a specific table (Table 3) which describes the classification performance as shown in Fig. 5.

(ii) Precision, recall: Precision is a ratio of TPs to the sum of TPs and FPs where recall is a ration of TPs to the overall TPs and FNs. These are the classification statistics on class predictions given in Figs. 6 and 7.

(iii) Test accuracy: is a trained model performance on unseen data. Model identifies a given random sample against 46 different classes with an accuracy of 99.8% as shown in Fig. 8.

Validation accuracy of the proposed model is compared with the original VGGNet16 as well as simple seven-layered CNN architecture with similar set of

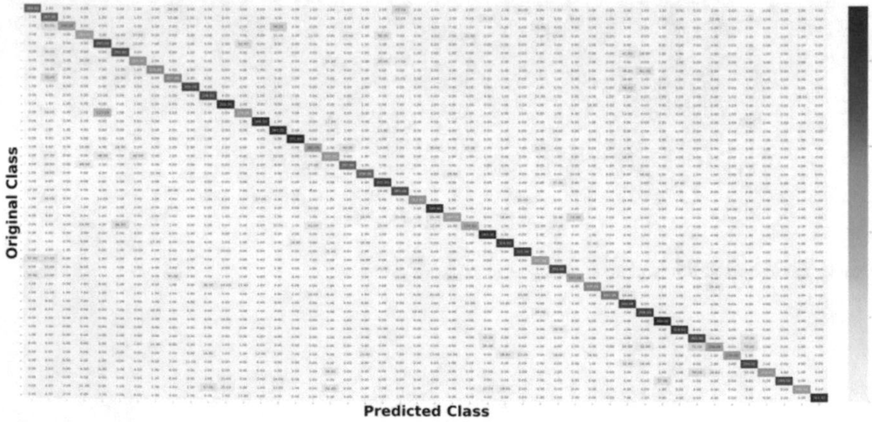

Fig. 5 Confusion matrix

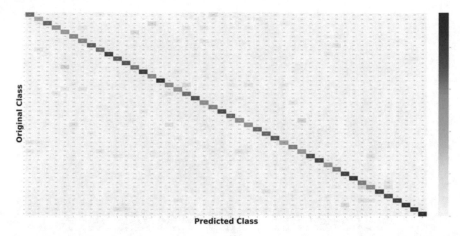

Fig. 6 Precision matrix

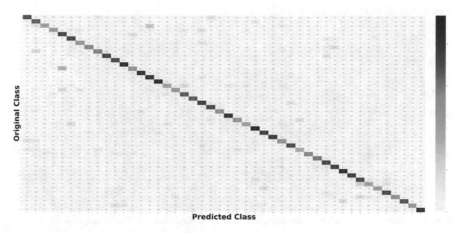

Fig. 7 Recall matrix

parameters. Proposed model converges better than VGG and outperforms CNN as shown in Fig. 9.

7 Conclusion

In this paper, an extensive study on offline recognition of characters in Tigalari documents is proposed. A database comprising of 92,000 Tigalari characters of Varnamala is prepared for the recognition task. A deep learning CNN architecture of 12 layers is developed for character classification. Depth of the convolutional layers is deliberately restricted as the features in text are less compared to the features of semantic

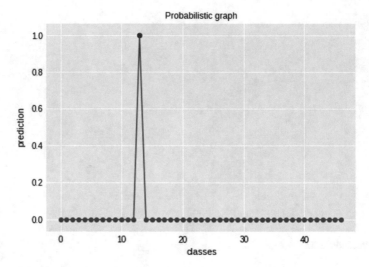

Fig. 8 Test accuracy on a randomly selected character

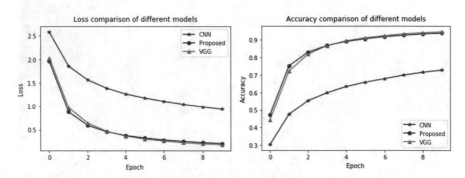

Fig. 9 Validation accuracy and loss comparison of proposed versus VGG versus DCNN

scene image. This model incorporates CLs, FCLs, activation function, optimizer, maxpooling layer, stride, dropout layer, and classifiers. These hyperparameters are fine-tuned with numerous permutations and combinations to attain maximum efficiency. Obtained result is ascertained with number of evaluation methods like train–validation accuracy, confusion matrix, precision–recall matrix, ROC of multiclass and misclassified classes. Train and validation accuracies are compared with the original model of VGG to establish the effectiveness of the proposed method.

References

1. R. Caldwell, *Comparative Grammar of Dravidian or South Indian Family of Languages* (Trübner &Co., London, 1856)
2. K.Y. Vaishnavi Murthy, Preliminary proposal to encode Tigalari script in Unicode (2017)
3. https://thetulufont.com/
4. S. Bhat, G. Seshikala, Preprocessing and binarization of Inscription images using phase based features. in *2018 Second International Conference on Advances in Electronics, Computers and Communications (ICAECC)* (IEEE, 2018)
5. S. Bhat, G. Seshikala, Restoration of characters in degraded inscriptions using phase based binarization and geodesic morphology. Int. J. Recent Technol. Eng. **7**(6), 1070–1075 (2019)
6. U. Pal, B.B. Chaudhuri, Indian script character recognition: a survey. Pattern Recogn. **37**(9), 1887–1899 (2004)
7. I.K. Sethi, B. Chatterjee, Machine recognition of constrained hand printed Devanagari. Pattern Recogn. **9**(2), 69–75 (1977)
8. N. Sharma et al., Recognition of off-line handwritten Devanagari characters using quadratic classifier, in *Computer Vision, Graphics and Image Processing.* (Springer, 2006), pp. 805–816
9. M. Hanmandlu, O. Murthy, Fuzzy model based recognition of handwritten numerals. Pattern Recogn. **40**(6), 1840–1854 (2007)
10. V. Bansal, R.M.K. Sinha, On how to describe shapes of Devanagari characters and use them for recognition. in *Proceedings of Fifth International Conference on Document Analysis and Recognition* (1999), pp. 410–413
11. K. Kale, P.D.S. Chavan, M. Kazi, Y. Rode, Zernike moment feature extraction for handwritten Devanagari (Marathi) compound character recognition. Int. J. Adv. Res. Artif. Intell. **3**(1), 68–76 (2014)
12. S.D. Connell, R.M.K Sinha, A.K. Jain, Recognition of unconstrained on-line Devanagari characters. in *IEEE International Conference on Pattern Recognition* (2000), pp. 368–371
13. M. Avadesh, N. Goyal, Optical character recognition for Sanskrit using convolution neural networks. in *2018 13th IAPR International Workshop on Document Analysis Systems (DAS)* (IEEE, 2018)
14. S.M.A. Sharif, N. Mohammed, N. Mansoor, S. Momen, A hybrid deep model with HOG features for Bangla handwritten numeral classification (IEEE, 2016), pp. 463–466.
15. M. Shopon, N. Mohammed, M.A. Abedin, Bangla handwritten digit recognition using autoencoder and deep convolutional neural network. in International Workshop on Computational Intelligence (IWCI) (IEEE, 2016), pp. 64–68
16. Halima Begum et al., Recognition of handwritten Bangla characters using Gabor filter and artificial neural network. Int. J. Comput. Technol. Appl. **8**(5), 618–621. ISSN: 2229-6093
17. Md. M. Rahman, M.A.H. Akhand, S. Islam, P.C. Shill, Bangla handwritten character recognition using convolutional neural network. Int. J. Image Graphics Signal Process. **8**, 42–49, (2015)
18. N. Das, B. Das, R. Sarkar, S. Basu, M. Kundu, M. Nasipuri, Handwritten Bangla basic and compound character recognition using MLP and SVM classifier. arXiv preprint arXiv:1002.4040
19. K. Sheshadri, S.K. Divvala. Exemplar driven character recognition in the wild. in *British Machine Vision Conference* (2012)
20. M. Thungamani, P.R. Kumar, K. Prasanna, S.K. Rao, Off-line handwritten Kannada text recognition using support vector machine using Zernike moments. Int. J. Comput. Sci. Netw. Secur. **11**, 128–135 (2011)
21. T.V. Ashwin, P.S. Sastry, A font and size-independent OCR system for printed Kannada documents using support vector machines. Sadhana **27**, 35–58 (2002)
22. M.M. Prasad, M. Sukumar, A.G. Ramakrishnan, Divide and conquer technique in online handwritten Kannada character recognition. in *Proceedings of the International Workshop on Multilingual OCR* (2009), pp. 1–7

23. R.S.R. Kunte, R.S. Samuel, Online character recognition for handwritten Kannada characters using wavelet features and neural classifier. IETE J. Res. **46**(5), 387–393 (2000)
24. R. Kunwar, P. Mohan, K. Shashikiran, A. Ramakrishnan, Unrestricted Kannada online hand-written Akshara recognition using SDTW. in *Proceedings of International Conference on Signal Processing and Communications* (2010), pp.1–5
25. G.K. Prasad, I. Khan, N.R. Chanukotimath, F. Khan, Online handwritten character recognition system for Kannada using principal component analysis approach: for handheld devices. in *World Congress on Information and Communication Technologies* (2012), pp. 675–678
26. R. Fernandes, A.P. Rodrigues, Kannada handwritten script recognition using machine learning techniques. in *2019 IEEE International Conference on Distributed Computing, VLSI, Electrical Circuits and Robotics (DISCOVER)* (IEEE, 2019)
27. R. Singh, M. Kaur, Ocr for telugu script using back-propagation based classifier. Int. J. Inf. Technol. Knowl. Manage. **2**(2), 639–643 (2010)
28. M.A. Rahiman, A. Shajan, A. Elizabeth, M. Divya, G.M. Kumar, M. Rajasree, Isolated hand-written Malayalam character recognition using HLH intensity patterns, in *Second International Conference on Machine Learning and Computing (ICMLC)* (IEEE, 2010), pp. 147–151
29. B.S. Moni, G. Raju, Modified quadratic classifier for handwritten Malayalam character recognition using run length count. in *International Conference on Emerging Trends in Electrical and Computer Technology (ICETECT)* (IEEE, 2011), pp. 600–604
30. R.G. Salomon, Brahmi and Kharoshthi. The World's Writing Syst. 373–383 (1996)
31. G. Raju, Recognition of unconstrained handwritten Malayalam characters using zero-crossing of wavelet coefficients. in *International Conference on Advanced Computing and Communications, ADCOM 2006* (IEEE, 2006), pp. 217–221
32. M. Manuel, S. Saidas, Handwritten Malayalam character recognition using curvelet transform and ANN. Int. J. Comput. Appl. **121**(6)
33. K. Manjusha, M.A. Kumar, K. Soman, Reduced scattering representation for Malayalam character recognition. Arab. J. Sci. Eng. 1–12 (2017)
34. N.V. Neeba, C.V. Jawahar, Empirical evaluation of character classification schemes. in *Seventh International Conference on Advances in Pattern Recognition* (IEEE, 2009), pp. 310–313
35. S. Bhat, G. Seshikala, Character recognition of Tulu script using convolutional neural network. in *Advances in Artificial Intelligence and Data Engineering* (Springer, Singapore), pp. 121–131

Handwritten Character Recognition Using Deep Convolutional Neural Networks

R. Shashank, A. Adarsh Rai, and P. Srinivasa Pai

Abstract The automatic detection and recognition of characters in images are an important problem in various applications. The traditional shallow networks in machine learning have limitations in image classification due to their inability to effectively utilize the spatial relationships between the pixels of the image. But the incredible advances in deep learning methods and deep architecture in the recent years have opened doors to the possibility of employing these techniques. In this paper, a deep convolutional neural network (CNN) with minimal preprocessing for the effective classification of handwritten characters has been proposed. The application of this network yielded an accuracy of 99.50 and 94.66% on the test data of MNIST and EMNIST datasets, respectively.

Keywords Convolutional neural networks (CNNs) · Character recognition · Deep learning

1 Introduction

Optical character recognition represents the process of extracting text embedded in images and converting it into a digitized text format. Text data embedded in images can generally be classified into two classes: printed text and handwritten text. While a lot of work has been done for the recognition of printed characters [1–4], including open-source systems like the Tesseract [5], these systems are not suited for general purpose text extraction and tend to perform poorly for handwritten text.

Handwritten character recognition (HCR) tends to be a harder problem, due to the sheer variability of text written by hand. HCR systems find extensive application in industries that generate large corpuses of handwritten documents. These systems

R. Shashank · A. Adarsh Rai · P. Srinivasa Pai (✉)
Department of Mechanical Engineering, NMAM Institute of Technology Nitte, Udupi, Karnataka 574110, India
e-mail: srinivasapai@nitte.edu.in

A. Adarsh Rai
e-mail: adarsh.rai@nitte.edu.in

© The Author(s), under exclusive license to Springer Nature Singapore Pte Ltd. 2022   253
P. Shetty D. and S. Shetty (eds.), *Recent Advances in Artificial Intelligence and Data Engineering*, Advances in Intelligent Systems and Computing 1386,
https://doi.org/10.1007/978-981-16-3342-3_21

could be used for automatic processing of cheques in banks, to read patient forms in healthcare, and to read grades in universities to name a few.

HCR can either be online or offline. Online HCR involves automatic processing of a message as it is being written, whereas offline HCR deals with text in scanned images [6]. These variants differ mainly in the way the input is fed into the classifier. Previous methods in classifying handwritten characters tend to use handcrafted feature extraction methods [7, 8]. These manually extracted features tend to be limited by the prior knowledge of the user [9], leading to a degradation of the inputs to the classifier, thereby yielding poor results.

Deep learning (DL) is a rapidly developing class of machine learning algorithms, which employ multiple nonlinear hidden layers, allowing it to capture complex relationships in the inputs. This is done by directly extracting features from raw data in a number of nonlinear hidden layers [10]. The features extracted by DL methods like convolutional neural networks (CNNs) tend to be superior to manual engineered features [11]. These techniques have performed better than the traditional neural network models like multilayer perceptron (MLP) [12]. The increase in processing power of processors and graphics processing units (GPUs) has made training a large number of parameters associated with these deep networks possible. This has led to the development of several deep learning frameworks. TensorFlow is one of the most popular one released by Google in 2015 [13]. High-level libraries like Keras, which are built on top of TensorFlow, are also available. These are easier for writing native TensorFlow codes. Thus, this paper proposes a deep convolutional neural network model with minimal preprocessing for the recognition of handwritten characters using Keras. This paper is divided into the following Sect. 2—Reviewing related research in handwritten character recognition, Sect. 3—Convolutional neural network, Sect. 4—Proposed CNN architecture, Sect. 5—Results of the CNN, and finally Sect. 6—Conclusions drawn from the application of the model.

2 Related Work

One of the earliest works of using CNNs for character recognition was probably the LeNet proposed by Yann LeCun et al., back in 1998 [14]. They found that CNNs specifically designed to deal with 2D shapes performed the best when it came to character recognition.

Later with the development of other machine learning and feature engineering processes, the application of methods like K-nearest neighbor and support vector machines were explored [15]. But most of these methods had limitations due to their shallow architecture and manual feature extraction.

The use of deep networks to solve computer vision problems with higher accuracies is increasing. It was found that the deep CNNs perform exceptionally well in problems such as the ImageNet classification [16].

Deep CNNs have shown a marked improvement over previous methods, when used for Arabic handwritten character recognition with minimal preprocessing, employing batch normalization and dropouts [17]. Techniques exploiting the hierarchical feature extraction process of CNNs have been used successfully in the recognition of handwritten characters in Tamil language [18]. Similar techniques have also been explored for Chinese characters [19] and Bangla character recognition [20].

3 Convolutional Neural Networks

Convolutional neural networks (CNNs) are a class of feed forward deep neural networks that are similar to MLPs, but mainly differ in the way that their weights are organized.

The weights in the convolutional layer have a matrix of weights called kernel or filters, which slide through the inputs performing the convolution operation at every stage. The outputs of these convolution operations generate a feature map, where each feature in the feature map corresponds to the local receptive field of the input, where the kernel performs the operation. The output of the nth convolutional layer with filter size $(M \times M)$ is computed as:

$$Y_{i,j}^n = B_{i,j}^n + \sum_{a=1}^{M} \sum_{b=1}^{M} W_{i,j}^n X_{(i+a)(i+b)}^{n-1} \tag{1}$$

where $B_{i,j}^n$ is the bias matrix and $W_{i,j}^n$ is the filter in the nth convolutional layer.

The learned weights in the kernel are common throughout the inputs of the given convolutional layer. This helps reduce the number of trainable parameters by many folds, thereby reducing the scope for over fitting, which is rampant in MLPs [21].

The use of subsampling or pooling layer further reduces the number of learnable parameters of the network. There are different methods of pooling, the most common method being max pooling. In this there is again a kernel sliding through the inputs of the pooling layer, where the most prominent element within the kernel is obtained. However, in the pooling layer, the local receptive field of the kernel does not overlap. This mechanism provides a method to extract the most prominent features within the kernel size allowing a reduction in overall number of parameters.

Convolutional layers are followed by a nonlinear activation function like the rectified linear unit (ReLU). This is used to overcome problems related to the plateauing of the loss function during backpropagation [22]. Batch normalization, a normalization technique is sometimes used between the convolutional layer and the activation function, and this technique is proven to reduce the training time of deep networks by reducing the internal covariate shifts [23].

A typical CNN architecture consists of multiple convolutional blocks, which consists of several convolutional and pooling layers. Each convolutional block extracts features from the inputs and the complexity of the features extracted increases deeper inside the model [24]. These convolutional blocks act as feature extractor, that extracts the most relevant features that is fed into a classifier, normally a fully connected layer [25]. The features that are extracted are fine-tuned by backpropagation during training.

The softmax function is applied to the outputs of output layer to determine probabilities for each class. The class with the highest probability would be the predicted class of the input. The probabilities are computed as:

$$x_i = \left(\frac{e^{x_i}}{\sum_{k=1}^{N} e^{x_k}} \right) \quad (2)$$

These CNN models have demonstrated to perform better with an increase in their depth, i.e., number of convolutional blocks [26]. But the– depth of the model is limited by the size of the data available, as overfitting becomes predominant as the depth increases.

4 Proposed CNN Architecture

For this task, Keras has been used to build a deep convolutional neural network and this network has been trained on two datasets, i.e., MNIST dataset [14] and EMNIST Letters dataset [27]. The MNIST dataset consists of handwritten digits from 0 to 9. It has 60,000 training samples and 10,000 test samples. The EMNIST Letters dataset consists of handwritten letters of the English alphabet both capital and cursive. It has 124,000 training samples and 20,800 test samples. The datasets used have letters in the form of binary 28 × 28 pixel images.

The pixels of the images have been normalized by dividing them by 255, i.e., the maximum pixel intensity before using them for training, thereby reducing the range of pixel values to 0–1. All the training samples of the datasets are retained for training and half of the test samples are used for validation of the model and the other half is used for determining the percentage accuracy of the model after the training process has been completed.

Figure 1 shows the basic architecture of the proposed model, and the particulars of the same are explained below.

Batch normalization and parametric rectified linear unit (PReLU) activation is applied after every convolutional and fully connected layer. This activation operation retains the positive pixel values and returns the product of the negative value and a term alpha, which is learnt from the data [28]. PReLU is used instead of ReLU as it can compute negative inputs and yields a lower error rate than the ReLU [29].

The first input layer is followed by three convolutional blocks with 256, 128, and 64 filter maps, respectively. The convolutional layers in these blocks have a kernel size of (3, 3), same padding and have a stride of 1. The max pooling layers that follow have a pool size of (2, 2) and a stride of 2. Thus, the feature map size reduces by half through every convolutional block, as every block has a max pooling layer. The feature maps reduce in size as 28-14-7-3 through these blocks. These blocks

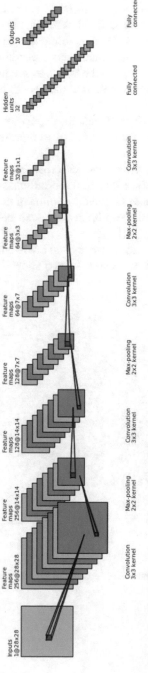

Fig. 1 Basic architecture of the proposed model

are followed by a single convolution layer without padding, so that it converts the two-dimensional array obtained from the convolutional blocks to a one-dimensional vector, so that it can be fed into fully connected layers for the final classification. This layer also has a kernel size of (3, 3). It has 32 filters and the output of this layer will be of the dimension (32, 1, 1). This one-dimensional vector would contain all the features extracted by the convolutional layers preceding it.

Dropout layers are used after some of the layers, as a regularization layer to reduce over fitting and improve the generalization of the networks, by setting a fraction of inputs to zero [30].

The output layer has either 10 or 26 neurons depending on the dataset used. The output layer uses a softmax activation function to represent the probability that a character belongs to a class.

During the training of the model, categorical cross-entropy [31] is used as the cost function for backpropagation, since it is ideal for multiclass classification problems. The AMSGrad optimizer was used to find the minima of the loss function. This optimizer is a modification of the Adam optimizer and deals better with the convergence issues associated with the latter [32].

Figures 2 and 3 show percentage accuracy and cross-entropy loss versus number of epochs for the MNIST dataset. Figures 4 and 5 show the same for the EMNIST Letters dataset.

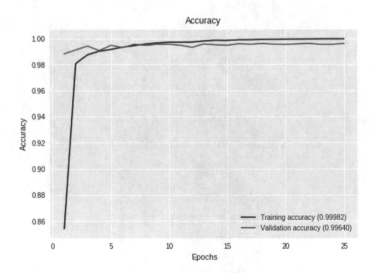

Fig. 2 Percentage accuracy versus number of epochs for MNIST dataset

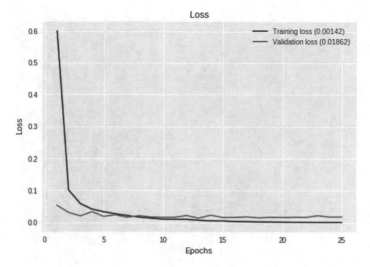

Fig. 3 Cross-entropy loss versus number of epochs for MNIST dataset

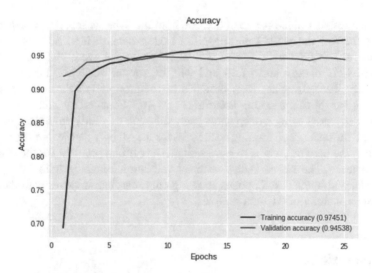

Fig. 4 Percentage accuracy versus number of epochs for EMNIST Letter dataset

5 Results and Discussion

The whole training section of both the datasets is used for training the network, whereas the testing samples provided in these datasets are divided into two halves: validation and testing. The validation set is used during the training process to know the performance of the model, during training.

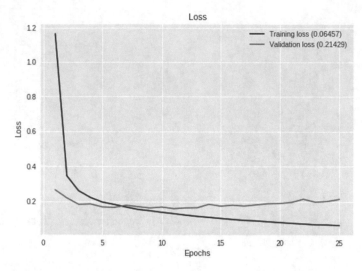

Fig. 5 Percentage accuracy and cross-entropy loss versus number of epochs for EMNIST dataset

The network is trained for 25 epochs on both the datasets separately. The testing accuracy for both the MNIST and EMNIST Letters datasets is 99.50% and 94.66%, respectively. The final training and validation accuracies were 99.98 and 99.64% for the MNIST dataset and 97.59 and 94.69% for the EMNIST Letters dataset, respectively. The testing accuracy obtained by the proposed method on the MNIST dataset is superior to the testing accuracy of 98.46% obtained by Younis et al. [33] on the same dataset using a deep neural network with five fully connected layers.

Though the testing and training accuracies for the MNIST dataset are very close to each other, the network seems to have a slight over fitting problem for the EMNIST Letters dataset. The network also reached validation accuracy within 1% of the final value, within the first 5 epochs showing only marginal improvement thereafter indicating a scarcity of the data samples.

6 Conclusions

The network described in this paper successfully classifies English alphabets and numbers using the EMNIST Letters and the MNIST datasets with test accuracies of 94.66% and 99.50%, respectively.

However, the performance of the CNN plateaus out after the initial epochs showing marginal improvement after. The performance of the CNN is limited by the size of the dataset available, as they require large datasets and are prone to over fitting on smaller datasets.

References

1. T. Pavlidis, Recognition of printed text under realistic conditions. Pattern Recogn. Lett. **14**, 317–326 (1993)
2. S.N. Srihari, Recognition of handwritten and machine-printed text for postal address interpretation. Pattern Recogn. Lett. **14**, 291–302 (1993)
3. R. Seethalakshmi, T.R. Sreeranjani, T. Balachandar, A. Singh, M. Singh, R. Ratan, S. Kumar, Optical character recognition for printed Tamil text using unicode. J. Zhejiang Univ. Sci. A **6**, 1297–1305 (2005)
4. H.A. Al-Muhtaseb, S.A. Mahmoud, R.S. Qahwaji, Recognition of off-line printed Arabic text using hidden Markov models. Signal Process. **88**, 2902–2912 (2008)
5. R. Smith, An overview of the Tesseract OCR engine, in *Ninth International Conference on Document Analysis and Recognition* (ICDAR 2007) 2007 Sep 23 (vol. 2, pp. 629–633). IEEE (2007)
6. R. Plamondon, S.N. Srihari, Online and off-line handwriting recognition, a comprehensive survey. IEEE Trans. Pattern Anal. Mach. Intell. **22**, 63–84 (2000)
7. N. Shanthi, K. Duraiswamy, A novel SVM-based handwritten Tamil character recognition system. Pattern Anal. Appl. **13**, 173–180 (2010)
8. N.A. Hamid, N.N. Sjarif, Handwritten recognition using SVM, KNN and neural network. arXiv preprint arXiv:1702.00723 (2017)
9. F. Jia, Y. Lei, J. Lin, X. Zhou, N. Lu, Deep neural networks, a promising tool for fault characteristic mining and intelligent diagnosis of rotating machinery with massive data. Mech. Syst. Signal Process. **72–73**, 303–315 (2016)
10. Y. LeCun, Deep learning & convolutional networks, in *2015 IEEE Hot Chips 27 Symposium* (HCS) (2015)
11. C. Bailer, T. Habtegebrial, K. Varanasi, D. Stricker, Fast dense feature extraction with convolutional neural networks that have pooling or striding layers, in *Proceedings of the British Machine Vision Conference 2017* (2017)
12. S.B. Driss, S. Ben Driss, M. Soua, R. Kachouri, M. Akil, A comparison study between MLP and convolutional neural network models for character recognition, in *Real-Time Image and Video Processing* (2017)
13. S.S. Girija, Tensorflow: large-scale machine learning on heterogeneous distributed systems. https://cse.buffalo.edu/~chandola/teaching/mlseminardocs/TensorFlow.pdf
14. Y. Lecun, L. Bottou, Y. Bengio, P. Haffner, Gradient-based learning applied to document recognition. Proc. IEEE **86**, 2278–2324 (1998)
15. P. Kumar, N. Sharma, A. Rana, Handwritten character recognition using different Kernel based SVM classifier and MLP neural network (a comparison). Int. J. Comput. Appl. **53**, 25–31 (2012)
16. A. Krizhevsky, I. Sutskever, G.E. Hinton, ImageNet classification with deep convolutional neural networks. Commun. ACM **60**, 84–90 (2017)
17. K. Younis, A. Khateeb, Arabic hand-written character recognition based on deep convolutional neural networks. Jordanian J. Comput. Inf. Technol. **3**, 186 (2017)
18. P. Vijayaraghavan, M. Sra, Handwritten tamil recognition using a convolutional neural network, in *2018 International Conference on Information, Communication, Engineering and Technology (ICICET)* (2014)
19. W. Yang, L. Jin, Z. Xie, Z. Feng, Improved deep convolutional neural network for online handwritten Chinese character recognition using domain-specific knowledge, in *2015 13th International Conference on Document Analysis and Recognition (ICDAR)* (2015)
20. S. Sen, D. Shaoo, S. Paul, R. Sarkar, K. Roy, Online handwritten Bangla character recognition using CNN, a deep learning approach. Adv. Intell. Syst. Comput. 413–420 (2018)
21. S. Lawrence, C.L. Giles, Overfitting and neural networks: conjugate gradient and backpropagation, in *Proceedings of the IEEE-INNS-ENNS International Joint Conference on Neural Networks. IJCNN 2000. Neural Computing: New Challenges and Perspectives for the New Millennium* 2000 Jul 27 (vol. 1, pp. 114–119). IEEE

22. K. Hara, D. Saito, H. Shouno, Analysis of function of rectified linear unit used in deep learning, in *2015 International Joint Conference on Neural Networks (IJCNN)* (2015)
23. S. Ioffe, C. Szegedy, Batch normalization: accelerating deep network training by reducing internal covariate shift (2015)
24. N. Eikmeier, R. Westerkamp, E. Zelnio. Development of CNNs for feature extraction, in *Algorithms for Synthetic Aperture Radar Imagery* XXV, vol. 10647, p. 106470C. International Society for Optics and Photonics (2018)
25. M.D. Zeiler, R. Fergus, Visualizing and understanding convolutional networks, in *European Conference on Computer Vsision*, pp. 818–833. Springer, Cham (2014)
26. S. Liu, W. Deng. Very deep convolutional neural network based image classification using small training sample size, in *2015 3rd IAPR Asian conference on pattern recognition (ACPR)*, pp. 730–734. IEEE (2015)
27. G. Cohen, S. Afshar, J. Tapson, A. Schaik, EMNIST: Extending MNIST to handwritten letters, in *2017 International Joint Conference on Neural Networks (ICJNN)*, pp. 2921–2926 (2017)
28. K. He, X. Zhang, S. Ren, J. Sun, Delving deep into rectifiers: surpassing Human-level performance on ImageNet classification, in *2015 IEEE International Conference on Computer Vision (ICCV)*, pp. 1026–1034 (2015)
29. B. Xu, N. Wang, T. Chen, M. Li, Empirical evaluation of rectified activations in convolutional network, in arXiv:1505.00853v2[cs.LG]
30. N. Srivastava, G. Hinton, A. Krizhevsky, I. Sutskever, R. Salakhutdinov, Dropout: a simple way to prevent neural networks from overfitting. J. Mach. Learn. Res. **15**(56), 1929–1958 (2014)
31. P. Golik, P. Doetsch, H. Ney, Cross-entropy vs. squared error training: a theoretical and experimental comparison, in *Interspeech*, vol. 13, pp. 1756–1760 (2013)
32. S.J. Reddi, S. Kale, S. Kumar, On the convergence of adam and beyond. arXiv preprint arXiv: 1904.09237. 2019 Apr 19
33. K.S. Younis, A.A. Alkhateeb, A new implementation of deep neural networks for optical character recognition and face recognition. *Proceedings of the new trends in information technology.* 2017 Apr 25 pp. 157–162

Implementing Face Search Using Haar Cascade

Ramyashree⑩ and P. S. Venugopala⑩

Abstract It is a known fact that human beings are able to distinguish distinct people primarily by observing the human face. Visual search engines are the ones that, when given the name, search for celebrity faces in a video file. This will be retrieval based on the data related to the file. The goal of this paper is to enhance the usability of video searches. Users would be able to input a video and a picture (passport size) of the person they are intended to search for, and then select the frames containing the given image from the uploaded file. This would be helpful for video capture purposes. This is especially useful for the people who are interested in a part of the video in which either they themselves or the person of their interest is available. This is achieved by removing all frames from the video and saving the extracted video in different directories. Facial detection method is used to recognize the faces in the frame, and these identified frames are used as training models for the next level of facial recognition. The facial recognition technique is used to differentiate the reference image from all the frames, to transform the selected frames to a video, and to play it for the user. This application faces many challenges like occluded face detection and recognition. Also, frontal view of the face will be necessary for this application to work successfully. These challenges can be taken into account as the future work for this particular application.

Keywords Face detection · Haar cascade · PCA · Viola–Jones technique

1 Introduction

In the current situation, facial identification and face recognition have become a significant area in machine perception science and thus the most effective applications

Ramyashree
Shri Madhwa Vadiraja Institute of Technology and Management, Bantakal, Udupi, Karnataka, India
e-mail: ramyashree.cs@sode-edu.in

P. S. Venugopala (✉)
NMAM Institute of Technology Nitte, Udupi, Karnataka 574110, India
e-mail: venugopalaps@nitte.edu.in

© The Author(s), under exclusive license to Springer Nature Singapore Pte Ltd. 2022   263
P. Shetty D. and S. Shetty (eds.), *Recent Advances in Artificial Intelligence and Data Engineering*, Advances in Intelligent Systems and Computing 1386,
https://doi.org/10.1007/978-981-16-3342-3_22

are comprehension and analysis of the image [1]. Face detection is the essential testing area with the most heterogeneity. The implementations include video conferencing, image retrieval based on information and automated authorization, etc. The classifier will isolate the face parts from the context and will be trained using a series of non-facial sample results. Positions and sizes are extracted and scaled to the same sample size when a replacement input image is given. This brute-force search technique is used in most face recognition methods [2].

The purpose of this work is to help those users who wish to view only those parts of the video in which the person of their interest is present. The application will result in a series of frames containing the picture of the requested person which is then converted into a video and presented to the use. In this system, the user has to select an image that has to be searched in a video [3]. The user must also select a video in which the search has to be performed. The important phases are:

Face detection: The first phase is converting the video into frames and then performing face detection on all these frames [4]. Implementation of face detection using Viola–Jones technique is one of the approaches. This technique of face exploration is designed to prepare faces extremely rapidly while maintaining high recognition speeds [5]. There are three main responsibilities. The first is the display of another image, called a fundamental frame, which enables the highlights used by the locator to be processed easily. The second is a quick and effective classifier that is generated using the AdaBoost learning equation to select a few specific visual highlights from a wide range of potential highlights [6]. The third pledge is a technique for entering classifiers in a route that requires image districts to be disposed of immediately while focusing time on promising face-like regions. The faces thus identified are used as training images for the purpose of facial recognition [7].

Face recognition: Principal component analysis (PCA) is one of the approaches that is used for the recognition purpose [8, 9]. The frames which contain the required face are then extracted and then converted back into a video [10]. The method of initialization comprises the following operations:

(i)   Get the initial set of face images referred to as a training set.
(ii) Calculate from the training collection the eigenfaces, holding highest eigenvalues only. These pictures of M describe the space faces. The eigenfaces, as new faces are met, it is possible to upgrade or recalculate.
(iii) In this M-dimensional space, calculate the distribution for by projecting his or her face pictures, each known individual into this space face [11].

The Haar cascade classifier is based on the Haar wavelet process of functionally analyzing pixels in the image into squares [12]. This utilizes "integral image" principles to measure the detected "features." Haar cascades use the AdaBoost learning algorithm to pick a small number of important features from a wide collection to give classifiers an effective result and then use cascading techniques to detect the face in an image [13].

2 Literature Survey

The face detection, face identification, and face recognition have been carried out by several researchers.

Sakle [1] has suggested the different strategies of face detection. Confidentiality and identity have become the core problem in the modern world today. Face recognition plays a role major function in verification and recognition. In this paper, multiple current facial recognition techniques are evaluated and it was addressed.

Prasad [4], a report on the identification of features in real images is explained and also provides a brief understanding of the issue in identification of features in real images. The aims of this paper are to describe the definition and scope of image processing, to address the different steps and methodologies involved in the typical processing of images, and to apply image processing tools and processes to research frontiers.

Chora [14], it describes the extraction of the feature and different techniques used for this. One of the main areas of technology that continues to evolve and expand is facial recognition. This paper, together with its pros and cons, provides the different strategies. Three major categories classified here as feature-based approach, hybrid approach and holistic approach, one of the key methods that many researchers may use.

Menser and Muller [7] suggested one algorithm to identify faces with dynamic background pictures of color. Use color details in a face detection system focused on study of key components (PCA). Using a color analysis, a skin color likelihood image is created and the PCA is performed on this new image, instead of the luminance image.

Perveen et al. [10], in this article, facial expression recognition (FER) framework supports estimated facial expression space by facial features and Gini indices for facial expression identification. Demeanor awareness also focuses on facial expression information regions, so that eyes, eyebrows, and mouth are extracted from the input image. When meeting a human face, an isolation is made of a feature that helps to distinguish the facial features. The values of the variable factors are defined in order to explain one of the six standard facial expressions. The clustering algorithm is applied to the JAFFE range of 30 images, each of which has six normal facial expression images.

Stojmenovic [15] explains about Viewdle concepts in face recognition. Viewdle feature for face recognition is explained in Video Search Engine, powered by Reuters Labs. It analyzes sequence of data clips and filters for names. Furthermore, by translating speech to text it adds contextual information to the facial recognition results.

Ghazi Mohammed Zafaruddin et al. [16] explained the applications of Face Search Engines. This search engine is as benign as it can be. It looks for faces that are based on picture tags. You will then ask the search engine for a name and return all matching images of faces marked with the name of the user. But this search engine does not appear to be very accurate at the retrieval of the images.

Nikolskaia et al. [2] about PicTriev function explained and go a step forward by actively looking for something similar faces. Then what you do is upload a portrait shot or other face picture in the format of.jpg or.jpeg, with a scale not exceeding 200 KB, and the search engine will return similar photographs found online. This will run a demo with a range of well-known images too. It fits particularly well with pictures of celebrities.

Pinto et al. [17] explain the approaches. There are several approaches to face recognition that are commonly known as systematic and feature oriented approaches. Currently there are relatively few studies that compare these two methods. Nowadays, there is a huge rise in facial recognition research, largely because of the numerous negative incidents taking place. In the areas of image processing and patter recognition, human face identification has become a difficult issue. This paper proposes a modern human face recognition algorithm using the primitive Haar cascade algorithm combined with three additional weak classifiers.

3 Methodology

The implementation is carried out using MATLAB. Viola–Jones algorithm is used in this work for face detection. The classic technique for image processing is to ramp the feature vector dynamically to varying lengths and then run the fixed-length detector through those images [18].

Although there are numerous face search engines, most of them are used for online purposes. Also, these search engines are used to identify the faces in still images rather than a video. So this led to the idea of detecting and recognizing faces in a video. This kind of application would be helpful in tracking a person and for video surveillance purposes [19].

In this work, there are four main steps. The first step is converting the given input video into frames. These frames are then stored in a separate folder. In this step, upload an image which has to be recognized in the video frames. The second step is face detection phase. Each of the frames is checked for the presence of face using face detection algorithm. The detected face from each of the frame is stored in a separate folder. The third step is face recognition on the frame from which the face has been detected. For recognition purpose, the face that are detected are used and the uploaded photograph is used as the test image. If there is a match then that particular frame is extracted and saved in a separate folder [16]. The final step is extracting all these selected frames from the folder and converting them into a video. The application then displays the final video which consists only the required frames of the given input video.

MATLAB script is used for converting video to frames. The script reads the input video in avi format [20]. The frame rate of the video is taken and using a counter the video is traversed till the end of file is reached. For each counter increment, a frame is extracted and saved in the desired output folder.

3.1 Face Detection

Here, the implementation is done using MATLAB. It includes control flow statements, features of object-oriented programming, etc. This will produce very speed and accurate result due to feature-based method [14]. When doing this work, several limitations were found out. To overcome this enhancement, techniques are applied using Viola–Jones algorithm.

3.2 Face Recognition

This is the third main step in this work. There are several algorithms for face recognition. Popular recognition algorithms include principal component analysis which uses eigenfaces, linear discriminate analysis, elastic bunch graph matching fisher face, the hidden Markov model, and the neuronal motivated dynamic link matching.

3.3 Converting the Frames into Video

This is the reverse process of converting video into frames. The stored frames are retrieved one at a time, and each frame is inserted into the video. The result is the video with only the required frames containing the person of your concern.

3.4 Algorithm

Step 1: Read the video
Step 2: Read the test image
Step 3: Convert the input video into a number of frames
Step 4: Perform face detection on each and every frame
Step 5: Select the frames in which the faces have been detected and discard the frames in which there has been no detection of the faces
Step 6: Perform face recognition on the selected frames
Step 7: Store only those frames in a different folder which has face match with the test image and discard the remaining frames
Step 8: Extract these frames from the folder and convert it back into a video
Step 9: Return the extracted video as the output

The steps that are used are shown in following Fig. 1.

Fig. 1 Flowchart of video
searching

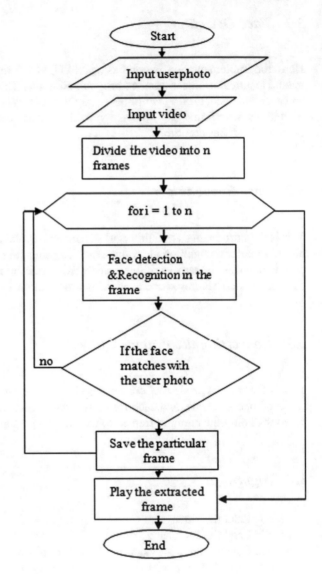

4 Implementation Details

Viola–Jones algorithm [15] is used in this work. The typical approach to image
processing will be to rescale the image data to various sizes, and then run the fixed
size detector through these images. This method turns out to be very time-consuming
due to the estimation of the various scale pictures. However, Viola–Jones developed
an invariant scale locator that needs a comparable number of counts regardless of
the distance. This finder is designed using a supposed important image and some
straight rectangular highlights reminiscent of the Hair wavelets.

Fig. 2 Integral image

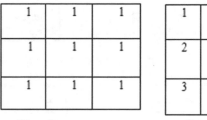

Given image   Output image

Fig. 3 Calculation of sum

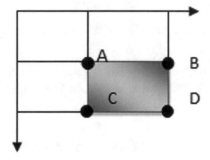

4.1 The Scale Invariant Detector

Converting a given image that is input to the integral image is the very initial stage of Viola–Jones algorithm. This can be done using some calculations that is initially calculated the sum of the pixel that is above and left of the particular pixel. Then set every pixel to the above calculated sum. This is shown in Fig. 2.

With the four values of any given rectangle gives the values like shown in Fig. 3.

$$\text{Rectangle sum} = D - (B + C) + A$$

From Fig. 3, it is noted that the rectangle A is covered with the two rectangle B and C. So the calculated sum of A should be added. Viola–Jones algorithm decided that, it gives the acceptable outcome when the constancy is about 576 pixel. So totally using this 160,000 features can be build.

4.2 The Modified AdaBoost Algorithm

AdaBoost algorithm is established by Freund Schapiro in 1996 [17]. So the revised adaptation of AdaBoost algorithm is used in the Viola–Jones technique. One of the popular machine learning algorithms is the AdaBoost algorithm. This has the ability

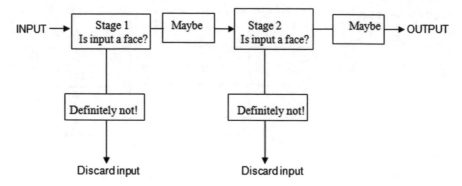

Fig. 4 Cascade classifier principal component analysis algorithm

to build a very active classifier. The threshold value and polarity are the main features of this type of algorithm. The simple brute-force approach is used in this technique.

4.3   The Cascaded Classifier

The very determinant in the Viola–Jones algorithm is scanning. For the given image, scan with various times with different size. This process should be done for a small number of faces in images too. Using this technique, the part which is not a face can be subtracted, and this method is very fast approach. Strong classifier is contained in every step of cascade classifier. The task of this step is to identify that whether the subwindow contains a non-facial part or any facial part. Non-face parts are eliminated during this stage, and image contains a face that is moved to next step. This is demonstrated in the following Fig. 4.

PCA [10] includes a scientific method that changes various perhaps co-related factors into few co-related factors called chief parts. The most relevant section, though, represents as much of the changeability of the information as could be predicted, and each subsequent segment reflects as much of the remainder of the fluctuation as could reasonably be anticipated. Steps involved in this are prepare the details, deduct the mean, calculate the covariance matrix, calculate the own vectors and own values, and choose the main part.

5   Results and Discussion

Recognition has recently attracted attention and is beginning to be applied in variety of domains predominantly for security. There are some image search engines that search for the image of a particular person and there are no such video search engines.

So this led to the idea of video searching. User interface is developed using MATLAB to provide the input. An image and a video are the inputs for this application. Output is the part of the video in which the interested person is present. This application is ideal for those wanting to see only portions of their interest in the stream. The experimental results are shown in following Fig. 5.

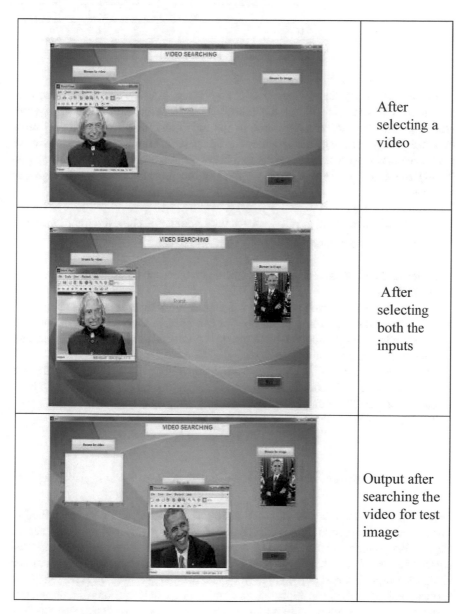

Fig. 5 Experimental results

Table 1 Performance of facial recognition

Number of samples	Recognition accuracy (%)	Execution time (s)
1	100	0.02
3	100	0.02
5	100	0.05
7	98	0.1
10	95	0.2
20	91	0.5

In the initial step, the user has to select the video, after selecting the video user have to select the test image from the dataset. After selecting both the inputs, system will show the presence of test image in the video and it will give suitable error messages when test image does not contain a face or test image is not present in the video. Accuracy percentage is calculated based on the classification accuracy rate (CAR). Based on the total number of samples and correct prediction, percentage was calculated.

$$\text{CAR} = (\text{Total Number of Faces} - \text{a total number of False Recognition})$$
$$/(\text{Total number of Faces}) * 100\%.$$

Performance of facial recognition is calculated based on the number of sample as shown in Table 1.

The bar graph in Fig. 6 shows the data segment and compares the data using solid bars to represent the quantity. From this we can observe that, when the number of samples are less then percentage of accuracy is more and execution time is also less compared to more number of samples.

The above experiment is compared with local binary patterns (LBP). By this we observe that quality of analysis is high and false positive rate is always less in Haar cascade approach. The comparison between this samples is shown in Table 2.

The graphical representation is shown in Fig. 7.

6   Conclusion and Future Work

Face recognition has recently attracted attention and is beginning to be applied in variety of domains predominantly for security. Although there are some image search engines that search for the image of a particular person there are no such video search engines. So this led to the idea of video searching. Input is an image and a video and output is a part of the video in which the interested person is present. This application is ideal for those appearances wanting to see only portions of their interest in the stream. For photojournalists and photographers, for surveillance systems, and

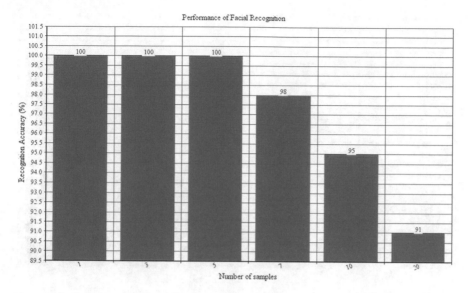

Fig. 6 Graphical representation of accuracy rate

Table 2 Comparison between Haar and LBP

Number of samples	Execution time (s) in Haar cascade	Execution time (s) in LBP
1	0.02	0.10
3	0.02	0.10
5	0.05	0.128
7	0.1	0.435
10	0.2	0.672
20	0.5	0.925

for tracking a person in a video, this program may also be essential. The above experiment is compared with local binary patterns (LBP). By this we observe that quality of analysis is high and false positive rate is always less in Haar cascade approach

Face detection works well with photographs or video with a very small audience, but it is not that good as the number of users rises. And face recognition operates on the probabilistic approach, because there is a possibility to identify a similar face to that of the input image. Thus, the structure can be generalized to ensure the precise matching of images by selecting threshold values, which is an art in itself. As if the task is now taking a long time to process and return the result, this drawback can be taken into account and the performance can be increased by using parallel processing techniques. This application faces many challenges like occluded face detection and recognition. Also, frontal view of the face will be necessary for this application to

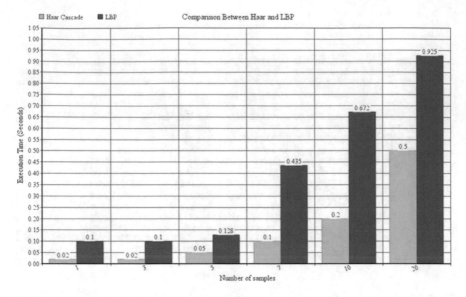

Fig. 7 Execution time using Haar cascade and LBP

work successfully. These challenges can be taken into account as the future work for this particular application.

References

1. M. Sakle, Study and analysis of different face detection techniques. Int. J. Comput. Sci. Inf. Technol. **5** (2014)
2. K. Nikolskaia, N. Ilya, A. Minbaleev, Development of the face recognition application using neural networks technology, in *2019 International Conference "Quality Management, Transport and Information Security, Information Technologies" (IT&QM&IS)* (Sochi, Russia, 2019), pp. 87–91. https://doi.org/10.1109/ITQMIS.2019.8928392
3. Z. Shaaban, Face detection methods, recent researches in applied informatics and remote sensing. Int. J. Comput. Appl. (0965–8687) **55**(1) (2010)
4. D.K. Prasad, Survey of the problem of feature detection in real images. Int. J. Image Process. (IJIP) **6**(6) (2012)
5. Y.K. Singh, V. Hruaia, Detecting face region in binary image, in *2015 IEEE Recent Advances in Intelligent Computational Systems (RAICS)* (2015, 10–12 December)
6. R. Zwiggelaar, B. Rajoub, M.H. Yap, H. Ugail, A short review of methods for face detection and multifractal analysis, in *2009 International Conference on CyberWorlds. 978–0–7695–3791–7.* https://doi.org/10.1109/CW.2009.47
7. B. Menser, F. Muller, Face detection in color images using principal component analysis. Institute for Computer Science, Social Informatics and Telecommunication Engineering (2012)
8. V. Bhandiwad, B. Tekwani, Face recognition and detection using neural networks, in *International Conference on Trends in Electronics and Informatics ICEI* (2017) 978-1-5090-4257-9
9. A. Robert Singh, A. Suganya, Efficient tool for face detection and face recognition in color group photos. 978-1-4244-8679-3

10. N. Perveen, S. Gupta, K. Verma, Facial expression recognition system using statistical feature and neural network. Int. J. Comput. Appl. (0975–8887) **47**(1) (2012, June)
11. I. Paliy, Face detection using Haar-like features cascade and convolutional neural network, in *2008 International Conference on "Modern Problems of Radio Engineering, Telecommunications and Computer Science" (TCSET)* (Lviv-Slavsko, 2008), pp. 375–377
12. V. Mohan, R. Deepa, M. Deepa, S. Somannavar, M. Datta, A simplified Indian diabetes risk score for screening for undiagnosed diabetic subjects. J. Assoc. Phys. India **53**, 759–763 (2005)
13. D. Tyas Purwa Hapsari, C. Gusti Berliana, P. Winda, M. Arief Soeleman, Face detection using Haar cascade in difference illumination, in *2018 International Seminar on Application for Technology of Information and Communication* (Semarang, 2018), pp. 555–559. https://doi.org/10.1109/ISEMANTIC.2018.8549752
14. R.S. Chora, *Image Feature Extraction Techniques and Their Applications for CBIR and Biometric System*
15. M. Stojmenovic, Mobile cloud computing for biometric applications, in *2013 16th International Conference on Network-Based Information Systems* (Melbourne, Australia, 2012), pp. 654–659. https://doi.org/10.1109/NBiS.2012.147
16. G.M. Zafaruddin, H. S. Fadewar, Face Recognition: A Holistic Approach Review, 4799-6629-5
17. N. Pinto, J.J. Di Carlo, D.D. Cox, How far can you get with a modern face recognition test set using only simple features? in *IEEE Conference on Computer Vision and Pattern Recognition (CVPR)* (2009)
18. A. Rastogi, A. Pal, B.S. Ryuh, Real-time teat detection using Haar cascade classifier in smart automatic milking system, in *2017 7th IEEE International Conference on Control System, Computing and Engineering (ICCSCE)* (Penang, 2017), pp. 74–79. https://doi.org/10.1109/ICCSCE.2017.8284383
19. Ramyashree, P.S. Venugopala, Proposal for enhancing face detection in group photos, in *2018 4th International Conference on Applied and Theoretical Computing and Communication Technology (iCATccT)* (Mangalore, India, 2018), pp. 113–118. https://doi.org/10.1109/iCATccT44854.2018.9001937
20. D.K. Ulfa, D.H. Widyantoro, Implementation of Haar cascade classifier for motorcycle detection, in *2017 IEEE International Conference on Cybernetics and Computational Intelligence (CyberneticsCom)* (Phuket, 2017), pp. 39–44. https://doi.org/10.1109/CYBERNETICSCOM.2017.8311712

Deep Learning Photograph Caption Generator

Savitha Shetty, Sarika Hegde, Saritha Shetty, Deepthi Shetty,
M. R. Sowmya, Reevan Miranda, Fedrick Sequeira, and Joyston Menezes

Abstract Generating caption for an image is challenging problem present in artificial intelligence. For a given photograph text, descriptions must be generated. So we need to have computer vision method for understanding the exact content present inside image and also natural language processing field to turn image understanding into proper words in right sequence. Methods of deep learning have successfully demonstrated the ability to solve this problem statement. The advantage of deep learning methods is that it is possible to define an end-to-end model for predicting the caption whenever picture is given, rather than needing a pipeline of specifically designed models or sophisticated data preparation. Our paper focuses on preparing the text and image data for training our deep learning model, we need to design and also train our deep learning caption generation model, and we need to evaluate the trained caption generation model, using it to caption entirely new photographs. Finally, we are hosting a webserver to which the user can upload images and fetch the generated captions.

Keywords Caption generation · Deep learning · End-to-end model · Webserver

S. Shetty (✉) · S. Hegde · R. Miranda · J. Menezes
Department of Computer Science and Engineering, NMAM Institute of Technology, Karkala, India
e-mail: shettysavi1@nitte.edu.in

S. Hegde
e-mail: sarika.hegde@nitte.edu.in

S. Shetty · F. Sequeira
Department of Computer Applications, NMAM Institute of Technology, Karkala, India
e-mail: shettysaritha1@nitte.edu.in

D. Shetty · M. R. Sowmya
Department of Computer Science and Engineering, NMIT, Bangalore, India
e-mail: deepthi.shetty@nmit.ac.in

M. R. Sowmya
e-mail: sowmya.mr@nmit.ac.in

1   Introduction

Caption generation requires both the computer vision methods to understand image content and language model, from natural language processing field to turn image understanding into words in the right sequence. The aim of computer vision problems at an abstract level is, we are using image data to infer some information about the world. It is considered as multidisciplinary field, which is also a subfield of machine learning and artificial intelligence. It involves the use of some special methods and usage of general learning algorithms. If input dataset is provided, then CNN has the ability of humans to do the classification of images [1].

Generation of natural language is the process of transforming structured data into natural language. It is also used to generate long-form content for automating customer reports for various organizations and also to produce the custom contents for web or mobile application. Description and recognition of image are fundamental challenge of computer vision [2].

Deep learning focuses on state-of-the-art results on various problems related to the generation of captions. Various deep learning algorithms try to exploit the unknown structure in distributing the inputs to further discover the good representations. At multiple levels, the features which are learned at higher levels define features at lower levels. The Internet is composed of images and text. Indexing and searching text are relatively straightforward but algorithm which is used should know what is the content of the image in order to index and search images. The images and video content have remained opaque for the longest time, and it is properly described using meta descriptions which are provided by person who uploads them. In order to get the proper information from the image, computer must understand content of image and generate descriptions in human understandable format. The paper focuses on utilizing the deep learning architectures, deep neural networks, recurrent neural networks, convolutional neural networks for generating captions for various images. The convolutional neural network is used for the computer vision. The recurrent neural network is used for language modeling, and the deep neural network is used to find proper mathematical manipulations to turn input into output. Web technologies are used to present generated captions to the client.

2   Related Work

Raffaella et al. [3] proposed a system which classifies existing approaches based on, how they conceptualize the problem, that is models that cast description as either a generation problem or visual or a multimodal representation space retrieval problem. It provides detailed review of various existing models and also highlights their pros and cons. It also gives an overview of various benchmark image datasets used and various evaluation measures which are developed to evaluate the quality of the machine-generated descriptions of images. Finally, it extrapolates future directions in the field of automatically generating proper descriptions of images.

Marc et al. [4] explained that in neural image caption system, and it is typically seen as a primary 'generating' component. It suggests that features of should be 'injected' to RNN. It is dominant view in case of the literature. RNN is used to encode all earlier generated wordings. RNN is used to encode various linguistic characteristics, and final representation is merged with features of image in next stage. After comparing two architectures, we came to know that late merging will outperform injection. RNN is seen as encoder than generator.

Marctanti et al. [5] explained that whenever a recurrent neural networking model is used for generating caption, information of the image is fed into neural network by including it in that RNN with a condition that our model by 'injecting' the image feature, layer following that RNN condition that particular language-related model by 'merging' all image feature. Both the options are attested in our literature, there are no systematic comparisons among two. The authors say it is not necessary to know whether performance first architecture will be used or not. This merge architecture has various advantages, since conditioning by merging will make RNN's hidden state vector will be shrunk by 4 times. Linguistic and visual features need not been coded together because that will result in very huge model generation.

Kelvin et al. [6] proposed that there is attention-based model which will learn automatically to describe contents of image. It also describes how a person is able to train this model in deterministic manner by the using backpropagation technique, stochastically by maximizing the lower bound. It describes that if we use visualization then model is able to fix its gaze by considering minute objects. But it generates proper words as in the output sequence. Validation is done by using three benchmark datasets, namely MSCOCO, Flickr30k, Flickr8k. Ramnath et al. [7] proposed auto-caption system which automatically suggests a caption for the photograph for the user.

3 Proposed System

We have used Flickr8k open-source dataset consisting 8092 photographs in JPEG format and five descriptions for each photograph to build the vocabulary and train the model. The dataset has a predefined training data with 6000 images, our development dataset with 1000 images, test dataset with 1000 images. Firstly, the features of images are extracted using VGG-16 model without its last classification layer. VGG-16 is a convolutional neural network model for recognizing and classifying images [5]. Since we only need the features and not classification, we simply remove the final layer in that model. Hence, each image will be represented as a 4096-element featured vector. Descriptions of images are cleaned, and a vocabulary of words is built. The description is encoded and will have to be split into many words. Model will provide with previous word, the image and next word as training. Then model is

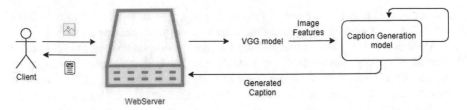

Fig. 1 High-level design architecture

used to generate descriptions and words which are generated are concatenated and provided recursively for generating caption of the image.

The input text is given to word embedding layer followed by LSTM. The photograph feature is given into dense layer to obtain 256 element representation. Both input models produce a vector with 256 elements. In addition, both input models use 50 percent dropout regularization. This will reduce overfitting of training dataset, since configuration of the model learns faster.

Decoder model uses an addition operation to merge vectors from 2 input models. This will be fed to 256 neuron layer which is dense, and after that, next to final output dense layer which will make softmax prediction for our very next word in that sequence over the entire output vocabulary. In this way, we obtain captions for any given image. The web server is hosted using express module of node JS to handle client requests. When the user submits a form by uploading an image, a python child process is spawned which generates the caption. Upon receiving the caption from child process, the caption is rendered on webpage to the user.

Figure 1 gives an abstract view of our work. The web server listens to client requests. When the client uploads an image to be captioned, it first uses the modified VGG-16 model to extract image features. The image features and a start token are fed to the caption generation model. The caption generation model generates the next word of caption, and it is appended and recursively fed to the caption generation model until it generates end token. Once the complete caption is generated, it is rendered on a web page to the client. The design of caption generation model and modified VGG-16 model is discussed in low-level design architecture.

4   Methodology

Figure 2 describes each layer of modified VGG-16 model. Final layer in originally used VGG-16 model classified the image into categories. The final layer is removed, and hence, output of fifteenth layer is taken as the final output. The output obtained is a 4096 feature elements vector of the image, which will be fed to caption generation model.

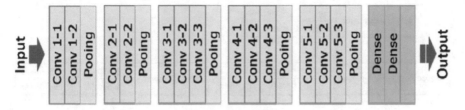

Fig. 2 VGG-16 model without final classification layer

Fig. 3 Merge model of caption generator

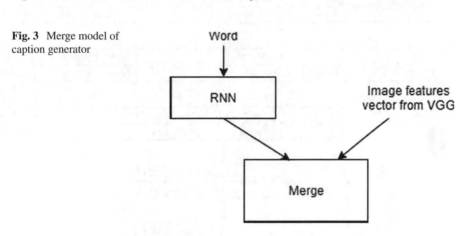

Caption Generation Model

Figure 3 explains the merged model used for caption generation [6]. The model can be described in three parts:

Photograph Feature Extractor: The photograph feature extractor expects features of the input photograph to have a vector of 4096 elements. A dense layer processes these to produce a photograph representation of 256 elements.

Sequence Processor: Here word embedding layer is used to handle our text input and it is followed by LSTM, recurrent neural network layer. Sequence processor is expecting input sequences with predefined length of thirty-four words to be fed into our embedding layer which is being masked to ignore associated padded values. It is followed by an LSTM layer which has 256, memory units. Both of input models which are used to produce a vector 256 elements. Both our input models are using regularization with 50% dropout. This is generally used for reducing the over fitting of our training dataset, since our model configuration learns too fast.

Decoder: Our decoder model uses an addition operation to merge the vectors from input models. It is fed to dense layer having 256 neurons and to one final output dense layer which will make a softmax prediction for next used word using sequence over our output vocabulary. Figure 4 visualizes structure of network which better

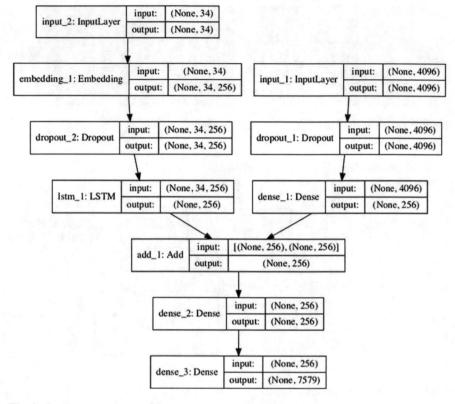

Fig. 4 Caption generation model

helps to understand two streams presenting input. It provides low-level structure of the model discussed in Fig. 3, specifically the shapes of the layers.

Model definition and training: The steps for defining the model and training using the dataset are:

Prepare image data:

For each image in the dataset:

1. Load the VGG-16 model provided by Keras: Remove the sixteenth layers of the model to make penultimate layer the output layer.

    (a) Load the image
    (b) Convert the image pixels into numpy array
    (c) Reshape data for the model
    (d) Prepare the image for input to model
    (e) Extract features using the model
    (f) Store the features into a dictionary with the corresponding image ID.

2. Serialize the dictionary into a file.

Prepare Text Data

1. Read the descriptions file and convert it into dictionary form. Image ID will be the key, and value will be an array of five descriptions each.
2. Clean the descriptions.

   (a)Make all words lower case
   (b)Remove all punctuations
   (c)Remove single characters
   (d)Remove all words which have numbers in them.

3. Store dictionary of image ID and descriptions into a file.

Loading the Data

1. Load the list of training and development image identifiers provided by dataset.
2. Load the image features and cleaned descriptions for corresponding identifiers inlist.
3. Add 'start seq' and 'end seq' tokens to the descriptions.
4. Encode the description to create a mapping of each word to unique integer values. Tokenizer class provided by Keras does the encoding of loaded description.
5. Serialize the Tokenizer for future use.
6. Create input–output pairs for training and validation.

Define and Fit the Model

1. Define the model using Keras
2. Train the model using input–output pairs on Google Colab [7] for 20 epochs [8].
3. Our model learns very fast, and it quickly overfits our training dataset. Hence, skill of trained model should be always monitored on our holdout development dataset. We have to save whole model to a particular file whenever model skill on our development dataset is improving at end of each epoch.

Image Captioning

The trained model and Tokenizer obtained can be used to generate new captions for clients. The steps are listed below:

1. Setup a Web server using express module in Node.js which displays an image upload page to the clients
2. Upon image upload by the client, save the image using express file upload module and spawn a python child process
3. The python child process will:

   (a) Extract image features using modified VGG-16model
   (b) Seed the generation process
   (c) Iterate:

      i.   Integer encode input sequence using Tokenizer
      ii.   Pad the input sequence
      iii. Predict next word using the model
      iv. Convert probability to an integer
      v.   Map integer back to word using Tokenizer
      vi. If word cannot be mapped then stop or else append it as input for the generation of next word
      vii. Stop the process if end of sequence is predicted.

(d)   Flush the generated caption onto standard output.

4.   Render the generated caption on a webpage to the client using Embedded JavaScript Template [9].

5 Results

Figure 5 shows the features of images in dataset being extracted using modified VGG-16 model to train the caption generation model. Figure 6 shows the descriptions in dataset before processing. Figure 7 shows the cleaned descriptions used to train the caption generation model.

Figure 8 shows the data being loaded for training and validation. The Tokenizer is built here, and vocabulary size indicates the number of unique mappings. Input–output pairs are created from the data for training and validation. Figure 9 shows the caption generation model being trained on training data and validated against development data. Figure 10 shows webpage rendered by web server upon client request.

Figure 11 shows the web page rendered by web server to the client with final generated caption.

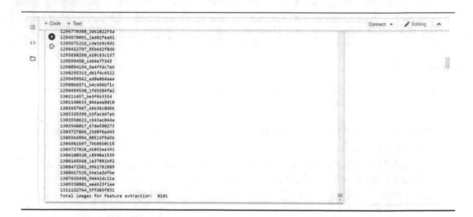

Fig. 5 Image feature extraction

Fig. 6 Descriptions in dataset

Fig. 7 Cleaned descriptions

Model evaluation is done by generating all descriptions for available photographs which present in test data and then evaluating each of these predictions using standard cost function. The predicted and actual descriptions are collected and then evaluated with corpus BLEU score which is used to summarize how close generated text is to text which is expected. Translated text is evaluated using bilingual evaluation understanding score. Generated description is compared against all reference descriptions for all photographs in the dataset. BLEU scores are calculated for 1, 2, 3, 4 cumulative n-grams.

The score can range from 0 to 1, where a score closer to 1 is always better and score close to 0 is always worse. The model scores fit perfectly within our expected range of skillful model on the problem.

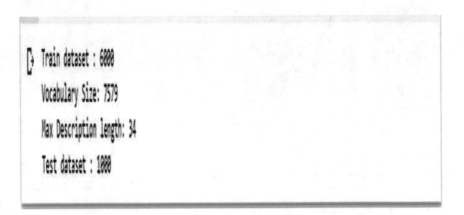

Fig. 8 Loading the data

```
[ ]  Epoch 1/5
[▸  Epoch 00001: val_loss improved from inf to 4.06052, saving model to model-ep001-loss4.519-val_loss4.061.h5
    9576/9576 - 700s - loss: 4.5187 - val_loss: 4.0605
    Epoch 2/5

    Epoch 00002: val_loss improved from 4.06052 to 3.91940, saving model to model-ep002-loss3.843-val_loss3.919.h5
    9576/9576 - 704s - loss: 3.8427 - val_loss: 3.9194
    Epoch 3/5

    Epoch 00003: val_loss improved from 3.91940 to 3.86909, saving model to model-ep003-loss3.638-val_loss3.869.h5
    9576/9576 - 701s - loss: 3.6377 - val_loss: 3.8691
    Epoch 4/5

    Epoch 00004: val_loss did not improve from 3.86909
    9576/9576 - 699s - loss: 3.5290 - val_loss: 3.8857
    Epoch 5/5

    Epoch 00005: val_loss did not improve from 3.86909
    9576/9576 - 676s - loss: 3.4672 - val_loss: 3.9112
    <tensorflow.python.keras.callbacks.History at 0x7fe6944bdc18>
```

Fig. 9 Training on Google Colab

6 Conclusion and Future Work

Our paper on training a deep learning model using text and image, and using it to caption new photographs. A web server was developed to listen to client requests, where the clients can upload images on the webpage and get them captioned. The convoluted neural network used here is a small 16-layer VGG model. Larger models that might offer better performance can be explored, such as ResNet. Word vectors pre-trained on training dataset or large corpus of text such as articles and Wikipedia

Fig. 10 Image uploading page

Caption ==> dog is running through the grass

Fig. 11 Generated caption

can be used rather than learning them as a part of fitting the model, to improve the performance. The webpage for the client can be made more user-friendly, and the model can be trained on a larger dataset.

References

1. K. Kafle, C. Kanan, Visual question answering: Datasets, algorithms, and future challenges. Comput. Vis. Image Underst. **163**, 3–20 (2017)
2. J. Donahue, L. Anne Hendricks, S. Guadarrama, M. Rohrbach, S. Venugopalan, K. Saenko, T. Darrell, Long-term recurrent convolutional networks for visual recognition and description. in *Proceedings of the IEEE Conference on Computer Vision and Pattern Recognition* (2015), pp. 2625–2634
3. Raffaella Bernardi Plank, et al., *Automatic Description Generation from Images: A Survey of Models, Datasets, and Evaluation Measures* (2016)
4. Marc Tanti Camilleri, Albert Gatt, P. Kenneth, *What is the Role of Recurrent Neural Networks (RNNs) in an Image Caption Generator?* (2017)
5. A. Marctanti, C. Kennethp, *Where to put the Image in an Image Caption Generator* (2018)
6. Kelvin Xu Bengio, et al., *Show, Attend and Tell: Neural Image Caption Generation with Visual Attention* (2015)
7. K. Ramnath, S. Baker, L. Vanderwende, M. El-Saban, S.N. Sinha, A. Kannan, ... A. Bergamo, Autocaption: automatic caption generation for personal photos. in *IEEE Winter Conference on Applications of Computer Vision* (IEEE, March 2014), pp. 1050–1057
8. K. Simonyan, A. Zisserman, *A Very Deep Convolutional Networks for Large-Scale Image Recognition* (2015)

Streaming of Multimedia Data Using SCTP from an Embedded Platform

E. S. Vani and Sankar Dasiga

Abstract With the Internet of things which is going to be happen in near future, the several of the functionality that currently being done by large computer systems are also going to become a requirement for embedded platforms. As we all know that nowadays, Internet is also being used for multimedia streaming. As such, the streaming of multimedia data is also envisaged to be a requirement for embedded platform. In our project, we have taken an embedded platform which is very cost effective and having an ARM processor and has got ARM A profile. With this platform, we are implementing an application using SCTP which we are attempting to transmit multimedia data both video and audio from an embedded platform, and we are comparing the performance of SCTP and TCP.

Keywords Streaming · ARM processor · SCTP · TCP

1 Introduction

For many of our day-to-day needs, embedded systems offer specific and cost-effective solutions and they have become an integral part of our lives. Nowadays, embedded technology is in prime, and the wealth of knowledge available in embedded technology is mind blowing. Also, nowadays for hands-on study, a number of development platforms are available along with tools, libraries, and peripheral devices.

Streaming is one of the data transfer technique [1]. Streaming means the end user can make use of the data when the data is on its way to the user. The user can view the data as and when he receives without waiting for the whole data to be received first. He

E. S. Vani (✉)
Department of Information Science and Engineering, Nitte Meenakshi Institute of Technology, Bangalore, India
e-mail: Vani.es@nmit.ac.in

S. Dasiga
Department of Electronics and Communication Engineering, Nitte Meenakshi Institute of Technology, Bangalore, India

© The Author(s), under exclusive license to Springer Nature Singapore Pte Ltd. 2022
P. Shetty D. and S. Shetty (eds.), *Recent Advances in Artificial Intelligence and Data Engineering*, Advances in Intelligent Systems and Computing 1386,
https://doi.org/10.1007/978-981-16-3342-3_24

can view it directly without downloading the whole thing [2]. Streaming is the real-time transmission of data. The data may be live or stored media. Compared to other data, multimedia data is typically large in size, and for its reasonable use it requires significant bandwidth. The multimedia data may be audio, video, slideshows, etc.

For inserted applications, the microcontroller having ARM processor has turn into a default processor. This assignment means to execute sight and sound spilling on ARM processor-based embedded platform.

The first objective of this work is to transfer multimedia data using TCP and also using SCTP protocols. Then, compare the performance of both. The second objective is to stream the multimedia data from an embedded platform using SCTP through multiple communication links simultaneously.

With IOT, the embedded systems are expected to be on the Internet and communicating with zero user intervention. These embedded systems have variants of 802.11 standards such as Wi-Fi, Ethernet, Bluetooth, and NFC [3].

The motivation for this project is to interconnect through multiple 802.11 when available, whereby stream the information such as multimedia with a protocol such as SCTP that support simultaneous communication through multiple links. These multi-channels should be selectable dynamically [4].

To establish connectivity through all, the available communication links to get maximum throughput [5].

2 System Specifications

2.1 Hardware Specifications

2.1.1 Raspberry Pi

With the intension of promoting the teaching of basic computer science, the Raspberry Pi foundation in the UK has developed the Raspberry Pi which is a series of credit card sized single board computer. The original Raspberry Pi includes an SRM1176JZF 700 MHz processor which is based on BroadcomBCM2835 system on chip. USB port is directly connected to SoC in model A and A+. Model B has two ports and model B+ has five-point hub.

2.2 Software Specifications

2.2.1 Linux

The Linux is one of the open-source operating system which is freely distributed, cross-platform based on Unix. It is similar to windows and OS X. The Linux kernel is the core of the operating system. Linux has begun as a server OS "From wristwatches to supercomputers" is the popular description of Linux capability. It brings up a streamlined Linux environment. It allows the user to turn on their devices within a matter of seconds. The natural fit for Linux is the growth of cloud computing.

2.2.2 The NOOBS Installer

The Raspberry Pi package comes with only Raspberry Pi board and nothing else comes with it. It does not come with OS. We have to install an OS on it. The SD card loaded with the OS from the computer is inserted in the Pi which becomes the primary boot device. To ease the installation of OS on Pi, Raspberry Pi foundation has made a software called NOOBS (New Out of Box Service). The NOOBS installer can be downloaded from the official Web site. To install NOOBS on the SD card, a user needs to connect the SD card to the computer and just run the setup file. Then, the card is inserted to the Raspberry Pi board. When booting it first time, a user can select an OS from a list of OS to install.

2.2.3 Raspbian

Raspbian is a free operating system which is based on Debian for Raspberry Pi hardware. It provides more than a pure OS. Raspbian comes with 35000 packages. For easy installation on Raspberry Pi, Raspbian comes with pre-compiled software bundled nicely.

3 Communication Protocols

3.1 Transmission Control Protocol (TCP)

TCP is one of the transport layer protocol and is a core protocol of the Internet protocol suite. It provides reliable, ordered message delivery. File transfer, email, and other Internet applications rely on TCP. It provides its service at the transport layer which is between application layer and Internet protocol layer. Host-to-host connectivity is provided by TCP. The transmission details and handshaking are handled by TCP at the transport layer. TCP detects the problems like packet loss, unordered message delivery. TCP requests retransmission of data and also rearranged the unordered message. It passes the data to the receiving application only when it reassembled the sequence of data transmitted originally. This protocol is unsuitable for real-time applications like video over. TCP provides a guaranty that the message received is identical and also is in correct order. In order to guaranty the reliability of data transfer, it uses technique of positive acknowledgment.

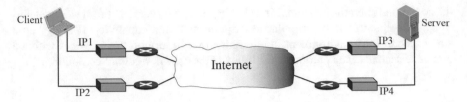

Fig. 1 Multi-homing

3.2 Stream Control Transmission Protocol (SCTP)

SCTP is one of the important transport layer protocol [6]. It provides the features of both TCP and UDP, i.e., it is message-oriented like UDP and reliable and congestion control like TCP. The message can be fragmented into a number of data chunks by SCTP. In SCTP, the message has been sent by the sender in only one operation. The receiver receives the exact message in one operation.

The two important features of SCTP are:

1. Multi-homing
2. Multi-streaming

3.2.1 Multi-homing

The multi-homing means an endpoint can have more than one IP address associated with it. The connection in SCTP is called as an association [7]. Among multiple IP addresses, one is taken as primary and the remaining is taken as secondary or redundant IP address. In the event of failure of primary path, the data transfer takes the secondary path. So, it provides redundancy. Figure 1 represents multi-homing.

3.2.2 Multi-streaming

Another important feature of SCTP is multi-streaming [8]. It means a stream of messages can be sent within a single connection. It is the capability of SCTP in parallel transmission of several independent messages. Rather operating on bytes, it operates on messages. The packet loss in one stream does not affect the packet loss of another stream. And, this feature of SCTP avoids head of line blocking (HOL). The Fig. 2 represents multi-streaming.

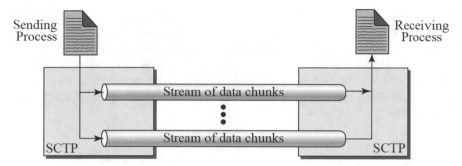

Fig. 2 Multi-streaming

4 Design and Implementation

4.1 Block Diagram

Figure 3 shows the experimental setup used for the project. One of the first steps to be done in the setup is to assign an IP address for the Ethernet interface as relevant for the configuration by using ifconfig. For example, ifconfig etho 192.168.200.200.

The system to the left is connected to the Raspberry pi using an Ethernet cable. The systems to the right interact with the board using Wi-Fi or Ethernet. Once the setup is created, the Raspberry Pi receives the multimedia files transmitted from the system

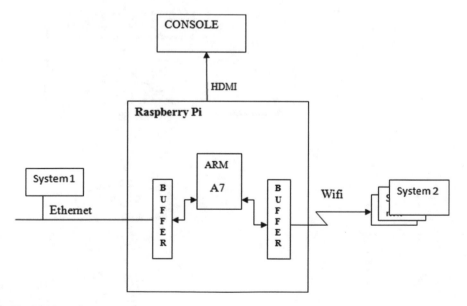

Fig. 3 Block diagram of the system

connected to its left using SCTP and it will stream it to the other side through Wi-Fi. So in this experimental setup, any number of systems can be connected simultaneously, and the streaming happens simultaneously. With this, we have demonstrated an embedded platform serving as a kind of router and also as a streaming device. Because we are using an embedded platform with limited memory and with limited CPU capability, the idea here is streaming of multimedia files simultaneously is of limited size.

Fig. 4 Flowchart for multimedia streaming

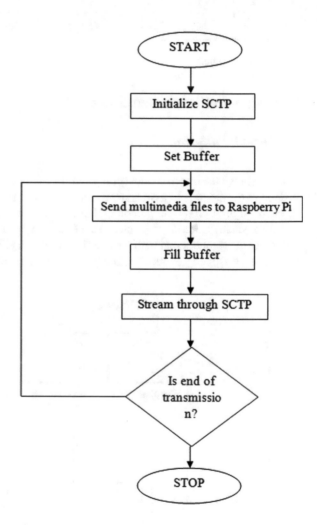

4.2 Flow Chart

Figure 4 is a flowchart for multimedia streaming which gives an overview of the sequence of steps involved in the multimedia streaming implementation. The first step is the initialization of SCTP connections, then setup the buffer, and then Raspberry Pi is provided with the multimedia data. Fill up the buffer and then stream the data to different systems.

5 Experimental Results

The multimedia data has been streamed using TCP and also SCTP. This has been done in both desktop and also in embedded platform. Also streaming is done through both wired and wireless. The time taken for the transmission of data of different sizes has been noted down for all the above-mentioned cases. The readings and the corresponding graphs are as below (Tables 1 and 2).

The corresponding graph for Tables 1 and 2 is shown below.

Table 1 Readings of streaming in desktop computers

(a) Using TCP through wired connection			
No. of bytes (MB)	15.7	14.4	13.8
Time (s)	3.953	3.585	3.457
(b) Using TCP through wireless connection			
No. of bytes (MB)	15.7	14.4	13.8
Time (s)	14.753	14.409	13.489
(c) Using SCTP through wired connection			
No. of bytes (MB)	15.7	14.4	13.8
Time (s)	1.388	1.26	1.208
(d) Using SCTP through wireless connection			
No. of bytes (MB)	15.7	14.4	13.8
Time (s)	12.383	11.46	10.868

Table 2 Readings of streaming in an embedded platform

(a) Using TCP through wired connection

No. of bytes (MB)	15.7	14.4	13.8
Time (s)	7.388	6.544	6.602

(b) Using TCP through wireless connection

No. of bytes (MB)	15.7	14.4	13.8
Time (s)	19.666	19.955	18.41

(c) Using SCTP through wired connection

No. of bytes (MB)	15.7	14.4	13.8
Time (s)	1.416	1.408	1.613

(d) Using SCTP through wireless connection

No. of bytes (MB)	15.7	14.4	13.8
Time (s)	13.260	16.376	16.214

1.   Streaming in desktop computers:

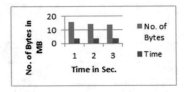

A) Using TCP through wired connection

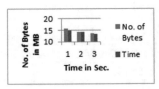

B) Using TCP through wireless connection

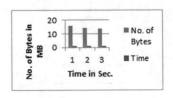

C) Using SCTP through wired connection

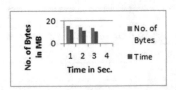

D) Using SCTP through wireless connection

2. Streaming in an embedded platform:

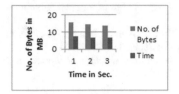

A) Using TCP through wired connection

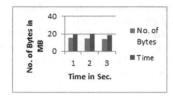

B) Using TCP through wireless connection

C) Using SCTP through wired connection

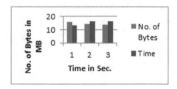

D) Using SCTP through wireless connection

6 Conclusion

In our project, we have compared the performance of both TCP and SCTP in the transmission of multimedia data. Also using an embedded platform, we have successfully transmitted multimedia data on multiple links simultaneously but with the restrictions for the reason that the available memory and computational power on the embedded platform is limited.

So, the conclusion is that the streaming of bigger and more complex multimedia data may not be suitable for embedded platform. Yet we envisaged that transmission of short and less resolution multimedia would be a necessity with IOT which is going to be happen. Embedded systems are becoming more powerful with computational power and available memory; they can also be a candidate for transferring large and more complex multimedia files.

References

1. K.-H. Kim, K.-M. Jeong, C.-H. Kang, S.-J. Seok, A transmission control SCTP for real-time multimedia streaming. Comput. Netw. 54, 1418–1425 (2010)
2. S. Ladha, P. Amer, Improving multiple file transfers using SCTP multistreaming. in *23rd IEEE and Communication Conference (IPCCC)*, pp. 13–22
3. R. Rajamani, S. Kumar, N. Gupta SCTP versus TCP: comparing the performance of transport. in *Protocols for Web Traffic* (Computer Sciences Department, University of Wisconsin-Madison)

4. A.L. Caro Jr., *SCTP: A New Internet Transport 1. Layer Protocol* (University of Delaware)
5. Y. Cao, C. Xu, J. Guan, H. Zhang, Receiver-driven SCTP-based multimedia streaming services in heterogeneous wireless networks
6. I.A. Rodriguez, Stream control transmission protocol—the design of a new reliable transport protocol for IP network
7. A. Jungmaier, E.P. Rathgeb, M. Schopp, M. Tüxen, SCTP—A multi-link end-to-end protocol for IP-based networks. Int. J. Electron. Commun. (AEU) **55**(1), 46–54 (2001)
8. S.J. Wee, J.G. Apostolopoulos, Secure video streaming for wireless networks. in *IEEE International Conference on Acoustics, Speech and Signal Processing*

A Fast Block-Based Technique to Detect Copy-Move Forgery in Digital Images

Vaneet Kour, Preeti Aggarwal, and Ravreet Kaur

Abstract Images have a use in almost each and every field of the world, from institutions to the court of law. This huge acceptance of the images in every field gave birth to numerous image editing tools and hence, various forgeries. Copy-move is a very popular forgery technique. Generally, keypoint-based and block-based methods are used for the copy-move forgery detection (CMFD). In this paper, a block-based approach for CMFD is implemented in which quantized discrete cosine transform (DCT) coefficients are used to detect forgery over an image. As, many forgers apply various post-processing like JPEG compression, blurring to make forged image traceless. Thus, these quantized DCT coefficients help in detecting the approximate match between the regions by setting user defined threshold parameter.

Keywords CMFD · Copy-move · Block-based · CoMoFoD

1 Introduction

Sophisticated tools for image manipulation have increased very much and are easily available. Editing is very easy to do because of such kind of tools but it is difficult to detect. Consider a scenario, an image needs to present at the court as an evidence. But before presenting, it has been edited which alters the original content of the image. Thus, this image can prove a criminal person as an innocent or an innocent as a criminal. Detecting forgery becomes a basic need of today. Therefore, forensic investigation is required for the images. Forensics can be mainly divided into two categories: analog forensics and digital forensics [1].

Analog forensics is all about physical evidences such as validating evidences and finding traces. It is also known by the name of classical forensics, e.g., forensic

V. Kour (✉) · P. Aggarwal · R. Kaur
University Institute of Engineering and Technology (UIET), Panjab University, Chandigarh, India

P. Aggarwal
e-mail: pree_agg@pu.ac.in

R. Kaur
e-mail: ravreetkaur@pu.ac.in

© The Author(s), under exclusive license to Springer Nature Singapore Pte Ltd. 2022
P. Shetty D. and S. Shetty (eds.), *Recent Advances in Artificial Intelligence and Data Engineering*, Advances in Intelligent Systems and Computing 1386,
https://doi.org/10.1007/978-981-16-3342-3_25

investigation of the murder weapon which was found at the crime scene comes under analog forensics.

As the name suggests, digital forensics is the investigation or exploration of the digital evidences like images, videos, audio clips, etc. It can be grouped into two categories: computer forensics and multimedia forensics. Computer forensics is used when the evidence need to be extracted from the computer through which the malicious or criminal code of conduct took place. Audio, video, and image forensics come under the multimedia forensics. Assuring the authenticity and integrity of the multimedia is the principle goal of the multimedia forensics. Hence, image forensics is a part of multimedia forensics. Thus, image forgery refers to changing the content of the original image and producing transformed image as the original image with illegal intentions. There are three popular ways of image forgery: copy-move, image splicing, and image retouching [2].

Copy-move is copying the content from the image and pasting within the same image to hide the original information or in order to representing the fake information. Figure 1 is the example of copy-move forgery in which bird is copied from the original image and very cleverly pasted into the forged image. The images shown in Fig. 1 are taken from CoMoFoD_small dataset. Image splicing refers to copying the multiple region from two or more images and pasting into the fake image which is to be created. Image splicing is also known as cut-paste image forgery [3]. Image retouching is basically the enhancement of an image. It is done in order to draw the attention of the viewers to a particular object in an image or make the image more attractive [4]. Forgers apply various attacks and post-processing after the forgery to make the forged image more natural which makes detection of the forgery more challenging.

(a)   **(b)**

Fig. 1 **a** Original image, and **b** copy-move forgery

In this paper, a method to detect copy-move type of forgery is implemented. The rest of the paper is organized as follows. Section 2 is about the state of the art on CMFD techniques. The method implemented in this paper is described in Sect. 3. Section 4 shows the copy-move forgery detection results based on proposed method. Section 5 is the conclusion and discussion section.

2 Related Work

Copy-move forged images contain duplicate regions. Thus, the main idea behind detection is to detect duplicate region within the image. CMFD method involves four main steps: preprocessing, feature extraction, feature matching, and taking decision [3]. Preprocessing steps generally involve color conversions. There are mainly two approaches followed to detect the copy-move forgery: keypoint-based and block-based. If the detection is block-based, then dividing the image into blocks also comes under preprocessing step. Afterward features vectors are computed by extracting the features from the image or blocks. Feature matching is done in order to check the similarity within the image. The final step is taking the decision whether the image is authentic or not. In keypoint-based methods, the keypoints are extracted using a particular feature like SIFT, SURF, KAZE, LBP, etc., and then the matching is performed between the keypoints extracted to detect the copied region. In 2010, the method for detecting copy-move forgery based on SURF features was proposed by Xu et al. [5]. KAZE and SIFT features are used for keypoint extraction in the method proposed by Yang et al. [6]. Roy et al. [7] proposed a method for the CMFD in the presence of SGOs using the RLBP features. In 2019, Lin et al. [8] presented a novel approach for the CMFD. They had used combined features in which they extracted SIFT and Local Intensity Order Pattern (LIOP) at first. For the feature matching purpose, they used transitive matching.

Blocks-based approaches first divide the images into blocks and then compare these blocks with each other to detect the copy-move forgery. Block-based approaches using DCT are presented in various works [9–12]. Hu et al. [10] proposed an improved lexicographical sort algorithm for copy-move forgery detection. Authors first divided the image into 8×8 blocks and then applied discrete cosine transform (DCT). Then, they cluttered DCT coefficients according to their frequencies. At the end, they used eigenvalues for the block matching [10]. Sondos M. Fadl et al. [12] proposed a fast and efficient method to detect copy-move forgery. This method is a block-based method which firstly divide the image into fixed size overlapping blocks. Then, authors had applied DCT to each block. After that, they used fast K-means clustering technique. This technique makes clusters by putting the blocks into different classes.

In the maximum block-based approaches, the square shape is used for the blocks. However, Zunliu Zhang et al. [13] used circular blocks to extract DRHFM in their method which shows good results in detecting forgeries under geometrical transformations. Dixit et al. used statistical image properties. They used mean and variance as features [14]. Fadl et al. [15] used frequency as a feature. The frequency is calculated using the Fourier transform over the polar system representation of the blocks of the image.

Keypoint-based methods are generally faster than the block-based methods. However, block-based approaches are known for producing good results. In this work, a block-based method using DCT is implemented which is described in the following section.

3 Proposed Methodology

The idea provided by Fridrich et al. proposed in [9] is applied with modification in this paper. The modification is done in order to reduce the false positives in the final output. The pairs of the matching blocks that create more false positives are discarded. When a region is copied and pasted within the same image, there is a correlation between the copied and pasted region. Due to the various post-processing on the forged image, these two regions, i.e., copied and the pasted, may not exactly match but approximately. Therefore, the method for the approximate match is implemented in the paper. The flowchart of the implemented algorithm is shown in Fig. 2.

The algorithm takes grayscale image for processing but if the image is RGB, then grayscale conversion for the image is done by using the following standard formula:

$$\text{Grayscale Image} = 0.299\text{R} + 0.587\text{G} + 0.114\text{B} \tag{1}$$

In Eq. (1), R, G, and B represent the red, green, and blue components, respectively.

Afterward, the image is divided into B × B overlapping blocks. For dividing the image into blocks, a sliding square window of size B is taken from the upper left corner of the image. The pixels coming under each window are considered as a single block. To make the blocks overlapping with each other, the window is shifted by 1 pixel toward the right and hence this procedure is followed till the last pixel of an image [9]. DCT coefficients are calculated and quantized for each block. The quantization steps depend upon the Q-factor which will be set by user. Q-factor is the quality factor which is similar to the JPEG compression's quality factor. Higher the value of Q-factor, finer will be the quantization and hence, much closer match between the blocks to be declared as similar. On the other hand, low value of Q-factor results in some false matches. After the quantization, a matrix 'A' is created for storing these quantized coefficients. These are stored in row-wise manner. Each row represents a single block of an image. It is to be noted that, here quantization is the key to find the approximate matching between the blocks. However, if there is a requirement to find only the exact matching between the blocks, then there is no need to do quantization and hence the step of doing quantization can be skipped.

As for now, the matrix 'A' is created in which each row represents a block. To find the correlation between the blocks, matching needs to be done. But before performing the matching, one more step is required which is lexicographical sorting.

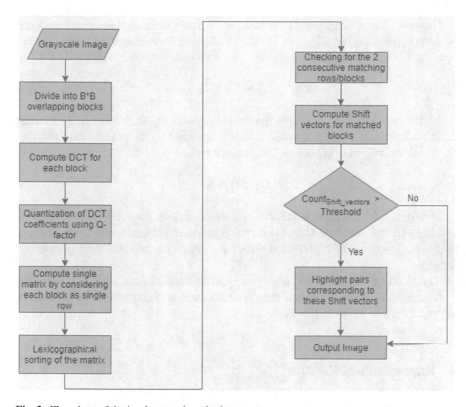

Fig. 2 Flowchart of the implemented method

It will create order between the rows which makes the matching process simpler and less time consuming.

The next step is to look for the matched blocks. As the matrix A is sorted, therefore, the algorithm only looks for the two consecutive matching blocks. But remember, algorithm can result into more false matches because DCT coefficients are used in this detection method instead of exact pixel values. For that, all the matching pairs of the blocks required to be remembered for further detection process. Hence, algorithm creates a separate list for storing the positions of the matching blocks. For each matching pair, a shift vector is calculated by using the locations of the blocks. These locations are actually the coordinates of the upper left pixel of the block. Let (M_1, M_2) and (N_1, N_2) are the locations of the two matching blocks. In other words, (M_1, M_2) are the coordinates of upper left (first) pixel of first block and (N_1, N_2) are the coordinates of the upper left (first) pixel of the other block in the matching pair. Then shift vector is calculated as shown in Eq. (2):

$$S = (M_1 - N_1, M_2 - N_2) \tag{2}$$

However, S or $-S$ represents the same shift. Thus, normalization for the $-S$ can be done by multiplying it by -1. The count of the shift vectors is stored in the shift counter 'C'. Counter is incremented by 1 for each matching pair as shown in Eq. (3):

$$C(S_1, S_2) = C(S_1, S_2) + 1 \qquad (3)$$

At the last, counter represents the frequencies of the shift vectors. For taking decision, threshold 'T' plays a role which is specified by user. Then the following condition shown in Eq. (4) is checked for all $r = 1, 2, 3, \dots k$:

$$C(S^{(r)}) > T \qquad (4)$$

$S^{(1)}, S^{(2)} \dots S^{(k)}$ are the normalized shift vectors calculated by the algorithm. If the condition represented in the Eq. (4) is met, then the matching blocks corresponding to that shift vectors are highlighted with the same color, hence detects copy-move forgery.

Threshold need to be chosen wisely; very small threshold can lead to high number of false positives whereas by setting very high threshold can result into missing of the some closely matched blocks.

4 Experimental Results

The above algorithm is implemented using MATLAB R2018a, RAM 4 GB, and Processor 2.50 GHz. The block size chosen is 16×16 for the testing of the algorithm. As the block size is quite large, the reason behind taking this block size is that there will be low chances of the false matches. This algorithm is tested on the several images of the CoMoFoD dataset. The results are shown in the Figs. 3, 4, and 5. These figures clearly show that algorithm correctly detects the copied regions which are highlighted in the resultant image. For example, Fig. 3b is the copy-move forged

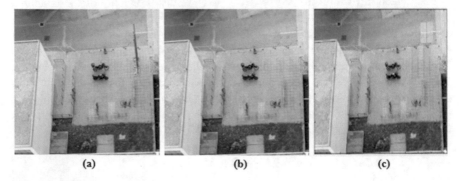

(a)   (b)   (c)

Fig. 3 a Original image, b copy-move forged image, and c forgery detection result

Fig. 4 **a** Original image, **b** copy-move forged image, and **c** forgery detection result

Fig. 5 **a** Original image, **b** copy-move forged image, and **c** forgery detection result

image in which the poles are missing. Figure 3c is the detection results which shows that two regions are duplicate in this image, thus proving Fig. 3b copy-move forged.

The quality parameter and threshold chosen were 0.25 and 25, respectively. The criteria behind choosing these parameters are already explained in Sect. 3. In short, the parameters settings are done in order to avoid the false matches. Although, the algorithm is tested by taking the other threshold and quality parameters. Not much differences are noticed except the images containing the flat surfaces like wall, plain background, etc. Figure 6 represents the time taken by the algorithm of 10 random images to show the results.

The average time taken by the algorithm to run on a single image is 36 s. However, it can vary according to system specifications.

5   Conclusion and Discussion

CMFD is a quite challenging task because the algorithm needs to put a label on an image as forged or original. Therefore, in this research, a method based on DCT is

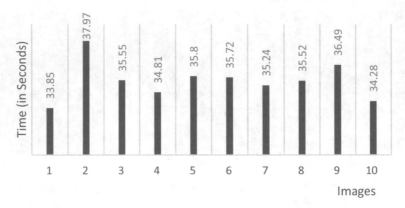

Fig. 6 Performance time

implemented. The foremost step of the algorithm is to divide the grayscale image into 16×16 blocks. DCT coefficients are calculated and quantized which are then stored in the matrix form. A lexicographical sorting is performed on the matrix before comparing the rows which represents the blocks of an image. After the matching, the copied blocks are highlighted in the output image. Although the proposed algorithm took very less time approximately 36 s to provide the result for one image, it can generate some false positives for the uniform areas for example flat surfaces like sky.

References

1. S. Sadeghi, S. Dadkhah, H.A. Jalab, G. Mazzola, D. Uliyan, State of the art in passive digital image forgery detection: copy-move image forgery. Pattern Anal. Appl. **21**(2), 291–306 (2018)
2. M.H. Alkawaz, G. Sulong, T. Saba, A. Rehman, Detection of copy-move image forgery based on discrete cosine transform. Neural Comput. Appl. **30**(1), 183–192 (2018)
3. L. Zheng, Y. Zhang, V.L.L. Thing, A survey on image tampering and its detection in real-world photos. J. Vis. Commun. Image Represent. **58**(December), 380–399 (2019)
4. M.A. Elaskily, H.A. Elnemr, M.M. Dessouky, O.S. Faragallah, Two stages object recognition based copy-move forgery detection algorithm. Multimed. Tools Appl. **78**(11), 15353–15373 (2019)
5. B. Xu, J. Wang, G. Liu, Y. Dai, Image copy-move forgery detection based on SURF in *Proceedings of 2010 2nd International Conference on Multimedia Information Networking and Security (MINES 2010)*, pp. 889–892, (2010)
6. F. Yang, J. Li, W. Lu, J. Weng, Copy-move forgery detection based on hybrid features. Eng. Appl. Artif. Intell. **59**(October), 73–83 (2017)
7. A. Roy, A. Konda, R.S. Chakraborty, Copy move forgery detection with similar but genuine objects. in *Proceedings of International Conference on Image Processing ICIP*, vol. 2017, (September 2018), pp. 4083–4087
8. C. Lin et al., Copy-move forgery detection using combined features and transitive matching. Multimed. Tools Appl. **78**(21), 30081–30096 (2019)
9. J. Fridrich, D. Soukal, and J. Lukáš, Detection of copy-move forgery in digital images. in *Proceedings of Digital Forensic Research Workshop* (Aug 2003)

10. J. Hu, H. Zhang, Q. Gao, H. Huang, An improved lexicographical sort algorithm of copy-move forgery detection. in *Proceedings of 2nd International Conference on Network Distribution and Computing ICNDC 2011* (2011), pp. 23–27
11. S.M. Fadl, N.A. Semary, A proposed accelerated image copy-move forgery detection (Dec 2014), pp. 253–257
12. A. Alahmadi, M. Hussain, H. Aboalsamh, G. Muhammad, G. Bebis, H. Mathkour, Passive detection of image forgery using DCT and local binary pattern. Signal Image Video Process. **11**(1), 81–88 (2017)
13. J. Zhong, Y. Gan, J. Young, L. Huang, P. Lin, A new block-based method for copy move forgery detection under image geometric transforms. Multimed. Tools Appl. **76**(13), 14887–14903 (2017)
14. A. Roy, R. Dixit, R. Naskar, R.S. Chakraborty, Copy-move forgery detection exploiting statistical image features. Stud. Comput. Intell. **755**, 57–64 (2020)
15. S.M. Fadl, N.A. Semary, Robust Copy-Move forgery revealing in digital images using polar coordinate system. Neurocomputing **265**, 57–65 (2017)

Bottlenecks in Finite Impulse Response Filter Architectures on a Reconfigurable Platform

Kunjan D. Shinde and C. Vijaya

Abstract Digital filters are primary and important part in digital signal processing, often used in the scope of separation or restoration of signals. Due to inherent properties and advantages of finite impulse response (FIR) filter, it is more preferred over infinite impulse response (IIR) filter. In the present work, bottlenecks of various FIR filter architecture are discussed to estimate and understand the impact of these architectures on reconfigurable platform. The RTL implementation of various architectures is focused while justifying the critical path and the computation complexity involved. Use of symmetric coefficients (SC) is a key in reducing the area constrains of filter, and further use of data representation for coefficient is addressed with CSD and its impact is discussed. Impact of pipeline and parallel architectures is discussed, and their performance is evaluated for reconfigurable platform.

Keywords FIR filter · FPGA · Reconfigurable platform · Bottlenecks · FIR filter architectures · Direct from transposed form · Multiplier-less architecture · Canonical signed digit (CSD) · Pipeline architecture · Parallel architecture

1 Introduction

Gadgets around us are more sophisticated than earlier, powered on battery and process real-world signals while interacting with the environment. Digital signal processing (DSP) is important as it improves the overall quality of signal that we are interested to study and evaluate. Nowadays, it is equally important to develop an effective algorithm for signal processing and efficiently implement the same without degrading the performance.

Most of the signal processing algorithms have explicit parallelism or the parallelism can be explored and used. In such cases, the implementation platform is unable to support the parallelism offered by the algorithms. Reconfigurable platform using FPGA is often preferred for exploring and implementing parallelism found

K. D. Shinde (✉) · C. Vijaya
Department of E&CE, SDM College of Engineering & Technology, Dharwad, Karnataka, India

© The Author(s), under exclusive license to Springer Nature Singapore Pte Ltd. 2022   309
P. Shetty D. and S. Shetty (eds.), *Recent Advances in Artificial Intelligence and Data Engineering*, Advances in Intelligent Systems and Computing 1386,
https://doi.org/10.1007/978-981-16-3342-3_26

in algorithms, prototyping in ASIC, lower time to market, low NRE cost, functional verification of the design at early cycle, designer works at RTL-level while optimizing and modifying the designed RTL to meet the requirements. Change in targeted hardware has no impact on RTL described, as the tool will regenerate the necessary net list describing the design. Upgrading the signal processing algorithms on FPGA platform is easy as they are reconfigurable.

In [1], the author has given detailed study on VLSI digital signal processing algorithms and their impact on VLSI platform. Several methods are presented for exploring various architectures. In [2], the authors have explained in detail the study on digital filter design using various methods and representation of FIR filter using DF and TF is provided. In [3], the authors have designed FIR filter and improved its performance using CSD representation and optimized using common subexpression elimination methods, the number of logical operators (addition units) required is reduced while preserving the logical depth (number of adder stages) required to implement constant multiplication. In [4], the authors have applied pipeline architecture to transpose for FIR filter of two-stage/level and has obtained better results in their work. Simulations on ECG signal is carried out and analysis on critical path is made, filter designed is of order 16, 32, and 64, and coefficients are represented in 16 bits for reconfigurable platform. In [5], parallel FIR filter structures using iterated short convolution is proposed by the authors; this architecture can be used while exploring the parallel architectures. In [6], FIR filter architecture is presented with high performance for fixed and reconfigurable applications. Data flow graph for TF is used to identify the replication of computations made in successive iterations and has presented a structure with pipeline adder unit. Authors quote that their work consumes less ADP and EPS than DF structure. In [7], the authors have designed FIR filter with odd length and used fast FIR algorithm for parallel architecture. The authors have obtained area-efficient results for the implementation. In [8], the FIR filter is designed for block memories and obtained better performance in computational speed and power consumption while comparing with conventional architectures for ASIC and FPGA applications. In [9], the block FIR is designed using multiple constant multiplication for direct form FIR structure as block processing cannot be used for transposed form. Improvement in sampling rate and EDP is observed while consuming less area. In [10], authors have investigated pipeline and parallel architecture for FPGA Platform and observed significant reduction in area by using fast FIR filter architecture.

2 Architectures for FIR Filter

Digital filters are important class of LTI system. The properties of such filters can be represented by their impulse response or frequency response or using difference equation [1]. In the present work, difference equation is used as it provides the implementation details needed for VLSI platform. Whereas, the impulse response and frequency response gives the filter properties in time and frequency domain, respectively. Digital filter can be represented as given in Eq. 1.

$$y(n) = \sum_{k=0}^{N-1} b_k x(n-k) - \sum_{k=1}^{M} a_k y(n-k) \tag{1}$$

If $a_k = 0$ for $1 \le k \ge M$, the Eq. 1 gets reduced to N tap of FIR filter as given in Eq. 2 [1]. The computations in FIR filter are non-recursive as there is no feedback from the output.

$$y(n) = \sum_{k=0}^{N-1} b_k x(n-k) \tag{2}$$

Expanding the above equation

$$y(n) = b_0 x(n) + b_1 x(n-1) + b_2 x(n-2) + \cdots b_{N-1} x(n-N-1) \tag{3}$$

where $y(n)$ is output sample, N is filter order, b_k is filter coefficients, and $x(n)$ is input sample.

Equation 3 gives the expanded version of Eq. 2 where the output sample is computed by multiplying filter coefficients with past and present values of input samples. The impulse response of the filter has finite duration and the output depends on past and present values of inputs samples, due to which the filter is causal and stable. In applications where the phase or shape of the signal is to be addressed, FIR filters are preferred.

Filter coefficients can be symmetric (SC) or asymmetric based on the type of application the filter coefficients are selected [2]. FIR filter requires higher order to obtained desired response, the area and computation complexity has direct impact on the filter designed, use symmetric coefficients is a key and plays a vital role on reconfigurable platform. Implementation of coefficients can be of various forms and the same can be explored for its impact on the filter design [8].

In general, Nth order FIR filter needs $N - 1$ delay elements, N multiplication, and $N - 1$ addition units. Architectures for FIR filter gives details on the arrangement of these blocks and improvement in the filter performance by rearranging/replicating the blocks. Critical path in a FIR filter is the longest path for data to travel from input to output node. In general, the critical path for Nth order FIR filter is $\mathbf{T_M} + (\mathbf{N} - \mathbf{1})\mathbf{T_A}$ where the computation time of adder unit to be $\mathbf{T_A}$ and multiplication unit to be $\mathbf{T_M}$. The clock cycle is represented by $\mathbf{T_{clk}}$.

The following are the FIR filter architectures.

2.1 Conventional Architecture (Direct Form and Transposed Form)

The realization of Eq. 2 by developing signal flow diagram in a simple and straight-forward way gives the direct form realization. Here the coefficients appear as gain units of the function, and number of delays are equal to filter order. Figure 1 gives the block diagram of generalized direct form structure of FIR filter.

Using Tellegen's theorem or Mason's gain formula, for a LTI system, its signal flow graph is inter-reciprocal with its transpose. It may be noted that in transposed form, the input sample gets multiplied with all the coefficients at a time [2]. This structure is later utilized to implement constant multiplication which results in new architecture (Fig. 2).

FIR filters are stable and can be practically realized to meet the requirements. The first bottleneck in the design process is the order of the filter as this increases the computation complexity and area constrain.

The critical path for tap 3 FIR filter represented in DF structure is $T_M + 2T_A$, and the critical path for TF is $T_M + T_A$ [1]. The critical path in transposed form is dependent on the effective realization of constant multiplication. Shift and add method is more commonly used along with its variants. Pipeline for TF can further increase the performance by reducing critical path [6]. As quoted in results section, the TF gives better performances as compared to DF, by the use of symmetric coefficients. Further optimization in computational complexity and arithmetic involved is seen.

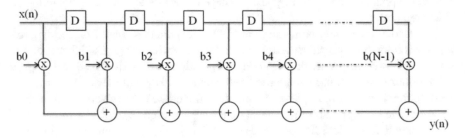

Fig. 1 Block diagram of generalized direct form structure of FIR Filter

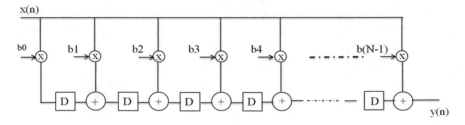

Fig. 2 Block diagram of generalized transposed form structure of FIR filter

Let $T_{M, max}$ be the term for maximum delay through the constant multiplication and $T_{A, Tree}$ be the maximum delay of adding the $N - 1$ products (or $(N - 1)/2$ partial products and one shift register delay in case of symmetric coefficients) of constant multiplication to generate the final output. This results in the critical path delay of $T_{M, max} + T_{A, Tree}$.

Use of fast/parallel adders can resolve the speed metrics but may increase in area and power metrics of the designed system. Precision of the input sample and coefficients also act as a bottleneck for the architectures. As the precession increases, more effective is the signal interpreted while further increasing the design complexity.

2.2 Multiplier-Less Architecture

The transposed form structure of FIR filter results in constant multiplication as the input sample getting multiplied with all the coefficients. The coefficients are constants for a given design and it can be implemented using a series of shift and addition blocks.

Use of common subexpression elimination can reduce the logical depth and number of adders required [3, 9] to implement the constant multiplication, and shift operations can be hardwired. The resultant is new architecture known as multiplier-less FIR filter architecture. The generalized structure for multiplier-less architectures is given in Fig. 3.

Canonical signed digit (CSD) is a special manner for encoding a value in a signed-digit ternary representation, which itself is non-unique representation and allows one number to be represented in many ways. Generally, digital signal processing algorithms require a large number of multiplications. Depending on the number of bits used to represent the input samples and the filter coefficients, these multiplications can become a time-consuming process. CSD is an interesting solution in implementing efficient multipliers. Figure 4 gives the generalized CSD-based FIR filter architecture.

The multiplier-less architecture resembles TF structure. The delay $T_{M, max}$ is replaced by T_{CSD}, where T_{CSD} is the delay for implementing constant multiplication using shift and add method with data represented in CSD notation. The critical

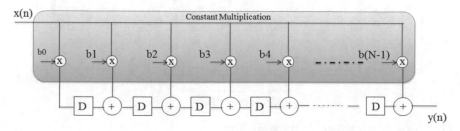

Fig. 3 Block diagram of TF structure with constant multiplication

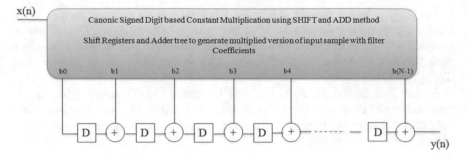

Fig. 4 Block diagram of TF structure with CSD-based constant multiplication

path for this architecture would be $\mathbf{T_{CSD} + T_A}$, tree and hence the bottleneck. The key to this is an efficient design of constant multiplication with CSD by eliminating common subexpressions and reducing its critical path.

2.3 Pipelined Architecture

Pipelining transformations lead to reduction in critical path [1]. In this architecture, pipeline registers are introduced in between paths as shown in Fig. 5, by adding pipeline register. The effective critical path is reduced with small increase in area due to pipeline register [4, 10]. Bottleneck of this architecture is that we cannot insert pipeline registers in every stage of FIR filter. Further by increasing the pipeline registers, the design may not reduce the critical path beyond some value. So it is important to identify the limit and effectively use pipeline registers.

A register is added at identified feed-forward cut; at every new interval, the data in this register will be updated and the previous data is forwarded within the network as shown in Fig. 5. In the present work, the pipeline is applied to the DF representation of FIR filter; the same concept can also be extended to transpose from FIR filter.

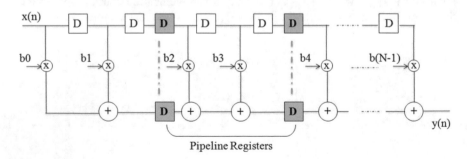

Fig. 5 Generalized structure of pipeline FIR filter

Lets consider a tap 3 FIR filter using DF structure with critical path of $T_M + 2T_A$. With this structure, two-level pipeline is the limit, and the computation this cannot be reduced beyond $T_M + T_A$ as bottleneck of this design, and hence, the minimum critical path achievable with pipelining.

$$T_{sample} \geq TM + TA \qquad (4)$$

where T_{sample} is the iteration period of pipeline architecture and $T_{sample} = T_{clk}$.

2.4 Parallel Architecture

Parallel processing architectures compute multiple outputs at a given clock cycle; hence, the effective processing speed is increased by the level of parallelism [1].

The conventional architectures are single input single output (SISO) systems. In parallel architectures, multiple input multiple output (MIMO) is created by replicating the functional block as indicated in Fig. 6, where the inputs are $x(3k)$, $x(3k + 1)$, and $x(3k + 2)$ and generating the outputs as $y(3k)$, $y(3k + 1)$, and $y(3k + 2)$. Most of the reconfigurable platforms have limited IO pins; it is difficult to support MIMO architecture, and hence, serial to parallel and vice versa is needed. Availability of large area and IO pins may be the bottleneck for the design. Whereas the design becomes faster by L fold as given in Eq. 5 [1] (Fig. 7)

$$\text{Tparallel} = \frac{1}{L}T_{clk} \geq \frac{1}{L}\text{CriticalPath} \qquad (5)$$

Fig. 6 Parallel FIR filter for tap 3 with DF structure

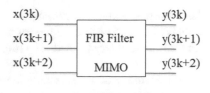

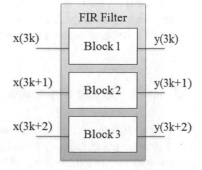

where $T_{parallel} \neq T_{clk}$ as multiple outputs are computed in single clock cycle.

3 Requirements of FIR Filter: A Reconfigurable Platform Perspective

In the present work, band pass filter is designed for base band signals of frequency range 300–3400 Hz using hamming window. As it is a voice signal band, the sampling rate is set to 48,000 Hz. The filter coefficients are obtained from FILTER DESIGNER tool available with MATLAB; the above specifications are used for the design of tap 3 and tap 11 FIR filter.

The following are the filter parameters that may have a major impact on the FIR filter design on reconfigurable platform.

3.1 Filter Order

The notation "N" represents the order of the filter, which defines the length of the sequence to be sampled at a given interval of time. FIR filter must have higher-order to achieve the ideal response; higher-order means more number of taps which in turn means more data to be processed.

3.2 Precession of Data to Represent the Sample/Coefficients

Real-world signals and filter coefficients are often fractions, and hence, any standard data type can be used to represent it. In the present work, factions are scaled, and its equivalent binary is used. The impact of scaling the filter coefficients is given in Table 1.

The accuracy of scaled coefficients is improved by increasing the scaling factor at a cost of increase in computation complexity and area, while reducing the round-off error due to scaling. A scaling factor of 2^{32} is preferred as it produces the least round-off error, and hence, the precession is set to 32 bits.

3.3 Data Type to Represent the Signal

VLSI and DSP platforms provide various methods to represent a signal-like fixed point, floating point, CSD, Q-Notation, signed integer, binary, and more. Every representation has its own limitations and strength of use. Effective method can be chosen

Table 1 Impact of scaling factor of tap 11 FIR Filter coefficient using Hamming Window (fs = 48,000 Hz, fc1 = 300 Hz, fc2 = 3400 Hz)

Coefficient Value	Scaling Factor	Scaled coefficient	Round off	Round off error	Retained coefficient after scaling & round off
0.00001713500058 — Minimum Value among coefficients	2^8	4.3865E-3	4E-3	0.38E-3	0.000015672
	2^{10}	0.01754624	0.018	0.45E-3	0.000017578
	2^{12}	0.07018496	0.070	-0.18E-3	0.000017089
	2^{14}	0.28073984	0.2	-0.0807	0.000012207
			0.3	0.0192	0.00001831
	2^{16}	1.12295939	1	-0.1229	0.000015258
			2	0.8770	0.000030517
	2^{18}	4.49183759	4	-0.4918	0.000015258
			5	0.5081	0.000019073
	2^{20}	17.9673503	17	-0.9673	0.000016212
			18	0.0326	0.000017166
	2^{24}	287.4776	288	0.5224	0.000017166
	2^{28}	4599.64	4600	-0.36	0.000017136
	2^{32}	**73594.26**	**73594**	**0.26**	**0.000017134**
0.12895671047709 — Maximum Value among coefficients	2^8	33.0129	33	0.0129	0.12890625
	2^{10}	132.05	132	0.05	0.12890625
	2^{12}	528.20	528	0.20	0.12890625
	2^{14}	2112.82	2113	-0.18	0.12896728
	2^{16}	8451.30	8451	0.30	0.128952026
	2^{18}	33805.22	33805	0.22	0.128955841
	2^{20}	135220.91	135221	-0.09	0.128956794
	2^{24}	2163534.58	2163535	-0.42	0.128956735
	2^{28}	34616553.36	34616553	0.36	0.128956709
	2^{32}	**553864853.8**	**55386484**	**-0.2**	**0.128956710**

based on precession and dynamic range of data to be represented. In the present work, binary notation is used whereas the signs associated with coefficient can be resolved by using an add/sub block as a need to fulfill the functionality of FIR filter.

3.4 Sample Rate

It is dependent on the type of signal being processed and signal reconstruction requirements; sampling rate for the case study is set to 48,000 Hz.

3.5 Filter Coefficients

Selection of filter coefficients is a very important task as the filter coefficients determine how effectively the filter can segregate the wanted signal from noise. Equiripple, least square, windowing, and more methods are available for the selection of filter coefficients.

4 Results and Discussions

The present work gives detailed analysis on the various architectures for FIR filter and its bottlenecks are as quoted above. The work is simulated on Xilinx 9.1i ISE and synthesized on Virtex device xcv600-5-hq240. Specification of filter designed is mentioned in early section, and to generate FIR filter coefficients, FDA/FILTER_DESIGNER tool on MATLAB is used.

4.1 Simulation Results

Digital FIR filter processes discrete samples as input and produces filtered version input samples. To observe this behavior and verify its functionality, a common set of input $x(n)$ samples are applied for various architectures of FIR filter, and it is observed that the same set of results are obtained as shown in Figs. 8, 9, 10, and 11 for tap 3 and tap 11 FIR filter

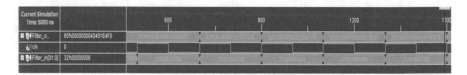

Fig. 7 Simulation result of tap 3 FIR filter (DF, TF, CSD, and pipeline architecture)

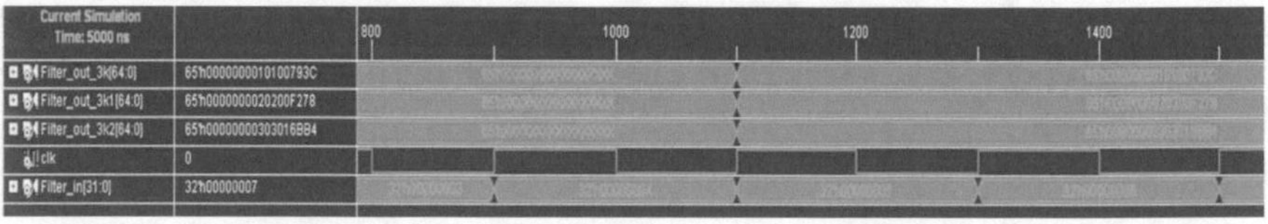

Fig. 8 Simulation result of tap 3 FIR filter for parallel architecture

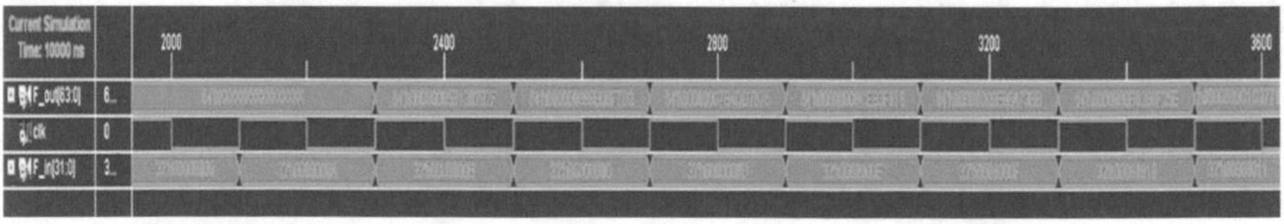

Fig. 9 Simulation result of tap 11 FIR filter (DF, TF, CSD, and pipeline architecture)

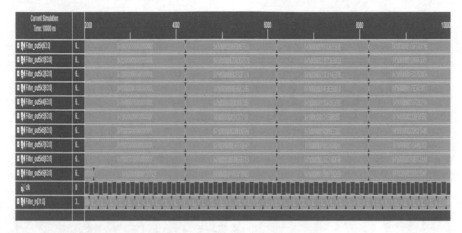

Fig. 10 Simulation result of tap 11 FIR filter for parallel architecture

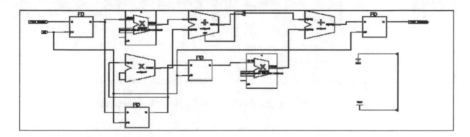

Fig. 11 RTL schematic of tap 3 FIR filter with all coefficients in DF representation

4.2 RTL Implementation

It gives the implementation details on FPGA hardware selected after the process of synthesis. On observing the top-level RTL schematic, the following deductions are made for the FIR filter architectures present in the work.

The RTL implementation of DF uses all the three coefficients of FIR filter, 2 adders and 3 multipliers are used with shift registers interconnected as seen in Fig. 11. The implementation resembles the DF structure (Fig. 12).

Using the symmetric coefficients of filter and revising the RTL code, the implementation details of DF FIR filter is changed to 2 adders and 2 multiplies. There is a reduction in number of multipliers required to implement the structure and increase in computational speed is observed.

Figure 13 gives the RTL implementation of TF structure of FIR filter; it uses 2 adders and 3 multipliers due to the TF arrangement, the computation is fast as its critical path is $T_M + T_A$.

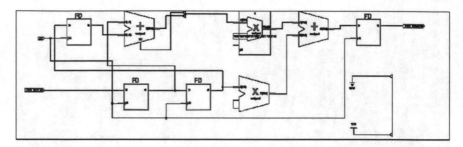

Fig. 12 RTL Schematic of tap 3 FIR filter optimized with symmetric coefficients in DF representation

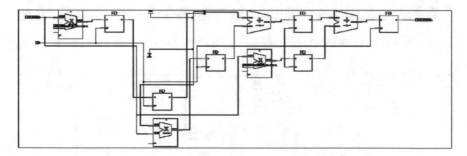

Fig. 13 RTL schematic of tap 3 FIR filter with all coefficients in TF representation

In TF RTL implementation, the input gets multiplied with the coefficients at a time, as $b_0 = b_2$, the required multiplier is reduced by one as shown in Fig. 14. It uses 1 adder and 2 multipliers.

With the use of CSD to implement the constant multiplications observed in TF architecture, the RTL is implemented by a series of shift and add units, thereby reducing the area consumed to a greater extent as seen in Fig. 15.

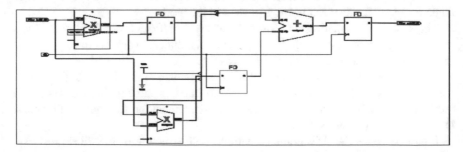

Fig. 14 RTL schematic of tap 3 FIR filter optimized with symmetric coefficients in TF representation

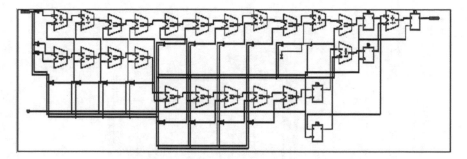

Fig. 15 RTL schematic of tap 3 FIR filter optimized with CSD for multiplier-less architecture

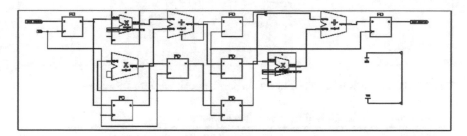

Fig. 16 RTL schematic of tap 3 FIR filter with pipeline architecture for DF representation

In the RTL implementation of FIR filter using pipelined architecture, we use feed forward cut and inserts pipeline registers to reduce the critical path of the FIR filter and to improve the performance in terms of speed. Pipelined register can be seen in the RTL implementation of Fig. 16.

The RTL implementation of FIR Filter using parallel architecture with block size of 3 is given in Fig. 17, it can be verified that there are 3 identical blocks of DF structure as we have used threefold the parallel structure generating 3 outputs. It uses adders and multipliers three times the DF structure producing with the highest computational speed. Fast FIR algorithms can be used to enhance efficiency and area utilization [7].

4.3 Comparative Analysis

In the present work, the comparison is made for tap 3 and tap11 FIR filters using direct form, transposed form, multiplier-less (CSD), pipeline and parallel architectures for our study and its performance is as shown in Tables 2 and 3, respectively.

Area metric in FPGA is number of LUT's used and intern the FPGA slices consumed. Delay metrice on FPGA is minimum period which is further subdivided

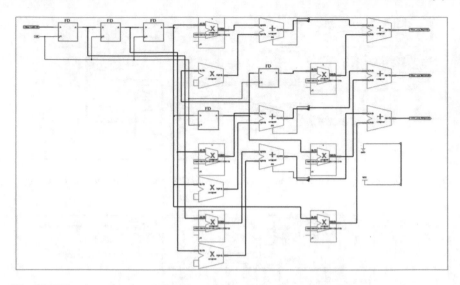

Fig. 17 RTL schematic of tap 3 FIR filter with parallel architecture for DF representation

Table 2 Comparative analysis of tap 3 FIR filter architectures

Parameters		Tap 3 FIR Filter architectures synthesis on device xcv600-5-hq240 Virtex						
FIR Architectures Vs **FPGA Status**		DF	DF with SC	TF	TF with SC	Multipl ier less TF (CSD)	Pipelin e DF 2 stage	Parallel DF L= 3
Multiplier	32*32	3	1	3	2	0	3	9
	33*32		1			0		
Adder	32 bit		1					
	64 bit	2					1	3
	65 bit		1	1	1	6	1	3
Subtractor	64 bit					16		
	65 bit							
Register	32 bit	3	3				4	5
	64 bit	1		3	2			
	65 bit		1	3	1	5	2	
No. of slices		1002	737	1061	716	666	1031	3009
Slice Flip Flop		183	161	309	183	314	246	224
4 input LUT		1926	1324	1903	1259	1128	1894	5618
Min period		26. 292	24.858	8.218	8.169	8.078	24.310	4.759
Logic delay		15.752	15.698	6.7381	6.6891	6.738	15.110	1.859
Route delay		10.540	9.160	480	480	1.340	9.200	2.900
Max Freq(M Hz)		38.034	40.229	121.684	122.414	123.79	41.135	210.128
Min I/P arrival time(ns)		2.676	2.676	22.622	21.872	36.802	2.676	2.676
Max O/P after clk(ns)		7.511	7.511	7.511	7.511	7.511	7.511	34.702
IOB		97	98	98	98	98	98	228

Table 3 Comparative analysis of tap 11 FIR filter architectures

Parameters		Tap 11 FIR Filter architectures synthesis on device xcv600-5-hq240 Virtex						
FIR architectures versus FPGA status		DF	DF with SC	TF	TF with SC	Multipl ier less TF (CSD)	Pipeline DF 2 stage	Parallel DF L= 10
Multiplier	32*32	11	1	11	6	0	10	110
	33*32		5					
Adder	32 bit		5					
	64 bit	10	5			32	6	100
	65 bit			10	10		4	
Subtractor	64 bit					27		
	65 bit					10		
Register	32 bit	11	11				12	21
	64 bit	1	1	11	11	6	1	
	65 bit		11	11	11	12	2	
No. of slices		**1992**	**3632**	**2341**		**1922**	**3376**	**33867**
Slice Flip Flop		416	1302	1247		1123	528	1427
4 input LUT		3730	6271	3818		3619	6495	65141
Min period		**35.436**	**9.087**	**8.737**		**8.637**	**31.554**	**4.759**
Logic delay		20.566	6.787	6.787		6.787	18.284	1.859
Route delay		14.870	2.300	1.950		1.850	13.270	2.900
Max Freq (MHz)		**28.220**	**110.05**	**114.46**		**115.78**	**31.692**	**210.13**
Min I/P arrival time (ns)		2.676	30.815	24.473		38.935	2.676	2.676
Max O/P after clk (ns)		7.511	7.511	7.511		7.511	7.511	53.357
IOB		97	98	98		98	97	673

into logic and route delay whereas the multipliers, adders, subtractors, and registers on FPGA gives the logic blocks utilized.

Tap 3 FIR filter, the impact of architectures, and use of symmetric coefficient are:

- DF with symmetric coefficients has slices reduced by 26.44% and speed is increased by 5.7% compared to DF without symmetric coefficients.
- TF with symmetric coefficients has slices reduced by 32.5% and speed is increased by 0.59% compared to TF without symmetric coefficients.
- Multiplier-less TF has slices reduced by 6.9% and speed is increased by 1.73% compared to TF with symmetric coefficients. It has slices reduced by 33.5% and speed is increased by 225.5% compared to DF without symmetric coefficients.
- Pipeline has slices increased by 2.8% and speed is increased by 8.15% compared to DF without symmetric coefficients.
- Parallel has slices increased by 320.2% and speed is increased by 71.6% compared to TF with symmetric coefficients. It has slices increased by 200.2% and speed is increased by 452.7% compared to DF without symmetric coefficients.

Tap 11 FIR filter, the impact of architectures, and use of symmetric coefficient is:

- DF with symmetric coefficients has slices reduced by 41.27% and speed is increased by 21.79% compared to DF without symmetric coefficients.
- TF with symmetric coefficients has slices reduced by 35.5% and speed is increased by 4% compared to TF without symmetric coefficients.
- Multiplier-less TF has slices reduced by 17.89% and speed is increased by 1.15% compared to TF with symmetric coefficients. It has slices reduced by 43.33% and speed is increased by 399.6% compared to DF without symmetric coefficients.
- Pipeline has slices reduced by 0.4% and speed is increased by 36.77% compared to DF without symmetric coefficients.
- Parallel has slices increased by 1346.6% and speed is increased by 83.58% compared to TF with symmetric coefficients. It has slices increased by 898.43% and speed is increased by 806.85% compared to DF without symmetric coefficients.

5 Conclusion

The present work provides a detailed study on FIR filter architectures like DF, TF, multiplier-less pipeline, and parallel architecture with their bottlenecks and possible solutions to overcome it for a reconfigurable platform. Comparative analysis and implementation details justify the performance of filter designed while providing the architectural impact. Inherent parallelism available in algorithms is explored using reconfigurable platform, and its performance is measured. Selection of effective architecture for an application can be made using Tables 2 and 3 and deductions made in comparative statements.

Acknowledgements We would like to thank Head of Research Center, Department of E&CE, SDMCET, Dharwad for proving us the tools, resources and platform to carry out the research.

References

1. K.K. Parhi, *VLSI Digital Signal Processing System-Design and Implementation* (Wiley Students Edition 2013). ISBN: 978-81-265-1098-6
2. C. Vijaya, U. Kumar, *Digital Signal Processing*. Elite Printers & Publishers. ISBN 1234567156882
3. A.P. Vinod et al., An improved common subexpression elimination method for reducing logic operator in FIR filter implementation without increasing logic depth. Elsevier, Integr. VLSI J. **43**, 124–135. https://doi.org/10.1016/j.vlsi.2009.07.001
4. P. Patali, S.T. Kassim, High throughput and energy efficient FIR filter architectures using retiming and two level pipelining, CoCoNet' 19. Elsevier, Procedia Comput. Sci. **171**, 617–626 (2020). https://doi.org/10.1016/j.procs2020.04.067
5. C. Cheng, K.K. Parrhi, Hardware efficient fast parallel FIR filter structures based on iterated short convolution, in *IEEE Proceeding ISCAS 2004*, pp. 361–364. https://doi.org/10.1109/TCSI.2004.832784

6. B.K. Mohanty, P.K. Meher, A high—performance FIR filter architecture for fixed and reconfigurable applications. IEEE Trans. VLSI Syst. **24**(2), 444–452 (2016). https://doi.org/10.1109/TVLSI.2015.2412556
7. Y.-C. Tsao, K. Choi, Area-efficient VLSI implementation for parallel linear-phase FIR digital filters of odd length based on fast FIR algorithm. IEEE Trans. Circ. Syst. II: Express Briefs **59**(6), 371–375 (2012). https://doi.org/10.1109/TCSII.2012.2195062
8. M. Pristach, V. Dvorak, L. Fujcik, Enhanced architecture of FIR filter using block memories. Elsevier IFAC-2015 **48**(4), 306–311. https://doi.org/10.1016/j.ifacol.2015.07.052
9. P.K. Meher, Y. Pan, MCM-based implementation of block FIR filters for high-speed and low-power applications, in *IEEE Proceeding IC-VLSI&SoC* (2011), pp 118–121. ISSN: 2324-8432. https://doi.org/10.1109/VLSISoC.2011.6081653
10. M. Kadam, K. Sawarkar, S. Mande, Investigation of suitable DSP architecture for efficient FPGA implementation of FIR filter, in *IEEE Proceeding ICCI&CT 2017*. ISBN: 978-1-4799-5522-0. https://doi.org/10.1109/ICCICT.2015.7045672

Copy-Move Image Forgery Detection Using Discrete Cosine Transforms

R. P. Vandana and P. S. Venugopala

Abstract Digital image forgeries have led us to a situation where no digital image obtained using computing devices like PCs, smartphones and laptops are trusted to be authentic. As there are plenty of tools available free and open source, digital image forgery is no more a sophisticated job. Here we have come up with a simple but an effective technique for copy-move forgery detection in digital images as this type of forgeries is hard to be identified visually. The proposed technique uses block-based forgery detection technique. The features are extracted for each overlapping block, and then, it is compared with the features of other blocks to identify if there is forgery existing in the image. Few different images were tried, and difference in the accuracies has been observed. Accuracy varies depending on the copied image.

Keywords Copy-move forgery · Block based forgery detection · Discrete cosine transforms

1 Introduction

In this digital era, we all have realized that images speak more than text. "A picture is worth thousand words" is found to be true. As the field of computer science grows, introduction of new tools and techniques for manipulating digital images also grows at rapid rate.

Digital image forgery can be explained as falsely and fraudulently altering the digital images. This malicious activity started may be a century ago. But until the advancement in image forgery tools, these types of activities were hardly used by common man. The free and open-source availability of image manipulation tools nowadays have made the task of image forgery a funny and easy task to perform.

R. P. Vandana (✉) · P. S. Venugopala
Department of Computer Science and Engineering, NMAM Institute of Technology, Nitte, Karnataka, India

P. S. Venugopala
e-mail: venugopalaps@nitte.edu.in

© The Author(s), under exclusive license to Springer Nature Singapore Pte Ltd. 2022   327
P. Shetty D. and S. Shetty (eds.), *Recent Advances in Artificial Intelligence and Data Engineering*, Advances in Intelligent Systems and Computing 1386,
https://doi.org/10.1007/978-981-16-3342-3_27

Broadly classified, there are four types of image forgeries possible. They are copy-move forgery, image splicing, resampling and retouching [1]. In copy-move forgery, a part of the image itself is cut and pasted on to another part. This may be used to conceal some information in the image or to add more objects onto the image. In image splicing, two or more images are used. The copied portions of different images are combined to form a new false image. Researchers have been concentrating on copy-move and image splicing than the other two methods as these techniques are widely used for illegal purposes by cyber criminals. Researchers who work on image forgery detection have always been trying to device an ideal technique for forgery detection, but that is the most challenging job as criminals are always trying to find new techniques for performing image forgery.

This paper proposes a simple but effective technique for detecting copy-move forgery in digital images. Copy-move forgery is the one where one portion of the image is copied and pasted onto another portion of same image. This method is said to be an active technique because you need no original image for detecting the forgery. As the copied portion is from the same image, it is hard to be identified as the properties like noise, color and texture do not change.

The detection techniques for this type of forgeries are classified into three types.

- Brute force technique
- Block-based technique
- Key-point-based technique.

Brute force technique follows exhaustive search auto-correlation techniques, whereas key point method is based on algorithm that captures image features and detects forgeries using the key points obtained [2–4]. Block-based method is the third method for forgery detection. In this method, we first divide the image into overlapping blocks. Features are extracted for these blocks and not for the image as a whole. Algorithms such as discrete cosine transforms, principal component analysis, signature value decomposition and discrete wavelet transform are used to perform forgery detection [5–8].

Any block-based technique for forgery detection in digital images must follow five steps given in Fig. 1 [9, 10].

2 Proposed Technique

The technique that we are proposing here is a block-based technique. In this technique, the given image is divided into blocks before the feature extraction is performed. The list of steps that we are following here is well described in Fig. 2.

- Preprocessing: This is the first step in the detection process. Here we convert the color image into grayscale image. When an image is converted to grayscale, its complexity decreases. Less number of values are enough to define a pixel. It is a reduction process.

Fig. 1 Steps to follow in block-based digital image forgery detection

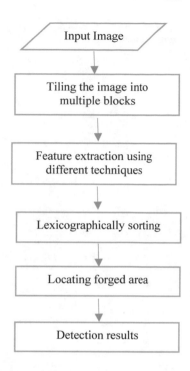

- Segmentation: The grayscale image obtained in the first step is divided into overlapping blocks of size 16×16 or 32×32 pixels depending upon the need. The further processing is done on individual blocks and not on the image as a whole.
- Discrete cosine transform (DCT): DCT usually expresses a finite sequence of data points in terms of a sum of cosine functions. DCT is applied to each and every block obtained in the segmentation process. It helps to separate the image into spectral sub-bands.
- Zigzag scan: Zigzag scan is applied onto the DCT values obtained in the previous step. Zigzag scan is preferred over raster scanning because zigzag scanning separates the DCT coefficients such that the ones with higher energy appear first and lower energy after that. Here we need not consider the lower energy regions. We can just neglect them. It is enough if higher-energy regions are considered. The vector hence obtained is converted to quantization matrix.
- Lexicographical sorting: The vector of DCT values obtained in the above step is lexicographically sorted. This type of sorting can be seen in dictionaries where words are sorted based on the letters constituting them.
- Euclidean method: This method is used for finding out if there are similarities between the blocks that are sorted lexicographically. This step is very crucial as it identifies the actual forgery in the image. If the similarity indices between two blocks are beyond the threshold value, then such blocks are said to be similar. A threshold value for distance and similarity must be decided by the designer before the coding is done. The threshold values have high impact on the accuracy of

Fig. 2 Steps followed in the proposed technique

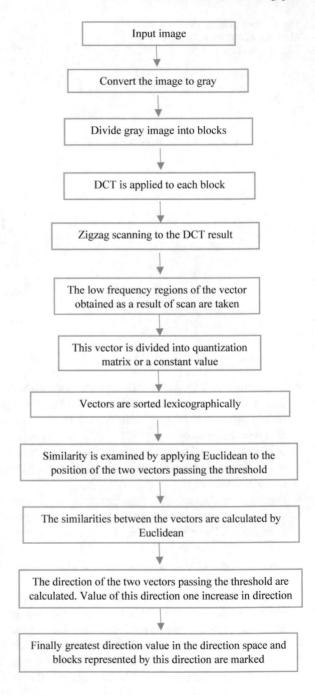

detection. The direction of the vectors which are found similar using Euclidean are also noted. As the last step, greatest direction value is marked and all the vectors in this direction are marked.

Presentation of the detection results is another challenge. A prediction mask can be the output of detection. This is an image containing only white and black pixels. The area with white pixels in prediction mask show the forged region. Fine-tuning the borders of the forged region is a biggest challenge in any type of detection. The accuracy also depends on this.

3 Results

The aim of this kind of study is identifying the forged region in the copied image as accurately as possible. This can be achieved by identifying the duplicated regions where the similarity index between them is less than a certain threshold and the duplicated regions are non-overlapping. Therefore, for forgery detection, two conditions are imposed over the duplicated block detection procedure: (1) The blocks should be non-intersecting and non-overlapping, and (2) the similarity index does not exceed a threshold. To meet the first requirement, the shift distance criterion is used. For this, let us consider that (x_i, y_i) and (x_j, y_j) are the top-left corner coordinates of the two blocks that are represented by the feature vectors fv_i and fv_j, then

$$\forall \sqrt{(x_i - x_j)^2 + (y_i - y_j)^2} \tag{1}$$

If the two feature vectors satisfy Eq. (1), then only they are considered for similarity index calculation in the next step where Euclidean distance method is used as given in Eq. (2). Here d_t is the minimum threshold distance.

$$d(fv_i, fv_j) = \sqrt{\sum_{k=1}^{10} (fv_{ik} - fv_{kj})^2} < d_t \tag{2}$$

3.1 Performance Evaluation

Accuracy of the detection is calculated by comparing the real and prediction masks. Confusion matrix method is a well-known method for visualization of performance of an algorithm. Here we make use of same method for calculating the following performance evaluation factors.

$$\text{Recall} = \text{True Positive} / (\text{True Positive} + \text{False Negative})$$

$$\text{Precision} = \text{True Positive (True Positive} + \text{False Positive)}$$

$$\text{Accuracy} = (2 * \text{Precision} * \text{Recall})(\text{Precision} + \text{Recall})$$

Precision denotes the probability which shows that a detected forgery is indeed a forgery, while recall denotes the probability where actually a forged image is detected [11]. Precision and recall are used to show the accuracy of a detection technique at image level and can be computed using the formulae given above. Images shown in Figs. 3 and 4 along with their real masks showing the forged regions are tested. The results obtained are individually shown in Figs. 5, 6, 7, 8, 9 and 10.

Observing Figs. 5, 6, 7, 8, 9 and 10, we can understand that there are differences in the real mask and prediction mask. The real mask shows which part of the image has been copied and pasted onto other part of the image. Prediction mask is prepared by the code that is detecting the forgery. Greater the similarity between the prediction and real masks, greater is the accuracy.

By looking at Table 1, we can make out that the value of accuracy can vary to a great extent when different forged images are used for testing. Figure 11 shows a chart that compares the different results. We can observe great variation in accuracy in the chart. Obtaining a good accuracy for any sort of image is a biggest challenge for researchers in this area. The chart given below compares the recall, precision and accuracy for the different tested images.

(a)   **(b)**   **(c)**

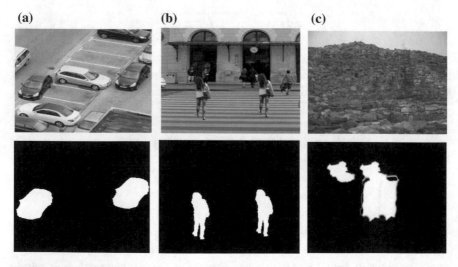

Fig. 3 **a** Copied Image1 and corresponding real mask, **b** Copied Image2 and corresponding real mask and **c** Copied Image and corresponding real mask

Fig. 4 **a** Copied Image4 and corresponding real mask, **b** Copied Image5 and corresponding real mask and **c** Copied Image6 and corresponding real mask

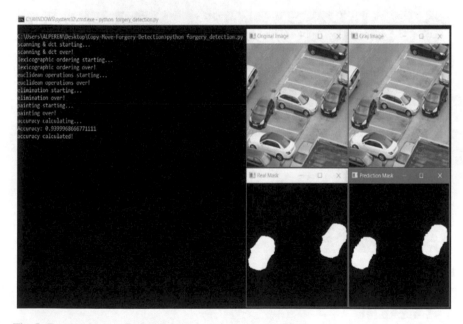

Fig. 5 Results obtained for Image1 producing an accuracy 93.99

Fig. 6 Result obtained for Image2 producing accuracy of 74.78

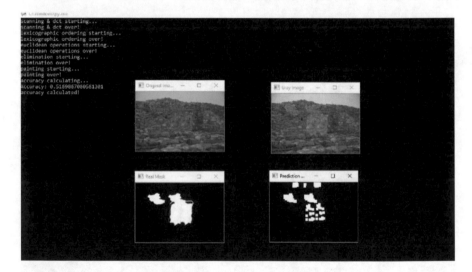

Fig. 7 Result obtained for Image3 producing accuracy of 51.69

4 Conclusion

Copy-move forgery in the field of digital images has become one of the most troubling activities. It is mainly used for illegal purposes which is creating a lot of problems related to cyber-crime. The researchers are taking this topic for their study as there is always need for a better technique as criminals keep coming with new methods for performing forging activities. The technique proposed here works well in identifying

Fig. 8 Result obtained for Image4 producing accuracy of 91.88

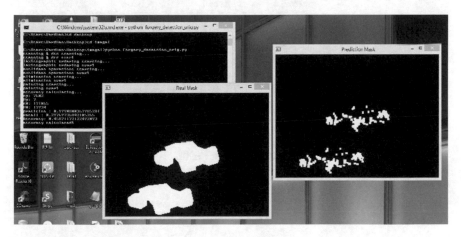

Fig. 9 Result obtained for Image5 producing accuracy of 45.87

the copied region in the image. But accuracy of the technique varies with the input images. There is scope for further studies in the same area for getting better and stable accuracy for all type of forged images. As we have artificial intelligence applied to all fields of image processing, the same can be tried with forgery detection also. Deep learning techniques are well suited for solving such kind of problems [12]. Convolution networks are used in the paper [12], and good results are obtained. When a neural network is trained for set of images, it can identify the forgery in the new input image. There are plenty of datasets available online for training and testing such models. So application of deep learning techniques for forgery detection can

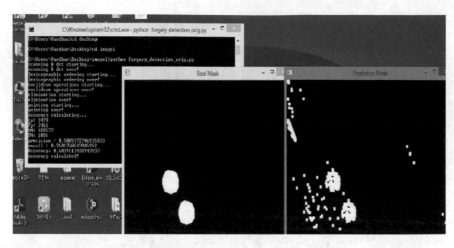

Fig.10 Result obtained for Image6 producing accuracy of 60.39

Table 1 Results for different images

Image	Recall	Precision	Accuracy
Image1	0.0905	0.9954	0.9399
Image2	0.6046	0.9801	0.7471
Image3	0.3941	0.7510	0.5169
Image4	0.9934	0.8546	0.9188
Image5	0.2976	0.9990	0.4587
Image6	0.7602	0.5009	0.6039

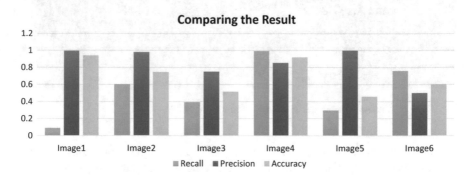

Fig. 11 Comparing the results obtained for various images

be a very good area for research. A technique alone cannot achieve the best result. Instead, combination of available techniques must be tried to obtain the best results.

References

1. K.B. Meena, V. Tyagi, *Image Forgery Detection: Survey and Future Direction* (2019)
2. N.K. Gill, R. Garg, A. Doegar, *A Review Paper on Digital Image Forgery Detection Techniques* (2017)
3. Y. Li, J. Zhou, *Fast and Effective Image Copy-Move Forgery Detection via Hierarchical Feature Point Matching* (2018)
4. H. Chen, X. Yang, Y. Lyu, *Copy-Move Forgery Detection Based on Key Point Clustering and Similar Neighborhood Search Algorithm* (2020)
5. R. Dixit, R. Naskar, *Review, Analysis and Parameterization of Techniques for Copy-Move Forgery Detection in Digital Images* (2017)
6. S. Alagu, K. Bhoopathy Bagan, *Copy-Move and Splicing Image Forgery Detection Using DCT and Local Binary Pattern* (2019)
7. B. Soni, D. Biswas, *Image Forensic Using Block-Based Copy-Move Forgery Detection* (2018)
8. M.A. Elaskily et al., *Comparative Study of Copy-Move Forgery Detection Techniques* (2017)
9. A. Kashyap, R.S. Parmar, M. Agarwal, H. Gupta, *An Evaluation of Digital Image Forgery Detection Approaches* (2017)
10. Y. Sun, R. Ni, Y. Zhao, *Nonoverlapping Blocks Based Copy-Move Forgery Detection* (2018)
11. V. Christlein, C. Riess, J. Jordan, C. Riess, E. Angelopoulou, *An Evaluation of Popular Copy-Move Forgery Detection Approaches* (2012)
12. Y. Liu, Q. Guan, X. Zhao, *Copy-Move Forgery Detection Based on Convolutional Kernel Network*

Sentiment Classification and Data Analysis

A Detail Analysis and Implementation of Haar Cascade Classifier

Gaurav Ghosh and K. S. Swarnalatha

Abstract Haar cascade is a one of the popular machine learning algorithm used for object detection. The Haar algorithm identifies objects in image as well as video. The Haar algorithm was initially used to identify the body parts; later, it was used to for identifying any kind of object. The Haar identify the objects based on the features provided. The Haar algorithm is divided into four-part/stages: Haar feature selection, creating integral images, AdaBoost training, and cascading classifiers; the detailed analysis and implementation of all the four stages along with statistical data have been carried out in this paper, and also the process/step to be followed to create the Haar cascade file has been discussed in this paper. We can conclude that the Haar algorithm is more effective to recognize the object based on the features.

Keywords Haar · Cascading · Features · Images

1 Introduction and Methodology

(i) Haar Feature and Feature selection: CNN can be used for identifying and selecting Haar features, consider an image in Fig. 1, and it is a color image and every color image is made up of three colors that are red, green, and blue. To identify similar kind of object, for example, car, you have to use the following image as a positive sample of your dataset, but having color images for training might not be a very wise choice, as:

- It increases the size of the overall sample/training set.
- It provides unwanted distractions for the model like: saturation, color value, and hue which would decrease the efficiency of the classifier [1].

So, to remove these problems, you must convert your image to gray scale image. Gray scale is the image which has its pixel value ranges from 0 to 255 ([0, 255]). This would not only reduce the size of every image but also remove the unwanted factors.

G. Ghosh · K. S. Swarnalatha (✉)
Department of Information Science and Engineering, NMIT, Bangalore 560064, India
e-mail: swarnalatha.ks@nmit.ac.in

© The Author(s), under exclusive license to Springer Nature Singapore Pte Ltd. 2022
P. Shetty D. and S. Shetty (eds.), *Recent Advances in Artificial Intelligence and Data Engineering*, Advances in Intelligent Systems and Computing 1386,
https://doi.org/10.1007/978-981-16-3342-3_28

Fig. 1 Sample color image

Fig. 2 Gray scale image

After converting the image to a gray scale image, it would look like as mentioned in Fig. 2

As humans can see the image like this, but computers see it in the form of a matrix of numerous pixels. As every image is made of some pixels, the computer reads the value of those pixels to understand the image. As this is a gray scale image, the pixel matrix would have multiple values between 0 and 255 [9]. It is impossible to show the pixel matrix of the whole image (as it will become unreadable), and therefore, Fig. 3 is an image of a sample pixel matrix of a part of an image.

In actual implementation, these matrices are enormously large, that is they can go to a size like 1900 × 2500 and still that is a small image. In the above image, you can see that every value inside the image is between 0 and 255, showing that it is a gray scale image. In the pixel matrix, we would even be able to see distinguish edges in it. Need to find the pixel values where you can see the contrast of two values by seeing the pixel values.

In Fig. 4 image within the two orange boxes, the two-pixel values have completely opposite values, that is one is dark and one is white. We can represent these boxes as separate colors like as shown in Fig. 5. The below image, i.e., Fig. 5, is called a Haar feature. Just like the above image, many other Haar features are created by different type of pixel values arranged in different ways. Some of them are shown in Fig. 6 [1].

Fig. 3 Sample pixel matrix

255	0	137	137	137	137	0
0	128	255	128	137	255	137
128	0	0	64	128	64	64
128	128	0	255	137	255	0
0	255	128	137	137	137	0
128	137	137	137	0	255	64
255	128	128	128	128	64	64

Fig. 4 Contrast of image pixels

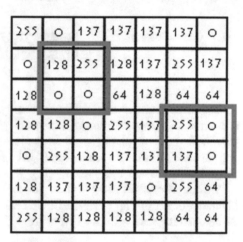

Fig. 5 Haar features

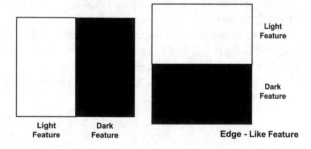

Fig. 6 Haar features with pixel values

1. Edge features

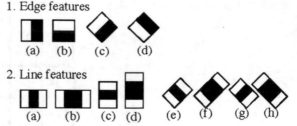

(a)　(b)　(c)　(d)

2. Line features

(a)　(b)　(c)　(d)　(e)　(f)　(g)　(h)

3. Center-surround features

(a)　(b)

4. Special diagonal line feature used in [3,4,5]

Fig. 7 Gray scale image

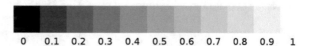

0　0.1　0.2　0.3　0.4　0.5　0.6　0.7　0.8　0.9　1

The Haar features are created internally, using inbuilt formula, in real-world scenarios; the grayscale image values are generally limited to a range of 0 and 1 [5]. Due to ease of calculation, the calculated values do not become very huge. This can be done by taking a normal grayscale image and normalizing it, so the values range between 0 and 1 as shown in Fig. 7.

We can also invert these values, that is 0 would mean white and 1 would mean black as shown in Fig. 8. In the future examples, we are going to use the inverted values, and the reason for using this is out of the scope, so we are not going to focus that.

If we put these values into the image pixel matrix, then our pixel matrix might look like as shown in Fig. 9.

To aim to detect any possible edges present, we consider any pixel with values less than 0.5 to be light or to be within the white group and any pixel values greater than 0.5 to be dark or within the black group.

According to Viola–Jones algorithm, the formula for detecting a Haar feature is:

$$\Delta = \text{dark} - \text{white} = \frac{1}{n}\sum_{\text{dark}}^{n} I(X) - \frac{1}{n}\sum_{\text{white}}^{n} I(X) \tag{1}$$

Fig. 8 Inverted image

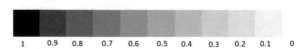

1　0.9　0.8　0.7　0.6　0.5　0.4　0.3　0.2　0.1　0

Fig. 9 Image pixel matrix

0	0	1	1
0	0	1	1
0	0	1	1
0	0	1	1

0.1	0.2	0.6	0.8
0.2	0.3	0.8	0.6
0.2	0.1	0.6	0.8
0.2	0.1	0.8	0.9

The above formula says that we have to subtract the average of all the white pixels from the average of all the dark pixels. Here, $I(x)$ is the intensity of the dark and the white pixels which are summed upon and divided by the total number of those pixels (or simply the average).

In the above pixel matrix, you can see the left one which describes an ideal case and the right one is depicted as a real-world scenario case. In the ideal case, the value of delta is always 1. According to the Viola–Jones algorithm, the real case scenario must be as close as possible to the ideal case to detect a Haar feature

Calculate the values of the right pixel matrix showing a real scenario:

$$\Delta = 0.74 - 0.18 = 0.56 \tag{2}$$

since the delta value is greater than 0.5, you can consider it as a Haar feature although not as a strong oneif this is to be used on a sample gray scale image as shown in Fig. 10.

The algorithm can find the features as shown in Fig. 11

(ii) Create Integral Images: consider Fig. 10 image. Run the algorithm; it would search for features by iterating through every pixel by using a particular sized block. You can visualize it as shown in Fig. 12.

Fig. 10 Gray scale image with haar feature

Fig. 11 Features finding

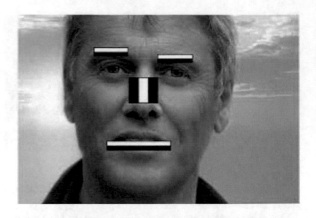

Fig. 12 Integral image

The image is being scanned by a block of some size. The blocks have been drawn very far away from each other, but in reality, the blocks are just a few pixels apart and in some cases, a new block is drawn by adjusting the box by just one pixel. Inside every single box, we can find thousands of pixels to be summed up and computed to compute a specific region for several number of times, which is very time expensive and this operation is executed in Big O of n^2 [O(n^2)]. To address this problem, we use integral images concept which is far more efficient than the one discussed. It provides a constant time complexity of O(1). These integral images are created as mentioned in Fig. 13, below you can see a pixel box for normal image and beside it another pixel box for the integral image.

Structure wise it looks the same but, if you have a closer look, you would see that the values are different from the original image and if you try a bit, you would be able to understand that every pixel value in the integral box is the sum of all the values to its left and all the values above it. So, let us see the following example shown in Fig. 14.

In Fig. 14 you can see that $1.2 = 0.1 + 0.1 + 0.2 + 0.3 + 0.1 + 0.4$ that is all the pixels to the left of 1.2 and above 1.2.

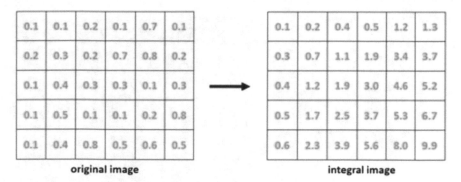

Fig. 13 Pixel box for normal image and integral image

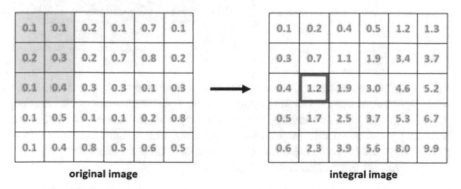

Fig. 14 Transformed pixel box

(iii)   AdaBoost

AdaBoost is one of the ensembles learning method (also known as "meta-learning") which was initially created to increase the efficiency of binary classifiers. It uses an iterative approach to influence the output of the weak learners, and by correcting the errors in the previous weak learners in our case study, we are using decision trees as the weak learners in the ensemble method. Decision trees are the ones which do not have a determined depth of its own, but if you restrict its depth only to 1, then every such tree would be having only 1 root and 2 leaves which are called stumps. AdaBoost technique has only decision trees in the form of stumps or better said as an ensemble of stumps. Since stumps have only one root node, so it can have only one variable to decide on and cannot have the advantage of having different weightage of different variables. Due to this reason, stumps are generally called weak learners.

Now, you might think that what would be the reason behind using something which is considered weak when we want to build something very strong and accurate.

So, to understand this, you must know that when you use Random Forest in an algorithm, it consists of a variety of trees which has the same weightage as the others

Fig. 15 Stumps in Adaboost

Chest Pain	Blocked Arteries	Patient Weight	Heart Disease	Sample Weight
Yes	Yes	205	Yes	1/8
No	Yes	180	Yes	1/8
Yes	No	210	Yes	1/8
Yes	Yes	167	Yes	1/8
No	Yes	156	No	1/8
No	Yes	125	No	1/8
Yes	No	168	No	1/8
Yes	Yes	172	No	1/8

but when you use stumps in AdaBoost you will note that each stump has a different weightage than the other depending upon the number of correct predictions. Learning from the mistakes of one stump, the other stump also influences its predictions accordingly and ultimately the accuracy level rises. The weightage of a stump is called "Say," due to which the order of these stumps also become important.

Therefore, the three ideas behind AdaBoost are:

- AdaBoost combines a lot of weak learners (mostly stumps) to make classifications.
- Some stumps have more weightage to their say in the classification, than others.
- Each stump is made by taking the previous stumps mistakes into account.

Let us take an example on how stumps are used in AdaBoost to make predictions. Let us assume you are given the following database as shown in Fig. 15.

Here you can see that the data has three columns and one output column. Make sure that the column "Sample Weight" is not a data column, but it is a column assigned because initially.

To give every row with same weightage

That is:
$$\frac{1}{\text{total number of samples}} = \frac{1}{8} \tag{3}$$

To create the stumps for each of the column, since the weight of every row is the same initially, so, at this stage, it does not matter which stump is created first.

Here you can see that three stumps are created due to three different columns present in the dataset as shown Fig. 16. Now, the Gini index of each stump is calculated to determine which one does a good job in classification. After calculating the Gini index,

Fig. 16 After creating stumps

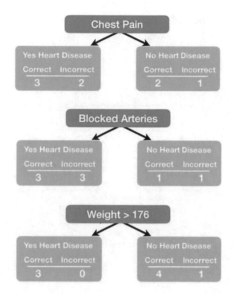

Chest Pain—0.47
Blocked Arteries—0.5
Weight—0.2.

The stump with column weight does the best job in classification as its Gini index is the least, and thus, this will be the first stump for our classification. Need to calculate the amount of say which stump will be having. To calculate the say, a formula is specified which is:

$$\text{Amount of Say} = \frac{1}{2} \log\left(\frac{1 - \text{Total Error}}{\text{Total Error}}\right) \tag{4}$$

Before calculating the amount of say of the stump, you must see the graph between amount of say and the error on the stump as shown in Fig. 17.

In the above graph, the say of a stump is high when it has a low error in it (that is close to 0) and the say of the stump will be low if the error is found to be high. Calculate the say of our stump. To calculate the say, you first need to calculate the error of the stump. Our stump incorrectly classifies one example out of total eight examples, and thus, the error becomes 1/8. Now let us put this error value in the formula for the amount of say. On calculating the values, you will find,

$$\text{Amount of say} = \frac{1}{2}\log(7) = 0.97 \tag{5}$$

By looking at the value, you can say that the stump does a pretty good job classifying and this value sits very near to 1 in the above graph. Now you need to update the weights of the rows, so that the next stump created learns from the mistake made

Fig. 17 Error on stumps

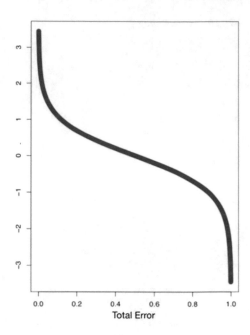

Total Error

by this stump and predicts a better output. To update the weights, you need to make sure that the weight of the incorrectly classified row is increased, so that the next stump gives more importance to it.

To increase the sample weight, the formula is as mentioned below i.e., Eq. 6:

$$\text{New sample weight} = \text{sample weight} \times e^{\text{Amout of say}} \qquad (6)$$

By the amount of say value in the formula with the previous sample weight, we get the new increased sample weight. Thus, on putting 0.97 in the amount of say and 1/8 in sample weight, we get new sample weight = 0.33 which is larger than 1/8 (0.125). Now let us decrease the weights of other samples, and for doing, so we have a very similar formula, i.e., Eq. 7:

$$\text{New sample weight} = \text{sample weight} \times e^{-\text{Amount of say}} \qquad (7)$$

The only difference between this formula and the previous one is a negative sign to the power of e. This would decrease the value exponentially as we increase the amount of say. On putting the amount of say as 0.97 and sample weight as 1/8 as previously used, you get the new sample weight = 0.05 which is less than 1/8 (0.125). Now your updated table with updated weights looks as in Fig. 18 [6].

The total sum of the new weight $\neq 1$, whereas previously we had the total sum of previous sample weights = 1. So, to make the sum of new sample weights = 1, you need to normalize the value by dividing each value by the current sum which is equal to 0.68. After dividing each value, you can see the sum = 1. Now we can

Fig. 18 Updated weights

Chest Pain	Blocked Arteries	Patient Weight	Heart Disease	Sample Weight	New Weight
Yes	Yes	205	Yes	1/8	0.05
No	Yes	180	Yes	1/8	0.05
Yes	No	210	Yes	1/8	0.05
Yes	Yes	167	Yes	1/8	0.33
No	Yes	156	No	1/8	0.05
No	Yes	125	No	1/8	0.05
Yes	No	168	No	1/8	0.05
Yes	Yes	172	No	1/8	0.05

replace the previous sample weights with the new one which changes the table as follows in Fig. 19.

As you started the first iteration by choosing any of the columns previously, you cannot do the same now because the weight of each row is not the same now. There are few different ways to start, but here you would be learning the method of making a new collection of samples that contains duplicate copies of the samples with the largest sample weights.

Fig. 19 Replaced sample weights

Chest Pain	Blocked Arteries	Patient Weight	Heart Disease	Sample Weight
Yes	Yes	205	Yes	0.07
No	Yes	180	Yes	0.07
Yes	No	210	Yes	0.07
Yes	Yes	167	Yes	0.49
No	Yes	156	No	0.07
No	Yes	125	No	0.07
Yes	No	168	No	0.07
Yes	Yes	172	No	0.07

So we start by making a new, but empty, dataset that is the same size as the original one. Then, we pick a random number between 0 and 1, and we see where the number falls when we use the sample weights like a distribution. If the chosen number is between 0 and 0.07, then we put the first row into the new collection of samples. If the number is between 0.07 and 0.14 (0.07 + 0.07 = 0.14), then we put the second row into the new collection of samples. If the number is between 0.14 and 0.21 (0.14 + 0.07 = 0.21), then we put the third row into the new collection of samples, and if the number is between 0.21 and 0.70 (0.21 + 0.49 = 0.70), then we put the fourth row into the new collection of samples [6].

Now, for example, the first number picked is 0.72, then the fifth row would be kept into the new collection of samples. If the second number picked is 0.42, then the fourth row would be kept into the new collection of samples. We continue to pick random numbers and add samples to the new collection until the size of the collection is the same size as the original. After completion, you would notice that the row with higher sample weight would be added to the new collection multiple times reflecting its larger sample weight.

Now this table, i.e., Fig. 19 is used further for second iteration and now each of the row, is again given the same sample weight as earlier (1/8); however, that does not mean that the next stump will not emphasize the need to correctly classify these samples. As four samples are the same, therefore they would be treated as a block, creating a large penalty for being misclassified.

The forest of stumps created by AdaBoost makes classification. Imagine four stumps are saying for a patient as having heart disease with their respective says as: 0.97, 0.32, 0.78, and 0.63. Alternatively imagine that two stumps are saying the patient does not have heart disease with their respective says as: 0.41 and 0.82. Now if we sum up the say values of positive stumps, it comes out to be 2.7 and summing the negative stumps comes out to be 1.23. Now, we can see that the sum value of positive stump is greater than the sum value of negative stump, and thus, the patient is ultimately classified as having a heart disease. This is how the errors that the first stump makes influence how the second stump is made and so on.

2   Cascading Classifiers

The cascade classifier have multiple stages as shown in Fig. 20, and within each stage, the boosting process is been performed, that is AdaBoost with multiple weak learners (ensemble method). The weak learners are the decision stumps, and boosting helps us to have highly accurate classifiers by taking the weighted average of the say value produced by the weak learners. You will be able to understand the workflow and the process of each segment while making an XML file by focusing on Fig. 20 [1, 4].

The classifier hovers over a specific region as per the current location of the sliding window on the image pixel matrix. The stage one of the classifier tries to detect any features within that window, and if any feature is found then that stage is said to be

Fig. 20 Larger sample weights

Chest Pain	Blocked Arteries	Patient Weight	Heart Disease
No	Yes	156	No
Yes	Yes	167	Yes
No	Yes	125	No
Yes	Yes	167	Yes
Yes	Yes	167	Yes
Yes	Yes	172	No
Yes	Yes	205	Yes
Yes	Yes	167	Yes

positive and the next stage in the same window starts searching for features. This continues until the last stage is reached, and if the last stage also returns positive then it is confirmed that a feature is present within the window and then the window slides to the next location. To the contrary, if any of the stages returns negative, then the window slides to the next location assuming that it has not encountered any promising feature for the detection. As mentioned above, the stages can neglect the negative features very quickly. Here, time complexity comes in to picture, where a big majority of the features are negative, whereas the object we want to detect is rarely found and must need time to get fully verified [5].

- A true positive is found when a positive feature is correctly classified.
- A false positive is found when a negative feature is mistakenly classified as positive.
- A false negative is found when a positive feature is mistakenly classified as negative.

Now to get an efficient result, the false negative rate of each stage should be as low as possible. This is because if a positive feature is classified as negative then, the mistake cannot be corrected in the subsequent stages and thus in such a case the classification stops. However, if it classifies a negative feature as a positive one, then it can be changed in the later stages, and this is the reason why a classifier can have a high false positive rate. Thus, if we keep on adding multiple stages, then the overall false positive rate decreases, but to the contrary, the true positive rate also decreases as the detected feature has to go through several levels of verification [7].

3   To Create the Haar Cascade Files

There are many ways to create a Haar cascade file, but the method that we focus here is the easiest and fastest way to create it with a high level of accuracy, as it offers a different UI to the user. Please do visit this link (http://amin-ahmadi.com/cas cade-trainer-gui/) to download the software. This software is named Cascade Trainer GUI which provides you with a very user-friendly interface to work with. This is developed by Amin Ahmadi (Tazehkandi) an Iranian author, developer, and computer vision expert [2, 3].

After downloading the software, it will look something like this as shown in Fig. 21.

Here, you can see that there are four partitions made:

- Input (used to provide the path for training data)
- Common (used for the settings of classifier)
- Cascade (settings used while cascading)
- Boost (settings used while boosting).

Before starting with the stages, we should first create a folder in any of your desired location and name (in my case let us name it cascade), and inside it, you must create two separate folders named 'p' (for positive image samples) and 'n' (for negative image samples). The positive image samples in the folder 'p' should strictly contain only the image that you want the algorithm to detect (e.g., if you want to detect faces, then the images in p folder should contain images which has only faces and no other objects in it). It would be even more helpful for your algorithm to train if you convert all the colored images in this folder to gray scale, so that the algorithm does not find any distractions in the image (in the form of hue, saturation, and brightness).

This technique is even more special because this software does not need to be fed a huge amount of samples for better accuracy. You can get decent accuracy even with a positive sample size of 50. Remember to provide at least double the negative number of samples than the positive ones. Similar to the positive samples, the negative samples can have anything related to the detecting object but should not contain any part of the object to be detected.

The Input Block:

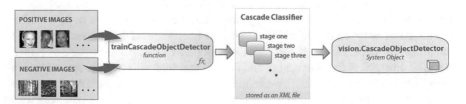

Fig. 21   Process of XML file

This stage contains only four parameters listed below.

(i) *Sample Folder*: You will be providing the folder directory/or the path of the folder in which you have kept the positive and negative samples.

(ii) *Positive Image Usage*: This is the percentage of positive images used for training in every stage of cascading as mentioned in page no. 16.

(iii) *Negative Image Count*: This is the number of negative images used for the training in every stage of cascading. Make sure not to make this as a high number, as this might result in memory shortage error.

(iv) *Force Positive Sample Count*: This is where you can create multiple positive images with a single positive image. This tab shows you the number of positives you want to generate from each positive sample (try to stick to 1 unless really required).

The common settings: common setting page as shown in Fig. 22.

Let us start with the **Common Settings Block**:

(i) *Number of stages*: This is the number of stages you want in every block of detection in the image. Generally, these stages are between 15 and 20 in number.

(ii) *Pre-calculated Buffer Size* (Mb): You can see two options by this name in your screen, and both of them are related to the speed of your training. It is on you how much memory you want to assign to these tabs, but always keep in mind that you should not assign more memory combining both, than you have in your system in total. To the contrary, assigning too less memory is also not good as it may not satisfy your algorithm's memory needs. Let us assume, if

Fig. 22 Cascade Trainer GUI

you have 8 GB ram in your system, then your combined allocated memory should not be more than that, or else you might encounter an algorithm crash.

(iii) *Number of Threads*: This is the highest number of threads you can use during your training. Make a note that the number of threads available may differ according to your machine. If you have TBB installed, then you can see that the total available threads are being displayed automatically, and which becomes a default option. It is recommended for you to have this software if you want to make changes to this parameter.

TBB is known as threading building blocks. It is a library in C++ which was introduced by Intel for purposes like parallel programming on systems having more than 1 core. It is used to break a problem into multiple sub-problems and compute each one simultaneously. Inside this library, you get functions which tell the threads to work in an organized way. To use TBB, you first need to install it in your system, and you can download it through this link https://github.com/oneapi-src/oneTBB (recommended not to use it unless you have a stronghold on it).

(iv) *Acceptance Ratio Break Value*: This is a parameter which shows how precise your algorithm must be. Generally this value is set to 10e−5, to ensure that the model gets perfectly trained. This value also shows that if the model crosses this threshold, then you might have a chance of getting a model which is over-trained. Now, if you want to disable this feature, then you can simply change the value to −1. This is a kind of threshold preciseness value that your algorithm must satisfy.

(v) *Base Format Save*: This parameter is related to the Haar features that you have learnt earlier. If you click the checkbox then you are actually telling the software to use the ancient version of the classifier because you do not possess the required specifications in your system to go forward with your training. Thus, do not put much importance on this option unless you have a really old configuration system.

The **Cascade Block**: the cascade block as shown in Fig. 23

(i) *Sample Width*: This is the width of the image that you are providing for training (in pixels).

(ii) *Sample Height*: This is the height of the image that you are providing for training (in pixels).

While setting the size of the image, it is recommended to keep the aspect of 24, and obviously do not put the size of the image very large or else it might affect the speed of your training. Regarding the aspect let us say you have an image of size 500 × 300, and on calculating its aspect, you get as 1.66:1. Now multiply both sides with 24 to keep the same aspect. On doing so, we finally get 40 × 24 for your image's width and height, respectively.

(iii) *Feature Type*: Type of features: Haar—Haar-like features, LBP—local binary patterns. HOG—histogram of oriented gradients.

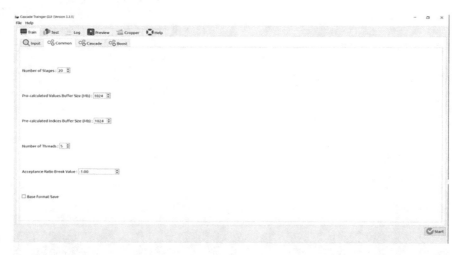

Fig. 23 Common settings in Cascade Trainer GUI

We use HOG only if we have OpenCV version of 3.1 or later. Since in this paper we are focusing on the details of Haar and its features, so please do keep this parameter value restricted to Haar.

(iv) *Haar Feature Type*: This parameter is defined if you want to use the basic features of Haar or all of the features. Basic includes only the vertical and the horizontal features described in page no. 4, whereas all include the features which are diagonal or even at an angle.

The Boost Block: the boost block is shown in Figs. 24 and 25

Fig. 24 Cascade block

Fig. 25 Boost block

(i)   *Boost Type*: You can see these options in the dropdown list {DAB, RAB, LB, GAB(default)}. These are the four different types of AdaBoost techniques that you can apply here. The different names of these techniques are: DAB (Discrete AdaBoost), RAB (Real AdaBoost), LB (LogitBoost), and GAB (Gentle AdaBoost). I recommend to use the GAB (default value), the one explained in the above sections.

(ii)   *Minimal Hit Rate*: Hit rate (HR) is the minimum threshold that any algorithm must pass through in each stage of the classifier. It can be defined as the number of positive objects that are correctly identified within the given stage. The default value is 0.995, and you can change this if you want your model to be lenient or strict accordingly. The total HR can be calculated using the formula as shown in Eq. 8

$$\text{minimum hit rate}^{\text{number of stages}} \tag{8}$$

(iii) *Maximal False Alarm Rate*: False Alarm (FA) is the maximum threshold that any algorithm must have in each stage of the classifier. It can be defined as the number of negative images that are incorrectly identified within the given stage. The default value is 0.5, and you can change this if you want your model to be strict or lenient accordingly. The total FA can be calculated using the formula as shown in Eq. 9

$$\text{maximum false alarm rate}^{\text{number of stges}} \tag{9}$$

(iv)   *Weight Trim Rate*: This parameter can have value between 0 and 1. It is used to save computational time. Samples with summary weight $\leq 1 -$

Weight Trim Rate do not participate in the next iteration of training. You can set the value to 0, to turn off this functionality. The default value to this parameter is 0.95.

(v) *Maximal Depth Weak Tree*: This is the maximum depth of the weak tree you want. Default value is 1 that is a stump as discussed earlier on page 10. You can even change to any value, but you must know the insights of the process when you use maximum depth more than 1.

(vi) *Maximal Weak Trees*: This is the maximum number of weak learners/stumps you want to have in each stage of the classifier.

4 Results and Conclusion

The detailed analysis and implementations proved that the Haar cascade algorithm can be effectively used to identify the features of an object, and the mathematical analysis is shown for all the stages of the Haar cascade algorithm, i.e., Haar feature selection, creating integral images, AdaBoost training, and cascading classifiers. Finally, an overview and step-by-step procedure has been shown to create a Haar cascade file.

References

1. http://www.willberger.org/cascade-haar-explained/
2. T.R. Phase, S.S. Patil, Building custom HAAR-Cascade classifier for face detection. Int. J. Eng. Res. Technol. (IJERT) **08**(12) (2019). ISSN (Online): 2278-0181
3. http://amin-ahmadi.com/cascade-trainer-gui/
4. https://docs.opencv.org/master/dc/d88/tutorial_traincascade.html
5. R. Yustiawati et al., Analyzing of different features using Haar cascade classifier, in *2018 International Conference on Electrical Engineering and Computer Science (ICECOS), Pangkal Pinang* (2018), pp. 129–134. https://doi.org/10.1109/ICECOS.2018.8605266
6. D.K. Ulfa, D.H. Widyantoro, Implementation of Haar cascade classifier for motorcycle detection, in *2017 IEEE International Conference on Cybernetics and Computational Intelligence (CyberneticsCom), Phuket* (2017), pp. 39–44. https://doi.org/10.1109/CYBERNETICSCOM.2017.8311712
7. É.K. Shimomoto, A. Kimura, R. Belém, A faster face detection method combining Bayesian and Haar cascade classifiers, in *2015 CHILEAN Conference on Electrical, Electronics Engineering, Information and Communication Technologies (CHILECON), Santiago* (2015), pp. 7–12. https://doi.org/10.1109/Chilecon.2015.7400344
8. I. Paliy, Face detection using Haar-like features cascade and convolutional neural network, in *2008 International Conference on "Modern Problems of Radio Engineering, Telecommunications and Computer Science" (TCSET), Lviv-Slavsko* (2008), pp. 375–377.
9. Q. Li et al., Multi-view face detector using a single cascade classifier, in *2016 10th International Conference on Software, Knowledge, Information Management & Applications (SKIMA), Chengdu* (2016), pp. 464–468. https://doi.org/10.1109/SKIMA.2016.7916267
10. http://uu.diva-portal.org/smash/get/diva2:601707/FULLTEXT01.pdf

Sentiment Analysis of Twitter Posts in English, Kannada and Hindi languages

Saritha Shetty, Sarika Hegde, Savitha Shetty, Deepthi Shetty, M. R. Sowmya, Rahul Shetty, Sourabh Rao, and Yashas Shetty

Abstract The World Wide Web has always been a huge repository for information. With this, an era of social networking came into existence that allowed people across the globe to connect and communicate with each other. The data present here is of various types like text, images, videos, audios, etc. The major part of the internet includes textual data which is written in different human-understandable languages. Recently, there has been tremendous increase in this data written in different languages over social media. This paper analyzes the semantic nature of a text, classifying them as either positive or negative for English, Hindi and Kannada languages. The following classification task is resolved with a CNN-LSTM-based neural network.

Keywords Twitter post · Semantic analysis · Machine learning · Natural language processing

1 Introduction

Social networking has evolved to become one of the major sources of information. Microblogging websites are one among these social networking websites. Twitter

S. Shetty (✉)
Department of MCA, NMAM Institute of Technology, Karkala, India
e-mail: shettysaritha1@nitte.edu.in

S. Hegde · S. Shetty · R. Shetty · S. Rao · Y. Shetty
Department of CSE, NMAM Institute of Technology, Karkala, India
e-mail: sarika.hegde@nitte.edu.in

S. Shetty
e-mail: shettysavi1@nitte.edu.in

D. Shetty · M. R. Sowmya
Department of CSE, NMIT, Bangalore, India
e-mail: deepthi.shetty@nmit.ac.in

M. R. Sowmya
e-mail: sowmya.mr@nmit.ac.in

being one of the very popular microblogging websites where the users have freedom to create and share short messages, also called as "tweets." Sometimes, these tweets are used to express the opinions as well as comments of the users on different topics that may be political, social, etc. Large number of businesses spend a fortune in these microblogging sites to engage with their customers and understand their opinions about different products and services. Sentiment analysis plays a major role in determining the emotions and attitudes of the user.

In this paper, we have applied sentiment analysis on Twitter to indicate if a tweet is of negative or positive sentiment. The text used for analysis can range from big document to small messages or tweets. A supervised learning algorithm is preferred for training a classifier that needs a hand-labeled suitable trained data. As the amount of data and the number of languages increases, it becomes very difficult to manually collect all data to do training for sentiment classifier in the case of tweets. Along with the already existing data sets, we added more training data to analyze the posts based on the ongoing trends and fashion. This helps the consumers to research about the products or services before making a purchase. The semantic analysis is used by organizations and businesses to get customer reviews and feedback about problems in newly released products. We present the experiment results on three different languages and also discuss on how to improve these results.

2 Literature Survey

2.1 Related Work

As we all know, there has been a vast amount of research in the field of sentiment analysis. This is true when we need to analyze product or movie reviews. Sentiment analysis of the tweets is much harder than the conventional documents we find. It is not only true because of the informal languages used but also because of the excessive use of short-length slang words. Pang and Lee [1] provide an up to date survey of the preceding work on sentiment analysis. Sentiment analysis started from the document-level classification which was then handled at the sentence level and now much recently is being carried out at the phrase level. Bruns et al. [2] discussed regarding communication patterns in Twitter tweets. Go et al. [3] explored making use of different n-gram features in conjunction with the parts of speech (POS) tags while training the supervised classifiers like naïve Bayes (NB), maximum entropy (MaxEnt) and support vector machines (SVMs). They found that MaxEnt trained to form a combination of POS tags unigrams by almost 3%. Also, a different finding was reported in Go et al. [3] that adding POS tag features in the case of n-grams gives better accuracy for sentiment classification on tweets.

Agarwal et al. [4] explained the POS features, lexicon feature along with microblogging features. After combining the different features, they also designed tree representation of the tweets to combine the many categories of features into one concise representation. Another significant word on this classification task is by Feng and Barbosa [5]. They used polarity prediction from three different websites and labeled these as noisy labels to provide training for a model and further they manually labeled 1000 tweets for further tuning and another 1000 for testing. They did not mention how they have collected their test data. They indicate the use of syntax features of tweet like retweet, hashtags, link, punctuation and exclamation marks along with features like prior polarity of words and parts of speech of words.

Instead of directly integrating the microblogging features into semantic classifier training. Speriosu et al. [6] constructed a graph which has all some of the microblogging features like hashtags and emoticons together with users, tweets, word unigrams and bigrams as its nodes which are connected based on the link that existed among them (e.g., Users are connected to the tweets they created and the tweets are connected to the word unigrams that they contain).

Existing work concentrates on the use of three main types of features: lexicon, POS features and microblogging features in case of sentiment analysis. Some emphasized the use of microblogging features, whereas some others argued about the importance of the POS tags with or without word priority involved.

In the paper written by Berend et al. [7], they used a supervised learning approach and text preprocessing and feature engineering to classify a Twitter post as negative, neutral or positive. Our approach to problem is a similar technique, but the feature engineering is embedded in the neural network that we use.

Shetty et al. [9] discussed about transliteration of input text from Kannada to Braille. Shetty et al. [10] discussed about the significance of anusvara and visarga in Kannada to English transliterations.

3 Proposed System

The proposed system classifies if a given text sample is of positive or negative sentiment. For this, we need to have the text ready for classification which has to be done by applying various preprocessing and transformations to make it suitable for our training model.

3.1 Model Training

We define three different models for English, Hindi and Kannada languages. The training inputs to these models are a preprocessed dataset. The preprocessing involves removal of stop words, unnecessary characters, lemmatization in the case of English language and finally tokenizing the sentence. A bag of words model is created which contains a list of all words from the dataset. A unique integer mapping is created for each token and we use this to map our sentence to a new integer vector. This vector is given to each model for training along with their corresponding positive or negative labels.

3.2 Classification

After the training process, we save the model weights and the bag of words model. The new test sample is preprocessed with the same steps and then given to our model. The model makes a prediction on a probability scale, saying if a given text is positive or negative.

4 Methodology

4.1 Dataset

In order to train our three different models, we needed datasets in English, Kannada and Hindi languages. It was quite easy to find the English dataset which was already classified as positive or negative in the sentiment 140 dataset. Here, they tried to gather data based on searching certain keywords. The work about this is mentioned in the paper written by Richa Bhayani, Lei Huang and Alec Go. This constituted about 1600 k samples of posts with already classified sentiment as neutral, negative or positive. For this paper, we took the positive and negative samples of English from this dataset. For the Hindi language, the dataset was taken from a website source where they used similar crawler to capture posts from Twitter that were written in Hindi [8]. Additional data were collected from Hindi news articles and were manually tagged as positive or negative. About 2 k positive and negative samples were made for Hindi. Table 1 shows the dataset used (Fig. 1).

Kannada text samples are quite less in numbers when compared to the remaining two languages. We found few samples by going through some article websites, and for the rest of the data, we used the Google translator API to convert these English and Hindi texts to Kannada. A better translation was occurring when English was converted to Kannada than Hindi. So initially, we took some 2 k pre-classified positive and negative samples from English dataset and converted them to Kannada.

Table 1 Dataset used

Dataset	Total samples
Sentiment 140 for English	1,600,000
Hindi dataset	2500
Kannada dataset	6300

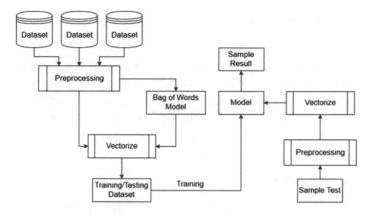

Fig. 1 Structure of the proposed system

Since the Twitter posts had some unnecessary or false tokens in English, the translator would straightaway keep them as it is. This issue was resolved in the preprocessing stage where the relevant tokens were only preserved. Later, we added in some 4 k extra samples by applying the same conversion process to increase the training accuracy and to increase the word count in the bag of word model.

4.2 Preprocessing

The dataset which was collected from Twitter posts will have unnecessary symbols like emoticons, foreign characters, hashtags or website links. The objective of this section is to remove all the noise from the data and prepare a normalized form that is suitable for our neural network. The following preprocessing steps were taken for all the three languages to normalize the data.

4.2.1 Removing Mentions

Mentions are common features for Twitter. They are used to specify or tag a particular individual person. All the mentions in the post always begin with '@' symbol. Since these mentions do not actually contribute to our sentimental classification, it is an unnecessary token. A regex string matching token r'@[a-zA-Z0-9]+' is used to identify a general set of mentions that occur in the post and are replaced with an empty character. For example, the text "happy @SparksCharity" is transformed to "happy."

4.2.2   Processing Hashtags

Hashtags are used to pin a Twitter post into a particular category and they always begin with "#" symbol. In most cases, people write a post and describe their emotions by writing in these hashtags. There the contents that come after the hash symbols are retained for further preprocessing. Consider, for example, the text "Feeling good #happy." This is transformed to "Feeling good happy" which retains the hashtag content.

4.2.3   Removing Links

People tend to post links with their post that points to a picture, video, gif or to some other website. These links do not contribute to the sentimental processing and are therefore removed. The links on the internet begin with a "http" or "https" and are followed by sequence of alphabets, digits and some of the punctuation symbols like "_" or "&". A regex expression is used to identify a general set of these from the Twitter post and is then replaced with null characters.

4.2.4   Processing Texts

For the English language, we have lower and upper case alphabets. To keep everything normalized, we convert any alphabet to its lower case form so that text tokenization and vectorizing become easier. Hindi and Kannada languages do not have this concept of lower and upper case letters. So no such transforms are applied in this case. Further unnecessary symbols, punctuation marks, numbers and foreign characters were removed by checking them against the corresponding alphabet sets for each language. As mentioned before, the Kannada dataset retained some English words, so these foreign characters are removed by the same procedure. Trailing whitespaces were removed from the text which were unnecessary for tokenizing.

4.2.5   Tagging Unknown Symbols

There are cases when some of the symbols do not match the alphabet sets or do not contribute to the other unknown symbols. For example, some of the Unicode characters contain a mixture of these alphabets and punctuations. We identified such symbols by the process of elimination and then marked them as "<UKN>" tag which specified that the token identified there is unknown.

4.2.6   Lemmatization

It is the process of converting a word or token to its actual form by removing its inflectional endings. The output of this is the word that conveys the meaning of the original entity. In any language, the words are in different forms. Just by considering the tenses, we can form it in three tenses like past, present and past participle.

In English, the word "studies," "studied" is related to "study." Here, "study" is the actual lemma transform of those two words. This transform was applied over English processing text using the WordNetLemmatizer provided by the natural language toolkit (NLTK) library. For the other language, we kept them as such and continued with the next stage.

4.2.7   Removing Stop Words

Stop words are the commonly used words in a sentence that do not contribute to its meaning or sentiment. For English, the words like "the, a, an, and, is" are some of the stop words. They are only used to specify a particular action or talk about an entity. Nltk library provides an English set of stop words which were used to eliminate the stop words in English dataset. As of Hindi dataset, a website source [10] provided with some set of stop words like " इस", " इसका", " इसकक" which was used to remove the stop words from Hindi dataset.

4.3   Bag of Words Model

The final model can only take in sequence of integer-based input for training and classification. For this, we need to have a mapping between each word to a unique number. We make use of bag of words model to collect and store all the words available in the dataset.

The first step for this is to tokenize our text into individual words. After applying all the preprocessing steps, we do this stage and collect all the words. Then we use a set data structure to store these words so that only unique words are collected. This model now represents a word to integer mapping where each integer specifies the location of that word in that bag (Fig. 2).

4.4   Dataset to Vector Transform

Using the bag of words, model and our tokenized set of texts, we create a new vector which is of some fixed length. This vector contains a sequence of integers where each integers maps on to a specific word. Since the neural network needs to take in a fixed-length input for training and classification, we limit the number of words in

Fig. 2 Structure of the
neural network along with
details about the number of
parameters and output shape
at each layer of the network

Layer (type)	Output Shape	Param #
embedding_2 (Embedding)	(None, 70, 128)	37282304
conv1d_4 (Conv1D)	(None, 70, 32)	12320
max_pooling1d_3 (MaxPooling1	(None, 35, 32)	0
conv1d_5 (Conv1D)	(None, 35, 64)	6208
max_pooling1d_4 (MaxPooling1	(None, 17, 64)	0
lstm_2 (LSTM)	(None, 128)	98816
dense_1 (Dense)	(None, 1)	129

```
Total params: 37,399,777
Trainable params: 37,399,777
Non-trainable params: 0
```

a text to 70 words. As of this paper, the tweet length is from 140 to 280 characters and based on this, we assume that a maximum length of words that might appear to be around 70. So a sentence is supposed to be less than or equal to 70 words.

In the case of texts with less word count, we pad in extra 0's to the list to make its length equal to 70 words and for larger length texts, the words after 70 are ignored. The procedure is used to convert English, Hindi and Kannada datasets to a suitable vector. For example, if our bag of word model has the following words: [a, is, the, apple, fruit], then the sentence "Apple is a fruit" is converted to [3, 1, 0, 4, 0...0].

4.4.1   Splitting the Dataset

The training process for our neural network needs two types of datasets. One for training and another for validation. The entire dataset is divided into 80–20% where the 80% contributes to the training steps and the remaining 20% for the validation. In most cases, the dataset tends to group positive or negative classified dataset in sequence of rows. Hence, shuffling was done before splitting in order to have the dataset normally distributed with both samples.

4.4.2   Training the Neural Network

The neural network model uses embedding layer as its first hidden layer where it takes in the size of vocabulary which in our case is the size of the bag of words and plus and also the length of each text sample which is 70 words per vector. Vocabulary size defines the number of unique words along with unknowns that are there in the entire dataset.

Following this, we introduce convolutional neural network (CNN) architecture layers which are found to produce promising results for classification problems. 32 filters are used at the first CNN layer and the outputs of these are downsampled by passing them through a maxpooling layer. The outputs of this layer are given as inputs to another CNN layer with 64 filters and are downsampled again.

A rectified linear activation function is used at CNN layers for triggering the neurons based on a particular threshold value. The outputs of this is given to LSTM and are mapped onto a single output layer where the result is normalized by using a sigmoid activation function. The normalized output lies within the range of 0–1. Here, the output specifies the sentiment of the given text vector. If the value is close to 0, then the sentiment is negative and if it is close to 1, then the sentiment of the text is positive. The binary cross-entropy is used as its loss function since our problem involves classifying results into two set of classes. Adam optimizer function is used to keep track of the model accuracy and adapt our model to classify the data samples into two categories.

The same architecture was used for all the three languages but trained with their corresponding dataset. Training was done for about 20 passes with giving in 128–256 data samples at a time. Validation dataset was used to compute the validation accuracy for the unknown samples.

5 Results and Discussion

5.1 Results for English Dataset

The model when trained for English dataset which had about 1600 k samples provided about 95% accuracy for 20 pass through the dataset. Initial model was created with lesser parameters to train and when it was trained for 5 epochs, it provided with 75% accuracy. Later it was redesigned and modified to give about 89% accuracy.

The current model was trained to 95% accuracy with 20 epochs. The graph in Fig. 3 shows how the model learned to improve its understanding about the dataset being positive or negative as it passed through the samples again and again. Then the model was tested against custom inputs as shown in Fig. 4. Here, for the input "this is bad," clearly, it depicts a negative context and due to which the model was able to identify that the sentiment is negative. Here, the numerical besides the results shows in the actual result of the model. The model generates a value of 0.15 which is quite close to 0, and it classifies the text as negative. As in general, we assumed that the values lesser than 0.5 are negative and higher or equal values are of positive sentiment.

The first example "this is a nice product" shows a positive sentiment as the keyword "nice" represents a good context. The model generates an output of 0.99 which is strongly close to 1 and shows that the text is of positive sentiment.

370   S. Shetty et al.

Fig. 3 **a** Graph of accuracy
versus time for the model
trained over English dataset.
b Graph of loss versus time
for the same model

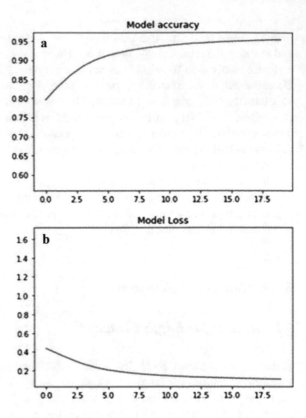

Fig. 4 Results for English
samples

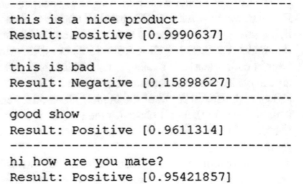

```
----------------------------------------
this is a nice product
Result: Positive [0.9990637]
----------------------------------------
this is bad
Result: Negative [0.15898627]
----------------------------------------
good show
Result: Positive [0.9611314]
----------------------------------------
hi how are you mate?
Result: Positive [0.95421857]
```

5.2 Results for Hindi Dataset

The model was about to achieve 99% accuracy with the Hindi dataset. Figure 5
represents how the model started off with 0.7 loss and as it trained over the data
again and again, it was able to reach a loss close to 0.

Fig. 5 **a** Graph of accuracy versus time for the model trained over Hindi dataset. **b** Graph of loss versus time for the same model

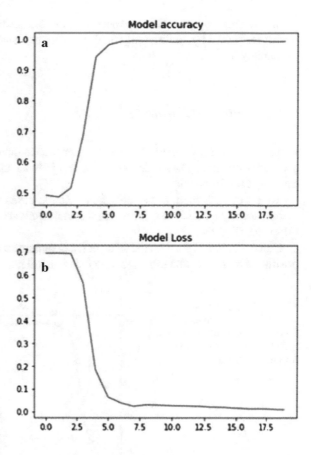

In Fig. 6, actual result represents the output from the model and expected result represents the real output.

Clearly, the expected and actual results are close to each other. The same network which was used for English dataset was used but with new dataset for training. Here,

Fig. 6 Results for Hindi samples

```
---------------------------------------------------------------
श्याम एक हठी बालक है, वह अपनी ज़िद्द के आगे किसी की नहीं सुनता
Actual Result: Negative [0.00028989]
Expected Result: 0
---------------------------------------------------------------
वापस लौटा मलेशियाई विमान, सभी यात्री सुरक्षित
Actual Result: Positive [0.9773548]
Expected Result: 1
---------------------------------------------------------------
वह कड़ी मेहनत करता है
Actual Result: Positive [0.9956928]
Expected Result: 1
---------------------------------------------------------------
वह बुरा बोलता है
Actual Result: Negative [0.00100877]
Expected Result: 0
```

the last two test samples were chosen out of the dataset and the results of them being positive is shown by getting the probability near to 1 and for negative case, we get a probability 0.

5.3 Results for Kannada Dataset

Figure 7 shows how the model trained over the Kannada dataset and improved its accuracy over time. The model reached about 99% accuracy with 6 k dataset just by training it for 20 epochs.

To test out the model, we took some sample cases from the dataset and some were manually typed. The expected and actual result accuracy matches with a small difference (Fig. 8).

Demonstration is done on the value of using semantics to classify a Twitter tweet as positive or negative. The model is built using neural networks consisting of different

Fig. 7 **a** Graph of accuracy versus time for the model trained over Kannada dataset **b** Graph of loss versus time for the same model

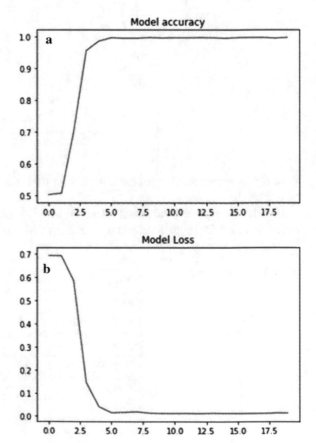

Fig. 8 Results for Kannada
samples

```
ಅವನು ಒಳ್ಳೆಯ ಮನುಷ್ಯ
Actual Result: Negative [0.1837786]
Expected Result: 1
---------------------------------------------
ವ್ಯಾಯಾಮ ಆರೋಗ್ಯಕ್ಕೆ ಒಳ್ಳೆಯದು
Actual Result: Positive [0.78110516]
Expected Result: 1
---------------------------------------------
ಅವನು ಕೆಟ್ಟವನು
Actual Result: Negative [0.21596132]
Expected Result: 0
---------------------------------------------
ವೇಗವಾಗಿ ಚಾಲನೆ ಮಾಡುವುದು ಅಪಾಯಕಾರಿ
Actual Result: Positive [0.7166374]
Expected Result: 0
---------------------------------------------
ಡ್ಯಾಮ್ಸ್! ನಾನು 11:11 ತಪ್ಪಿಸಿಕೊಂಡಿದ್ದೇನೆ.

Actual Result: Negative [0.00055438]
Expected Result: 0
---------------------------------------------
```

layers of CNN and LSTM, where the given input is mapped down to a single output of classification being positive or negative. The results discussed before show how well the English model behaves even though with 95% accuracy when compared to other two languages with 99%. Here, the accuracy mainly specifies how well the model behaves for the unknown validation dataset. Since the English dataset had 1600 k samples, the bag of word model it created consisted of huge words when compared to that of Kannada and Hindi language. So the English dataset trained model can take in a large set of values from the domain and produce a better result. The words appearing in the test run which are not in the bag of words would be replaced with an unknown or a padding symbol. This causes a problem where the entire words in the text become unknown and at that time, the model produces bias result.

To resolve this problem for English language, the NLTK library provides with WordNet which is a graph of how different words are connected. Words with similar meaning are close to each other. By using this and analyzing the text, we can replace these unknown words with the ones that match close. Again, there is a problem where the words never appear on the WordNet graph are used in the sentence. Usually, in Twitter posts, there could be cases where people use words that depict emotion like "Haha" or "huh." These words do not correspond to any means as they just represent an emotion. For these cases, our model produces bad results when the sentences have no other contexts. Solution for this includes creating or adding our custom words to WordNet which depicts these emotions. There are many models that still make use of n-gram feature extraction for classification. Here, they create a list of these n-grams which are words of length n. So a text "hello" can be reduced down to their bi-grams ($n = 2$) as "he, el, ll, lo" and use these to create the vector. This vector can be later used to classify the sentences. But the problem seen here is that the words

Table 2 Comparison of accuracies for various datasets

Dataset	Samples	Accuracy (%)
Hindi dataset	2500	99
Sentiment 140 for English	1600 k	95
Kannada dataset	6300	99

with prefix or suffix which changes the meaning of the word make the model produce ambiguous result. Considering the word "unhappy," the prefix "un" means that the word in actual is a negative meaning but the model needs to see the word as a whole than just its grams. Table 2 indicates accuracies for various samples used.

6 Conclusion and Future Work

We indicated how neural networks are used to classify a Twitter post as positive or negative sentiment for English, Hindi and Kannada languages. With the bag of words model, the network produces good results if the words from this bag are present in the sentence. More research is needed for techniques where parts of speech tagging is used to identify the relevant words and then perform other preprocessing operations and for identifying similar words. Using Google translator to produce the dataset for other languages helped us create more samples and produce a better result. Adding in the CNN and LSTM layers in the neural network helped in increasing the accuracy of the model in a significant way. When the model was evaluated against sarcastic posts where the intended meaning of the sentence is different than the post, here the model produced the wrong result due to the negative contexts being used. Even adding more samples to the dataset did not improve the output of these sentences. For handling such posts, we might need to make use of different networks to classify if the sentence is sarcastic or not and then apply sentimental analysis.

The future work for this involves adding in a word similarity check with WordNet model. Our current model only classifies the posts into negative or positive classes. We can add in more classes like neutral, anger, sad, happy, etc. and classify a text based on these. A post can also belong to multiple categories but not just two classes. Hence, work is needed on this to improve the sentimental analysis process.

References

1. B. Pang, L. Lee, Opinion mining and sentiment analysis. Comput. Linguist. **35**(2), 311–312 (2009)
2. A. Bruns, S. Stieglitz, Quantitative approaches to comparing communication patterns on Twitter. J. Technol. Hum. Serv. **30**(3–4), 160–185 (2012)
3. A. Go, R. Bhayani, L. Huang, Twitter sentiment classification using distant supervision. CS224N project report. Stanford **1**(12) (2009)

4. A. Agarwal, B. Xie, I. Vovsha, O. Rambow, R.J. Passonneau, Sentiment analysis of twitter data, in *Proceedings of the Workshop on Language in Social Media (LSM 2011)* (2011), pp. 30–38
5. L. Barbosa, J. Feng, Robust sentiment detection on twitter from biased and noisy data, in *Coling 2010: Posters* (2010), pp. 36–44
6. M. Speriosu, N. Sudan, S. Upadhyay, J. Baldridge, Twitter polarity classification with label propagation over lexical links and the follower graph, in *Proceedings of the First Workshop on Unsupervised Learning in NLP* (2011), pp. 53–63
7. V. Hangya, G. Berend, R. Farkas, Szte-nlp: Sentiment detection on twitter messages, in *Association for Computational Linguistics* (2013)
8. Github Sentiment-Analysis-of-Hindi-Tweets [Online]. Available https://github.com/Negibabu/Sentiment-Analysis-of-Hindi-Tweets
9. S. Shetty, S. Shetty, S. Hegde, K. Pandit, Transliteration of text input from Kannada to Braille and vice versa, in *2019 IEEE International Conference on Distributed Computing, VLSI, Electrical Circuits and Robotics (DISCOVER)* (IEEE, 2019), pp. 1–4
10. S. Shetty, S. Shetty, S. Hegde, K. Pandit, English transliteration of Kannada words with Anusvara and Visarga, in *Advances in Artificial Intelligence and Data Engineering* (Springer, Singapore), pp. 349–361

An Efficient Algorithm for Fruit Ripeness Detection

Sharath Kumar⊙ **and Ramyashree**⊙

Abstract The assessment and ranking are now carried out on the basis of observations and practice. A major computing concern is the identification of various fruit images. This thesis describes the grading approach applied to the fruit. The identification of good and bad fruits in this analysis illustrates the approaches used by MATLAB. The created approach starts the procedure by taking the image of the fruit using a normal digital camera. The image is then sent to the processing level, where MATLAB is used to isolate, identify, and rate the elements

Keywords Ripeness · Preprocessing · MATLAB · Pixel

1 Introduction

The primary industrial field is agriculture, and it plays a major role in India's economic growth. In India, 70% of the population relies on agriculture, but there is no robotic form of automation in agriculture for Indian industries. Service workers are particularly oriented on the processes of manual labeling and ranking that are used to differentiate between different fruit varieties [2]. As these methods are driven by human activity, they are vulnerable to some kind of error. Because humans are exposed to fatigue and lack of energy, it is necessary to adopt an automated system to remove labor, and the automated system also helps to minimize the time wasted on manual techniques [9].

University researchers are designing several new agricultural automation technologies, posing concerns about the efficiency and effectiveness of current agricultural operations. Fruits are delicate materials, but they can be checked by nondestructive processes. Classification is important for the evaluation of an agricultural

S. Kumar · Ramyashree (✉)
Shri Madhwa Vadiraja Institute of Technology and Management, Bantakal, Udupi, Karnataka, India
e-mail: ramyashree.cs@sode-edu.in

S. Kumar
e-mail: sharathkumar.cs@sode-edu.in

© The Author(s), under exclusive license to Springer Nature Singapore Pte Ltd. 2022   377
P. Shetty D. and S. Shetty (eds.), *Recent Advances in Artificial Intelligence and Data Engineering*, Advances in Intelligent Systems and Computing 1386,
https://doi.org/10.1007/978-981-16-3342-3_30

Fig. 1 Set of fruits

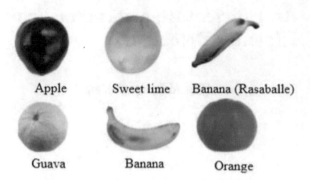

commodity [1]. The most significant physical property is fruit scale, while color resembles a visual property. It is therefore important to distinguish fruits in the assessment of agricultural product, to meet quality requirements and to improve market demand. In preparation, packing, transportation, and marketing activities, it is also useful. The process can be too cumbersome and sometimes error-prone if classification and rating are achieved by way of manual procedures. We may use various algorithms to classify the fruit and discover the quality of the fruit by using MATLAB software as an image processing tool [4]. Finally, after receiving a lot of information, we identify whether the fruit is good or poor (Fig. 1).

2 Literature Survey

Kavdir et al. [1] have built up a framework which grades apples by utilizing shading and size. Fluffy rationale framework is intended for this reason. Reviewing results acquired from fluffy rationale demonstrated 89% general concurrence with the outcomes from the human master, giving great adaptability in mirroring the master's desires and evaluating principles into the outcomes. This utilization of apple evaluating can be completely computerized by estimating the necessary highlights by methods for innovative sensors or machine vision and settling on the reviewing choice utilizing fluffy rationale.

Feng and Qixin [2] proposed a system in which color image preparation-based smart organic product arrangement structure has been investigated in which the shading proportion of the natural product, which was calculated with HSI shading room, was chosen as an arrangement. The organic commodity structure was recognized by an outstanding Bayes classifier, the limits of which were obtained by an inspection panel.

Mustafa et al. [3] developed an automatic grading system that was suggested to classify four fruit and vegetable types, namely apples, bananas, oranges, mangoes, and carrots. Qualities, for example, field, fundamental pivot, little hub, and edge, have been utilized to distinguish the examples. In this article, methods, for example, the assist vector with machining, are utilized to distinguish the examples and to rate the reviewed tests with fluffy rationale.

Chang et al. [4] have built up a dream-based framework for arranging the organic products. It depends on proportions of fluffiness and level of coordinating. Fluffy methodologies were utilized to decide ideal thresholding estimations of natural product's pictures, and fluffy level of coordinating was applied to order the shading and size of organic products.

Nagganaur and Sannanki [5] introduced the arranging and evaluating of organic products utilizing picture handling methods. The framework begins the cycle by catching the organic product's picture. At that point, the picture is communicated to the MATLAB for highlight extraction, arrangement, and evaluating. Both grouping and evaluating are acknowledged by fluffy rationale approach.

The shading and shape attributes of the organic product are definitive for visual examination. An incredible self-ruling natural product arranging framework must have the option to effectively order all boundaries. Natural product structure can precisely be delivered from an advanced picture utilizing customary strategies for picture preparing. There are a few techniques, for example, fuzzy rationale, neural network, in light of color histogram, genetic algorithm, etc. [6].

Wajid et al. [7] gave an overview to address the difficulties in natural products. Farming has a significant function in the financial improvement of our nation. Beneficial development what is more, high return creation of natural products is fundamental and required for the business. Use of picture handling has helped farming to improve yield assessment, infection discovery, natural product arranging, water system, and development evaluating. Picture preparing strategies can be utilized to lessen the time utilization and has made it cost effective. In this paper, they gave an overview to address these difficulties utilizing picture preparing methods.

Vaviya et al. [8], proposed the framework getting the picture of organic product under the test and contrast it and the highlights of normally aged leafy foods aged foods grown from the ground the yield with the likelihood. This method uses the cell phone that runs the Android application and the convolutionary neural organization to identify the natural product that is misleadingly aged.

Pramod and Devalatkar et al. [9] provide the techniques for preservation of fresh food, with the gains made in preserving fresh food items such as fruits and vegetables. The advanced visualization techniques for fruit and vegetable protection are now important parameters. Based on its different features, it involves the implementation of a fixed rule for individual items as a safe keeping algorithm. This research paper provides an overview of the tomato age factor focused on one of the color-based visual features of the tomato.

Iqbal et al. [10] explained about the classification of fruit diseases using ANN algorithm. Efficient growth and improved yield are so critical and important. As the agricultural industry is in high demand, it is important to use successful smart farming techniques in order to minimize more time and substitute manual sorting and does not offer significant results. The neural network are built for the classification of diseases and pattern matching and the network (ANN) algorithm are used. This methodology decreases human effort and can provide reliable results of 90%. The work to be done in the future is to suggest proper work treatments accordingly until the condition is compromised.

3 Methodology

The architectural depiction is concerned with setting up a basic layout of a system. System design mainly deals with recognizing major components of the system being developed and how the interaction takes place among these components. The entire process includes various steps. The flowchart of this process is given as follows (Fig. 2).

Step 1: Image acquisition
Step 2: Preprocessing
Step 3: Pull out the characteristics of fruit selected
Step 4: Feature training
Step 5: Feature matching
Step 6: Displaying the results.

Image acquisition

Image acquisition deals with getting the image of the fruit of the interest, and it will be the first step in the entire process [3]. The success rate of the entire process purely depends upon the quality of the image being captured, and in order to achieve more accuracy, digital camera with high resolution is preferred. The captured image will be stuffed into the MATLAB with the help of built-in function imread, and with appropriate parameter, entire image is represented in the form of two-dimensional matrix which consists of rows and columns [13].

Preprocessing

As mentioned before, the quality of image is of prime concern. The image being captured will be preprocessed, so as to eliminate the noise components present or elimination of unwanted background or to enhance the resolution of image that is

Fig. 2 System design

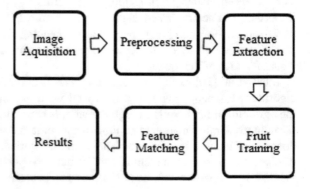

Fig. 3 Preprocessing

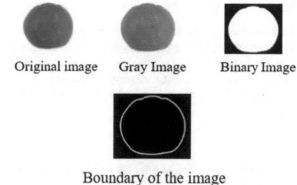

Original image   Gray Image   Binary Image

Boundary of the image

being captured by using camera [5]. Preprocessing mechanisms involves following steps

- RGB image: Entire image is represented by using three color components: red, blue, and green. In this format, the image is represented in the form of matrix, and the values within the matrix represent each pixel intensity value.
- Background subtraction: Area of interest is separated by removing the background objects with help of background diminution process. Background diminution is a process extricating of foreground objects [11]. One more advantage of removing unwanted background will reduce data processing overhead
- Gray image: The gray-level images will have only one color component, which has an intensity of gray color in abundant ranges in between.
- Binary image: In binary image, each pixel is represented as bit which will be in the form of 0 or 1 (Fig. 3).

Feature extraction:

Assembling of input data entities into group of characteristics is called as feature extraction. The process of feature extraction begins with the extricated boundary of the sample [6]. The function employed to discover the features is "regionprops." The key characteristics extracted are major axis, area, and minor axis. In this yellow, green, and red colors are used for categorization. As we know that, fruit's skin color will be different. Hence, these colors are useful for arranging the fruits. The green and red constituents are computed by summing up the pixel values which belong to the green and red colors, and yellow constituent is computed by first transforming the original image into CMY [12].

Feature training:

Feature training is a database which includes plenty of trained characteristic values for bad and good fruit [8]. As the number of trained characteristics increases, more accuracy can be obtained. These trained values are compared with that of the original fruit image to get the results.

Feature matching:

Feature matching mechanisms basically consist of determining the characteristics in image which can be matched with characteristics in supplementary images from which a transformation prototype can be determined [7]. Feature matching is a key task in the area of image processing. With extracted characteristics, individual values are correlated with other and based on that decision can be taken about good or bad fruit.

3.1 Detailed Methodology of Fruit Ripeness Detection

The mechanism comprises of the following steps:

1. From the menu provided choose the fruit name and the necessary image from the group of images exhibited.
2. Rescale the image selected and project it.
3. Ascertain the different color components red, blue, and green in the selected image.
4. Represent the selected image in the form of matrix.
5. From the matrix determine the greatest pixel value.
6. Differentiate all the color components (r, g, b) with greatest value selected.
7. Based on the color component, color of the fruit and ripeness parameter is decided; if the red constituent is more in the image being tested, then fruit may be red, orange, or yellow, and we can conclude saying that fruit is ripe.
8. If blue constituent is more in the image being tested, then we can conclude saying that fruit is ripe, and it is blue in color.
9. Fruit will be of color green if green constituent is more and fruit is not ripe.
10. A dialog box that shows the output will be displayed (Fig. 4).

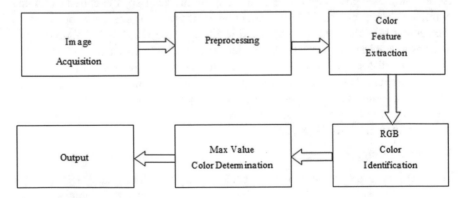

Fig. 4 Architecture of fruit ripeness detection

3.2 Detailed Methodology of Fruit Defect Detection

The mechanism comprises of the following steps:

11. From the menu provided, choose the fruit name and the necessary image from the group of images exhibited.
12. In window provided exhibit the input image.
13. Select the Analyze button from the menu provided.
14. Metamorphose the input image selected into HSV format and extricate all three components separately (H, S, and V).
15. Masking of the chosen image is done by comparing each color bands to the corresponding threshold value.
16. Once all three color components are masked, all the three components will be combined into a single image, and it will be displayed as a thresholded image
17. Thresholded image will be further processed and that will be represented in other forms such as grayscale and binary image and that will be displayed on the window provided.
18. The deficiency on the fruit is investigated and decision is taken; if the ripens quotient is more than 25, then rejected stamp will be displayed, otherwise approved stamp will be displayed (Fig. 5).

Thresholding

Thresholding includes various steps; out of these steps, the dominant one is background subtraction. Background subtraction is a mechanism to bring out foreground objects in a specific locus of an image [10]. A foreground object is interpreted as an object of scrutiny which assists in decreasing the quantity of data to be handled.

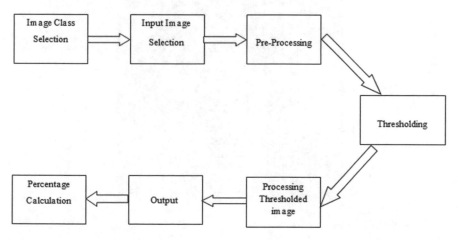

Fig. 5 Architecture of fruit defect detection

4 Results and Discussion

In main menu, there will be three functions: check ripeness, defect detection, and exit. Check ripeness function will check fruit is ripe or unripe; defect detection will analyze fruit and provide defected proportion in terms of percentage; and exit function is used to come out of the application (Fig. 6).

Once check ripeness function is selected, type of fruit selected from dropdown menu and fruit image must be uploaded for further processing (Fig. 7).

Once defect detection function is selected, select image from the drive and analyze the image. Based on image quality, approved or rejected stamp will be applied (Figs. 8 and 9).

Results on dataset:

As there is no relevant dataset available, we have gathered 76 images from few of the standard dataset and other Internet resources and analysis is done

Truly detected (TD)—The defective fruit which was concluded as defective one or healthy fruit which was declared as healthy fruit.

Falsely detected (FD)—Wrong decision taken by the algorithm either healthy fruit was detected as defective or defective fruit was concluded as health fruit.

Detection rate (DR)—TD/no. of images.

False detection rate (FDR)—FD/no. of images (Table 1).

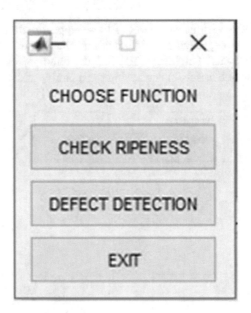

Fig. 6 Main menu

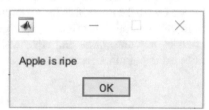

Fig. 7 Selecting an input image from the dataset and displaying the corresponding output

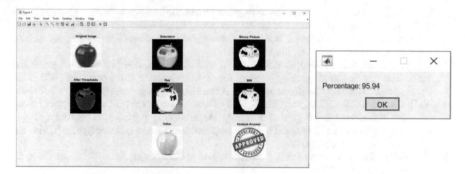

Fig. 8 Final output showing fruit is approved

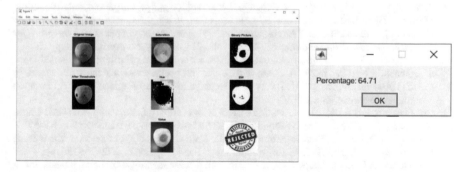

Fig. 9 Final output showing fruit is rejected

Table 1 Efficiency of proposed system

No. of images	DR	TD	FD	FDR
76	90.78	69	7	9.21

5 Conclusion and Future Work

This work confers mechanisms for arranging and grading of fruits based on quality. This process begins with recording the fruit's image using normal digital camera; the shape, size, and color of the fruit are used to categorize and estimate fruit's edibility. This automated approach can solve the problems of manual operating procedure, and we can make use of this in Agriculture Produce Corporation and Marketing, etc.

In this method, it is difficult to judge the quality of fruits that will not alter color whether they are unripe and ripe, while classifying one extra feature size can be included to get more accurate results.

References

1. I. Kavdir, D.E. Guyer, Apple grading using fuzzy logic. Turk. J. Agric. **2003**, 375–382 (2003)
2. G. Feng, C. Qixin, Study on color image processing based intelligent fruit sorting system, in *Proceedings of the 5 World Congress on Intelligent Control and Automation, Hangzhou, P.R. China*, June 15–19 (2004), pp. 4802–4805
3. N.B.A. Mustafa, S.K. Ahmed, Z. Ali, W.B. Yit, A.A.Z. Abidin, Z.A.M. Sharrif, Agricultural produce sorting and grading using support vector machines and fuzzy logic, in *2009 IEEE International Conference on Signal and Image Processing Applications* (2009), pp. 391–396
4. W.-H. Chang, S.C.S.-C. Lin, P.-Y. Huan, Y.-Y. Chen, Vision based fruit sorting system using measures of fuzziness and degree of matching. Department of Agricultural Machinery Engineering & Department of Electrical Engineering, National Taiwan University, Taipei, Taiwan, R.O.C. (1994), pp. 2601–2604
5. Naganur and Sannakki, Fruits sorting and grading using Fuzzy Logic. Int. J. Adv. Res. Comput. Eng. Technol. (IJARCET) 1(6) (2012)
6. Sagare, S.N.K., Fruits sorting and grading based on color and size. Int. J. Emerg. Technol. Comput. Appl. Sci. (IJETCAS), 12–333
7. A. Wajid, N.K. Singh, P. Junjun, M.A. Mughal, Recognition of ripe, unripe and scaled condition of orange citrus based on decision tree classification, in *2018 International Conference on Computing, Mathematics and Engineering Technologies, Invent, Innovate and Integrate for Socioeconomic Development, iCoMET 2018*, vol. 2018, January, no. April (2018), pp. 1–4
8. H. Vaviya, A. Yadav, V. Vishwakarma, N. Shah, Identification of artificially ripened fruits using machine learning, in *2nd International Conference on Advances in Science & Technology (ICAST)* 2019 on 8th, 9th April 2019 by K. J. Somaiya Institute of Engineering & Information Technology (Mumbai, India, 2019)
9. P.G. Devalatkar, S.R. Koli, Identification of age factor of fruit (tomato) using MATLAB-image processing. Int. J. Recent Trends Eng. Res. (IJRTER) 02(07) July-2016. ISSN: 2455-1457
10. S. Iqbal, A. Gopal, P.E. Sankaranarayanan, A.B. Nair, Classification of selected citrus fruits based on color using machine vision system. Int. J. Food Prop. **19**(2), 272–288 (2016)
11. S.R. Dubey, A.S. Jalal, Adapted approach for fruit disease identification using images. Int. J. Comput. Vis. Image Process. (IJCVIP) **2**(3), 44–58 (2012)
12. J.D. Pujari, R. Yakkundimath, A.S. Byadgi, Grading and classification of anthracnose fungal disease of fruits based on statistic texture features. Int. J. Adv. Sci. Technol. 52 (2013)
13. S. Maheswaran, Identification of artificially ripened fruits using smart phones, in *International Conference on Intelligent Computing and Control (I2C2)* (IEEE, 2017)

Kannada Document Classification Using Unicode Term Encoding Over Vector Space

R. Kasturi Rangan and B. S. Harish

Abstract Today, there is a great demand for extracting useful information and ability to take actionable insights from heaps of raw textual data. The processing of regional language texts, notably with low resources is challenging. Especially for Indian Regional Languages (IRL), language text processing is in huge requirement with regard to the applications of natural language processing (NLP) and text analytics. As a part of regional language processing and text analytics text documents classification for IRL are yet to be explored. Kannada is one of the official Indian regional languages. In this paper, the new benchmark Kannada document's dataset is created and analyzed using machine learning algorithms. This paper proposes an explicit Unicode term encoding based Kannada document classification, using the vector space model. Both term frequency (TF) and term frequency-inverse document frequency (TF-IDF), statistical measures are used for classification of Kannada documents using K-NN and SVM classifiers. SVM (linear) classifier performs better than the K-NN classifier in classifying the Kannada documents with 98.67% mean accuracy over K-Fold experiments.

Keywords Language processing · Text classification · Unicode encoding · Vector machine

1 Introduction

We are in the age of information, where the proliferations of data in the form of online documents, e-books, digital libraries, journal articles, etc., on the web, on corporate intranets, on news wire and elsewhere is overwhelming. However, ability to absorb

R. Kasturi Rangan (✉)
Department of Information Science & Engineering, Vidyavardhaka College of Engineering, Mysuru, India
e-mail: rkrangan@vvce.ac.in

B. S. Harish
Department of Information Science & Engineering, JSS Science and Technology University, Mysuru, India
e-mail: bsharish@jssstuniv.in

and process this large amount of data remains static. Text mining tries to solve this information explosion using various techniques like machine learning, natural language processing, information retrieval, and knowledge management. The textual information can be in any natural language. UNICODE for various natural language characters made easier in analyzing and handling multi-lingual data. India is a multi-lingual nation, where English is not widely understood. People prefer their regional language over English in many contexts. Hence computability of Indian regional languages overpasses many societal gaps in education, economy, and healthcare sectors.

Automatic text categorization is one among many tasks in natural language text analysis. It is the process of approximating an unknown category assignment function $F : D \times C \rightarrow \{0, 1\}$, where D represents the set of documents and C is the set of predefined categories. The value of $F(D, C)$ is 1 if the document D belongs to the category C and 0 otherwise [14]. Text analysis can be made at documents level, paragraphs level or at sentences level. The automatic text categorization is enforced in applications where there is a need of categorizing the dynamic flow of text information.

Kannada is one of the 22 official languages of Indian constitution. According to 2011 census, there are around 56 million Kannada speakers in India. In Kannada Document Classification (KDC), Kannada text documents are categorized into different domains based on the similarity. As there is lack of standard Kannada document's dataset, new dataset is created and presented in this paper. Section 2 of this paper consists of review of related work. In Sect. 3, preprocessing of data, explicit Unicode term encoding, feature extraction and selection and vector space model is explained. Dataset description, K-fold experimentation with K-NN and SVM classifiers are described with tabulation of results and graphs in Sect. 4. Concluding remarks are presented in Sect. 5.

2   Related Work

The abundant research works had carried out on text classification task in English and other non-Indian regional languages. Spam detection, documents categorization, named entity recognition, tagging contents or products in platforms like E-commerce, blogs, news agencies and so on are few applications of text categorization. Indian regional languages are less resourced, morphologically rich and agglutinative in nature [6]. Hence, it is quite difficult to develop intelligent systems for real time applications when compared to English natural language.

For Tamil which is one of the Indian regional languages, Rajan et al. in 2009 [11] presented automatic Tamil document classification system using most established and well-known vector space model (VSM) and also experimented using artificial neural network. The authors in Rajan et al. [11] created their own Tamil documents dataset. They experimented on both CIIL (Central Institute of Indian Languages) developed Tamil corpus and their own dataset. Rajan et al. [11] achieved around

90.33% and 93.33% accuracy, respectively, for VSM and ANN models. Similarly in VSM, Dhar et al. [1, 2, 5, 8] extensively experimented on Bangla text document categorization. The author in Dhar et al. [2] used TF-IDF features and distance metrics like Euclidean and Cosine similarity to categorize the dataset of 1000 Bangla documents. The results achieved are 95.20% and 95.80%, respectively. In [1], additional to term frequency (TF) and inverse document frequency (IDF) they added inverse class frequency (ICF) values and obtained good results for the newly created dataset of 4000 Bangla documents. In their experimentation, they incorporated K-fold (K = 5) cross validation. The proposed approach achieved a 98.87% of mean accuracy using Naïve Bayes Multinomial classifier. Further [5] presents the categorization task using multilayer perceptron (MLP) classifier with same TF-IDF-ICF frequency selection technique. Accuracy of 98.03% is achieved. Authors in Dhar et al. [8] use rule based classification algorithm called PART classifier for classifying 8000 Bangla documents into 8 categories. Term association and term aggregation are used in feature extraction process. Further K-fold cross validation is applied (K = 5) and obtained a mean accuracy of 96.83%. The experiments concluded that the PART classifier performs better when compared to other rule based classifiers (like RIPPER, Decision table, OneR) and Bayesian classifiers (like Naïve Bayes Multinomial, Naïve Bayes). Puri and Singh [13] experimented on Hindi documents using support vector machines (SVM). But the documents taken for experimentation are only four. Other than statistical approach, based on availability of Wordnet lexical database, few experiments are conducted on text classification. Mohanty et al. [3] presented Sanskrit language text classification based on semantic WordNets. This method constructs lexical chain of significant words from source document using semantic network of WordNet.

Text classification can be performed at document level, paragraph level or at sentence level. Jayashree et al. [12] focused on sentence level text classification in Kannada language. Both CIIL (Central Institute of Indian Languages) developed corpus and custom built dataset are used for experimentation. They found SVM classifier performed better than Naïve Bayes, and results are concluded by opining that the text classification can be extended to paragraph or document level. Similarly, Tummalapalli and Mamidi [10] presented sentence level classification based on sequence of syllables for English, Hindi, and Telugu languages. Further, based on sentence level classification, tasks like sentiment classification and question classification is experimented on 5 datasets. Following to this work, Tummalapalli et al. [4] applied neural network methods for sentence classification. Characters or syllables play a major role in understanding the morphologically rich languages. Hence in this work, the sentences are considered as input in both representations, i.e., sequence of syllables (or Characters N-gram) and sequence of words representations. Tummalapalli et al. [4] proposed MultiInput-CNN model that performs better with the representation of sequence of syllables over word vectors, in sentiment and questions classification tasks for Hindi and Telugu language datasets.

The clustering techniques are also used for document classification of Indian Regional Languages (IRL) by researchers. In 2017, Narhari et al. [7] worked on Marathi documents categorization using modified label induction grouping (LINGO) clustering algorithm. They created their own dataset of 200 Marathi documents. They claimed that their results (Precision = 88.89%) were better than the former LINGO algorithmic experimental results. As there is a lot more research experiments yet to be carried out on Indian regional languages in various domains, this work motivates to do experiments more on Kannada document level text classification task using machine learning algorithms.

3 Proposed Method

As there is a lack of publicly available standard Kannada document's dataset, Kannada document classification is performed on newly created Kannada document's dataset. From the literature survey, document classification in Kannada language found undiscovered and researchers suggested experimentations at paragraph and document level classifications [12]. In newly created, Kannada document dataset there are total 300 documents, spread across five categories. Details of the dataset are discussed in Sect. 4. The Kannada document classification is the task of assigning the category (c_i) for the given document (d_j), where $C = \{c_1, c_2, c_3, ...c_m\}$ are set of predefined (m) categories and $D = \{d_1, d_2, d_3, ...d_n\}$ are set of (n) documents. The block diagram of proposed methodology is shown in Fig. 1. The block diagram depicts the steps followed for the Kannada Document Classification (KDC).

3.1 Preprocessing

At first, the raw Kannada text documents are preprocessed to remove unwanted texts and clean the data as per the computational requirement. The removal of unwanted information in the texts like punctuations and numbers from the dataset are carried out based on the regular expressions. Removal of these, makes data clean and avoids

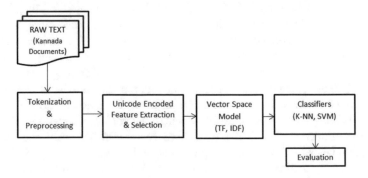

Fig. 1 The proposed block diagram of Kannada Document Classification (KDC) model

the ambiguity during the categorization. Later, tokenization is applied to obtain the tokens. Tokens are the basic unit of the documents. Each term of the document forms a token because tokenization is done at Word/Term level. It can also be achieved at sentence, character, and at N-gram levels [15]. Further, the documents are analyzed by the terms/tokens identified during tokenization. Tokenization requires delimiters to identify the tokens in given document. Here, whitespace is taken as delimiter and identified each Kannada word separated by whitespace as a token in the document. Natural language tool kit (NLTK) library (RegexpTokenizer) is used for the tokenization [20].

3.2 Unicode Encoded Feature Extraction

3.2.1 Unicode Encoding

Unicode (Universal coded character set) is designed to overcome the incompatibility of ASCII encoding, for the non-English language characters. Encoding is the process of revamping the Unicode characters into the sequence of bytes. In contrast, decoding is the process of reversing a sequence of bytes into a set of Unicode characters. In Unicode encoding, unique integer called code-point is used to encode every character uniquely. Unicode is a standard encoding system and Unicode string consists of zero or series of code-points. Code-points are written in the form \uXXXX, where XXXX is the four-digit hexadecimal value and \u indicates that the value is Code-point. Hence, each character is assigned with a standard value called code-point in a string.

E.g.: Normal String (Kannada language): " ಕನ್ನಡ".

In programming, Unicode strings can be handled like normal strings. However, if Unicode strings are to be part of file operations like storing and retrieving or to display in terminal, it has to be encoded as a stream of bytes [19]. Hence to perform any file accessing operations (like Count-Vectorizing, Retrieving, etc.) on the raw text files containing natural languages should be encoded. Unicode transformation format (UTF) is a way to encode the code-points. Based on the size of the bits used to store the code-points, there are three variants of UTF and they are UTF-8 (8 bits for a Code-point), UTF-16 (16 bits for a Code-point), and UTF-32 (32 bits for a Code-point).

The machine transliteration to English language or Romanization can be achieved for the Kannada terms. Later, default ASCII encoding for these Romanized Kannada terms makes language processing tasks easy. But the Romanization process itself is computationally tedious, it requires both languages' huge corpus and it has to maintain phonetics of source languages. Adding to this issue, if the normal text processing operations are performed directly on Kannada text then empirically we found that some agglutinative characters are faded in terms. This leads to the loss of valuable information of the features. Hence, we propose Unicode encoded number for each Kannada term which is not tedious, keeps term intact and not required additional language corpus. Encoded number can be decoded back to original term

Table 1 An example for Unicode string encoding and unique number generation

Characters	ಕ	ನ	ೕ	ನ	ಡ
Unicode standard values	\u0c95	\u0ca8	\u0ccd	\u0ca8	\u0ca1
Encoding (UTF-16)	b'\xff\xfe\×95\x0c\xa8\x0c\xcd\x0c\xa8\x0c\xa1\x0c'				
Decimal representation	390851694318730235535 8318335				

later with same encoded byte address order either Big-endian or Little-endian byte order.

In experimentation, Kannada terms are Unicode encoded (UTF-16). Further encoded hexadecimal values are converted to decimal numbers to retain the Kannada term intact during various operations. These numbers are unique and represent unique terms in the document. From these decimal numbers, the original term can be retrieved back by decoding in same byte order. Table 1 shows how unique decimal number is generated for a Kannada term. Steps to attain unique number generation for terms are shown in Algorithm 1.

Algorithm 1: Unicode encoding and numerical representation of terms
Input: Kannada Text Documents $D = \{d_1, d_2, d_3, ...d_n\}$.

Data: k = total number of tokens in a document, x = total number of characters in a term, n = total number of documents, h = encoded sequence of hexadecimal code points of a term.

Output: Unicode encoded numerical representations of Kannada terms in documents.
    STEP1: **for** $i \leftarrow$ to n by n do (each document in total documents).
    STEP2: Tokenization and Preprocessing.
    STEP3: **for** $j \leftarrow$ to k by k do (each term/token in a document).
    STEP4: **for** $l \leftarrow$ to x by x do (each character).
    STEP5: $(U + XXXX)_{i,j}^{l}$ Unicode character encoding based on system byteorder (UTF-16).
    STEP6: **END** for each character.
    STEP7: $z = 2 * len(h)$.
    STEP8: $DecimalValue = \sum_{s=0}^{z} (h_s * 16^s)$.
    STEP9: **END** for each term.
    STEP10: Every term's numerical value is appended in a document with a space.
    STEP11: **END** for each documents.

3.2.2   Feature Extraction

In text documents, the terms or words of the document are considered as features. These features represent the respective documents. Each term can be considered but it will lead to high-computation time, excess memory and more over it may create more ambiguity for machine to make decisions. All documents will be consisting of discriminative terms or features and even it could be a combination of features which will be discriminative factor for a document. Hence judicious selection of terms is preferred instead of all terms [16]. If given document is (d_j), then $T = \{t_1, t_2, t_3, ...t_k\}$ where (T) represents the (k) unique terms representing the document (d_j) so $T \in d_j$.

After the feature extraction, stop words are removed from the documents. Removing the stop words reduces the feature size and enhances the efficiency of the learning algorithm. Stop words are removed in two ways either to have a dictionary of the stop words or remove based on frequency of terms across the documents. In KDC, stop words are removed based on the frequency of terms across the documents. The lower term frequency threshold is kept at 2 and higher term frequency threshold is kept at 1150, where these thresholds are considered based on empirical evaluations.

3.2.3 Vector Space Model

Before applying any machine learning classifier, the digital text should be represented by its features in a more compact and computationally appropriate form. From the survey, vector space model (VSM) is most established and well-known method for document-term weighting. In VSM, sequential order information is not maintained. Each term or feature is weighted by its count across the documents is called as term frequency (TF). Another statistical measure TF-IDF (Term Frequency-Inverse Document Frequency) is also used for term weighting [17]. TF-IDF weight represents the relevance of the term to a document and specifically IDF (Inverse Document Frequency) is a good index of a term in classification [18]. In the document (d_j), set of terms $T \in d_j$ and $T = \{t_1, t_2, t_3, ...t_k\}$ contains (k) unique terms. These unique terms represent the respective text document. Each (t_i) has weight (wt_i) and calculated using $\text{TF}(d_j, t_i)$ or $\text{TF-IDF}(d_j, t_i)$ statistical measure. TF-IDF consists of $\text{TF}(t_i)$ and $\text{IDF}(t_i)$.

$$\text{IDF}(t_i) = \log\left(\frac{N}{d_f(t_i)}\right) \tag{1}$$

$$wt(t_i, d_j) = \text{TF} * \text{IDF} \tag{2}$$

In Eq. (1), N is the total number of documents and $d_f(t_i)$ is the frequency of documents containing term (t_i). In Eq. (2), $wt(t_i, d_j)$ represents the TF-IDF weight for each term. Later document (d_j) can be represented as a k-dimensional vector as shown in Eq. (3).

$$\vec{d_j} = (wt_1, wt_2, ...wt_k) \tag{3}$$

These k-dimensional vectors of all documents form the document-term matrix. Here, rows are documents and terms are represented in columns. Weights are the values assigned to the terms of the documents in the matrix. This document-term matrix is used further by machine learning algorithms for the classification task.

3.3   Classification

3.3.1   K-Nearest Neighbor (K-NN) Classifier

K-Nearest Neighbor (K-NN) classifier is a non-parametric method. This algorithm is an instance-based learning algorithm or lazy learning algorithm. It searches entire dataset for the 'K' most similar instances (documents) for the prediction of unseen data instance. Later among 'K' most similar instances, majority instance's label will be considered for the prediction. In Kannada Document Classification (KDC), the similarity between the instances is measured by one of the most widely used statistical measures called Euclidean distance. Equation (4) gives the Euclidean distance measure between instances.

$$\left| \vec{X} - \vec{Y} \right| = \sqrt{\sum_{i=1}^{k} (X_i - Y_i)^2} \tag{4}$$

In Eq. (4), $\vec{X}$ and $\vec{Y}$ are the document vectors and (k) represents the unique terms in the respective document vectors.

3.3.2   Support Vector Machine (SVM) Classifier

Support vector machine (SVM) is an efficient supervised algorithm for high-dimensional space classification task [9]. It is also used for regression tasks. This algorithm aims to find a hyper plane in an N-dimensional space which distinguishes the data points into its respective categories. Further, this hyper plane must have the maximum margin, i.e., maximum distance between the hyper plane and its closest data points of both classes. The data points nearer to the hyper plane are called as support vectors and they influence the orientation and position of the hyper plane. Maximum margin of hyper plane depends on these support vectors.

In linear SVM, distinguishing samples to their classes is done based on the output of linear function (Eq. 5). This linear function draws the plane between the classes. The optimization of positioning this plane in the sample space for better classification is the aim of the SVM classifier. As formerly discussed, margins helps in positioning this plane as far as possible from support vectors of classes. If there are two class and labeled as -1 and $+1$, then output of the linear function greater than '1' ($y \geq 1$) will be identified with one class and if output is '-1' or lower ($y \leq -1$), then it belongs to another class. In Eqs. (6) and (7), this condition is represented in terms of vectors. Hence $[-1, 1]$ pair acts like a margin between classes.

$$y = f\left(\overrightarrow{d_j}\right) = w_0 + \sum_{i=1}^{k} w_i t_i \tag{5}$$

From Eq. (5), (w_i) is the hyper plane parameter which is adjusted over the experiments in order to find the maximum margin between the support vectors. (w_0) is the bias parameter.

$$w^T x + b \geq +1 \text{ for class} + 1 \tag{6}$$

$$w^T x + b \leq -1 \text{ for class} - 1 \tag{7}$$

In Eqs. (6) and (7) w^T is the transpose of weight vector, x is the input feature vector of samples and b represent the bias. Equations (6) and (7) can be combined as $y(w^T x + b) \geq 1$ and hence the hyper planar equation becomes as shown in Eq. (8). Further to maintain maximum margin and less misclassification, i.e., for better generalization, weights are balanced with minimum error as shown in Eq. (9). From Eq. (9), C is the regularization parameter and ζ is the slack variable which stores the deviation of samples from margin.

$$y(w^T x + b) \geq 0 \tag{8}$$

$$\min \frac{\|w\|^2}{2} + C \sum_{n} \zeta^n \text{ with respect to } y(w^T x + b) \geq 1, \forall n \tag{9}$$

For the new test document (d_j) with (k) unique terms, if $f\left(\overrightarrow{d_j}\right) > 0$ then it belongs to a class else it belongs to other class in two class problem. Further, it will be same logic implied for the multi-class classifications also. If samples are not able to classify linearly then various Kernel tricks (e.g., polynomial, radial basis function, sigmoid) can be used, which intern transform the data into new space and eases the classification task.

Table 2 Details of Kannada documents dataset

Categories	Total number of documents
Space	40
Politics	95
Crime	55
Sports	60
Economics	50
Total	300

4 Experimental Results and Discussion

4.1 Kannada Documents Dataset

The Indian regional languages are less resourced [5]. Hence, there is a lack of publicly available standard datasets (in particularly for Kannada) for the various language processing tasks. For the Kannada document classification task, we have created our own Kannada documents dataset. This dataset consists of 300 Kannada documents; each document is comprised of two or three news articles. There are 5 categories like sports, politics, economics, space, and crime. Table 2 presents the details of the available documents at each category.

The articles in the documents are collected from the online publicly available news. Each document consists of more than 500 words and labeled with respect to its category. The source links are (a) https://www.kannadaprabha.com, (b) https://kannada.oneindia.com, and (c) https://kannada.asianetnews.com/india-news.

4.2 K-Fold Cross Validation

The cross validation techniques assess how the results of statistical analysis will generalize to an independent dataset. Cross validation address the problem of selection bias for training and testing division of samples and gives insight of model's ability of prediction for unknown data. In K-fold cross validation, given dataset samples are randomly divided into 'K' equal divisions. Among these divisions, $K - 1$ divisions will be taken for training and remaining will be used for testing the model. This process continues for 'K' number of experiments. In each experiment among all divisions one untested division will be part of testing. Finally, it generates result of each experiment and average accuracy of all experiments.

In KDC (Kannada Document Classification), the dataset is of multi-class labels hence the Stratified K-Fold cross validation is performed. In stratified K-Fold, the samples are divided into 'K' divisions unlike the normal K-Fold each division will have minimum representative samples from all labels. There will be 'K'

number of experimental results from which average accuracy is calculated. In this experimentation, we have considered 3, 5, and 10 values for 'K'.

4.3 Experimentation

The KDC experiments are carried out on 300 Kannada text documents. This dataset is comprised of 5 different categories and documents are unevenly distributed among these categories. The preprocessing, tokenization and Unicode term encoding are performed as formerly mentioned. Hence, each unique term is represented with unique number. Each Unicode encoded term is given weights with respect to the term frequency (TF) and term frequency-inverse document frequency (TF-IDF) statistical measures. This forms the document-term matrix with term weights. Experiments are carried out on both weighting measures. Later encoded terms with the highest and the lowest frequencies (stop words) over the documents are removed. Further, stratified K-Fold train-test division is applied, where we consider 3, 5, and 10 folds for experimentation. For $K = 3$, among 300 documents 3 stratified splits of 100 documents are created and each split of 100 documents undergo testing in any of the 3 experiments. Thus, at each experiment, the train and test division ratio will be 70:30, where 200 documents for training and 100 for testing. Similarly for $K = 5$ it is 80:20 train-test division ratio and for $K = 10$ it is 90:10 train-test division ratio. K-NN and SVM classifiers are used for classification task. In case of K-NN classifier, the nearest neighbors 3 and 5 are considered for classifying the Kannada test document. In SVM, the linear kernel is used to classify the Kannada test documents. The K-Fold mean accuracies obtained for KDC experiments for both weighting measures are presented in Figs. 2 and 3.

Evaluation of the classification results of the classifiers is calculated by the accuracy metric. From Figs. 2 and 3, classification using SVM classifier does better in all the K-Fold experimentation. It attained mean accuracy of 98.67% for 3-Fold experiment ($K = 3$) with TF-IDF as term weighting. It comparatively performs

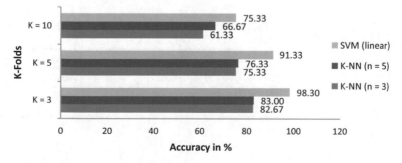

Fig. 2 K-Fold mean accuracies for TF based classification

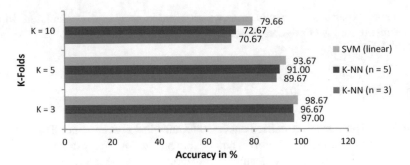

Fig. 3 *K*-Fold mean accuracies for TF-IDF based classification

better in TF-IDF term weighting measure and attains 93.67% and 79.66% accuracies in 5-Fold and 10-Fold experimentations, respectively. Table 3 presents the SVM classifier's performance in *K*-Fold experiments.

Table 4 shows the performances of K-NN and SVM classifiers at the 3-Fold cross validation experiments. At 3-Fold experimentation, classifiers do better when compared to all other K-Fold experimentations. SVM attains 98.67%, K-NN ($n =$ 3) attains 97.00% and K-NN ($n = 5$) attains 96.67% for TF-IDF Unicode encoded term weighting measure.

The combination of Unicode Encoded term's TF-IDF weighting measure and SVM classifier, in 3-Fold cross validation experiments performs better in KDC (Kannada Document Classification) for the newly presented dataset.

Table 3 Performance of SVM classifier

K-Fold cross validation	TF (accuracy) (%)	TF-IDF (accuracy) (%)
$K = 3$	98.30	**98.67**
$K = 5$	91.33	93.67
$K = 10$	75.33	79.66

Table 4 Performance of classifiers in 3-Fold cross validation

Machine learning classifiers	TF (accuracy) (%)	TF-IDF (accuracy) (%)
K-NN ($K = 3$)	82.67	97.00
K-NN ($K = 5$)	83.00	96.67
SVM (linear Kernel)	98.30	**98.67**

5 Conclusion and Future Work

Kannada is one of the Indian regional languages. As it is a low-resource language, it lacks in publicly available standard corpus for document text categorization. Hence, we have created a benchmark Kannada documents dataset consisting of 300 documents collected from online Kannada news articles. The document texts are in Kannada scripts. During experimentation, the Kannada terms are represented in Unicode encoded number because it aids in efficient, unambiguous and easier text processing task. These terms or features are represented in vector space model after the removal of term frequency based stop words. The conventional K-NN and SVM machine learning classifiers are used for document classification in K-Fold experiments. The SVM (linear) performs better when compared to K-NN classifier and gives 98.67% mean accuracy for the 3-Fold experiments, where term weighting is based on TF-IDF measure.

Here, the Kannada document classification is accomplished using supervised knowledge. Hence, still more research experiments are yet to be carried out under unsupervised condition. In using the similarity metrics, other than the Euclidean distance various measures like cosine similarity measures can be explored. In future, the corpus size should be increased to build complex and standard dataset. Analysis to be done for cohesive, complex and bigger datasets using better kernels, which handle the classification tasks efficiently. The various machine learning and neural classifiers can be explored using N-gram based features for large Kannada documents dataset. Deep learning networks handle complex and huge data efficiently. They concern over semantic issues, required for language understanding and are yet to be experimented on large Kannada dataset. In recent, pre-trained networks are developed and are used for arduous language tasks, which are well-trained with large corpus. These pre-trained models are to be explored on Kannada language corpuses for various language tasks.

Acknowledgements This work is supported by Vision Group on Science and Technology (VGST), Department of IT, BT and Science and Technology, Government of Karnataka, India [File No.: VGST/2019-20/GRD No.:850/397].

References

1. A. Dhar, N.S. Dash, K. Roy, Classification of bangla text documents based on inverse class frequency, *in 2018 3rd International Conference on Internet of Things: Smart Innovation and Usages (IoT-SIU)* (IEEE, 2018), pp 1–6
2. A. Dhar, N. Dash, K. Roy, Classification of text documents through distance measurement: An experiment with multi-domain bangla text documents, in *2017 3rd International Conference on Advances in Computing, Communication & Automation (ICACCA)(Fall)* (IEEE, 2017), pp 1–6

3. S. Mohanty, P. Santi, R. Mishra, R. Mohapatra, S. Swain, Semantic based text classification using wordnets: Indian language perspective, in *Proceedings of the 3th International Global WordNet Conference, South Jeju Island, Korea,* (Citeseer, 2006), pp. 321–324

4. M. Tummalapalli, M. Chinnakotla, R. Mamidi, Towards better sentence classification for morphologically rich languages, in *Proceedings of the International Conference on Computational Linguistics and Intelligent Text Processing*

5. A. Dhar, N.S. Dash, K. Roy, Categorization of bangla web text documents based on tf-idf-icf text analysis scheme, in *Annual Convention of the Computer Society of India* (Springer, 2018), pp. 477–484

6. P.K. Panigrahi, N. Bele, A review of recent advances in text mining of Indian languages. Int. J. Bus. Inf. Syst. **23**(2), 175–193 (2016)

7. S.A. Narhari, R. Shedge, Text categorization of Marathi documents using modified lingo, in *2017 International Conference on Advances in Computing, Communication and Control (ICAC3)* (IEEE, 2017), pp. 1–5

8. A. Dhar, N.S. Dash, K. Roy, An innovative method of feature extraction for text classification using part classifier, in *International Conference on Information, Communication and Computing Technology* (Springer, 2018), pp. 131–138

9. L. Wang, *Support vector machines: Theory and applications,* vol. 177. (Springer Science & Business Media, 2005)

10. M. Tummalapalli, R. Mamidi, Syllables for sentence classification in morphologically rich languages, in *Proceedings of the 32nd Pacific Asia Conference on Language, Information and Computation* (2018)

11. K. Rajan, V. Ramalingam, M. Ganesan, S. Palanivel, B. Palaniappan, Automatic classification of Tamil documents using vector space model and artificial neural network. Expert Syst. Appl. **36**(8), 10914–10918 (2009)

12. R. Jayashree, K. Srikantamurthy, B.S. Anami, Sentence level text classification in the Kannada language—A classifier's perspective. Int. J. Comput. Vis. Robot. **5**(3), 254–270 (2015)

13. Puri, S., Singh, S.P., An efficient Hindi text classification model using SVM, in *Computing and Network Sustainability* (Springer, 2019), pp. 227–237

14. B.S. Harish, D.S. Guru, S. Manjunath, Representation and classification of text documents: A brief review. IJCA, Spec. Issue RTIPPR **2**, 110–119 (2010)

15. J.J. Webster, C. Kit, Tokenization as the initial phase in nlp, in *COLING 1992 Volume 4: The 15th International Conference on Computational Linguistics* (1992)

16. M. Revanasiddappa, B. Harish, A new feature selection method based on intuitionistic fuzzy entropy to categorize text documents. IJIMAI **5**(3), 106–117 (2018)

17. G. Salton, C. Buckley, Term-weighting approaches in automatic text retrieval. Inf. Process. Manage. **24**(5), 513–523 (1988)

18. T. Tokunaga, I. Makoto, Text categorization based on weighted inverse document frequency, in *Special Interest Groups and Information Process Society of Japan (SIG-IPSJ, Citeseer* (1994)

19. S. Bird, E. Klein, E. Loper, Natural language processing with Python: analyzing text with the natural language toolkit (O'Reilly Media, Inc., 2009)

20. Project N (2020) https://www.nltk.org/_modules/nltk/tokenize/regexp.html. Last updated on 13 Apr 2020

Determining Stock Market Anomalies by Using Optimized z-Score Technique on Clusters Obtained from K-Means

Bibek Kumar Sardar, S. Pavithra, H. A. Sanjay, and Prasanta Gogoi

Abstract K-means has always been the most efficient technique to detect anomalies on any kind of dataset. It would be interesting to explore whether the algorithm could do marvels when used on a stock market dataset. Given the state-of-the-art methodologies, stock market data is prevailingly the most challenging data to work on, since the data values increase at a fast pace. Additionally, data analysis performed on time series data, taken from stock markets, has gained lot of popularity in recent past. Identification of any kind of anomaly in such dataset could be compelling; since this information can pave the way of growth for companies and investors hoping for higher returns and higher profits at lower risk. The manuscript aims to facilitate detection of such volatility by ascertaining outliers in the stock market data, without any prior knowledge of possible abnormalities. Though, z-score has been used extensively for determining deviations associated with data values for a given distribution, we strive to formulate a similar scoring formula, dev-score, that computes deviation for two-dimensional data (can be extended for more than 2D data as well), after generating clusters using K-means. The manuscript plots clusters for stock market data and identifies those stocks that deviate from their normal value on a particular trading day. It is important to note that the deviations are computed only for specific features of stock market data (volume and fluctuations), and this model can be easily extended on large number of features.

Keywords Machine learning · Anomaly detection · K-means clustering algorithm · z-score

1 Introduction

Dissemination of scientific knowledge, across the researcher and practitioners plays a key role in technology advancement in both industry and academia. Machine learning has become an important subject of exploration and an unavoidable tool to solve problems related to classification, clustering, and regression. Machine learning

B. K. Sardar (✉) · S. Pavithra · H. A. Sanjay · P. Gogoi
Nitte Meenakshi Institute of Technology, Yelahanka, Bangalore, India

© The Author(s), under exclusive license to Springer Nature Singapore Pte Ltd. 2022   401
P. Shetty D. and S. Shetty (eds.), *Recent Advances in Artificial Intelligence and Data Engineering*, Advances in Intelligent Systems and Computing 1386,
https://doi.org/10.1007/978-981-16-3342-3_32

domain basically is a subset of a larger domain—Artificial Intelligence that provides machines an impeccable ability to train, evaluate, and later improve its experience with the help of explicit programming. Around the globe, there have been an infinite number of articles which have explored powerful machine learning tools and their usage in solving problems of different domains. Essentially, machine learning domain solves problems of three broad categories that are designated under the umbrella of supervised, unsupervised, and reinforcement learning algorithms. Supervised algorithms make use of labeled dataset, observe the under lying patterns in these datasets, and predict a new sample by making use of the discovered patterns. Alternatively, unsupervised machine learning algorithms does not need any class-labels from training samples. These algorithms identify a mathematical relationship across the features and solve problem associated with regression and generation of clusters. On the other hand, reinforcement algorithms interact with their environment, predict future actions to solve the problem in hand, and calculate errors at every step and simultaneously, assigning reward points to these actions.

Anomaly, also known as abnormality, irregularity, or inconsistency, is a value that has deviated from the normal trend. An anomalous data can be a real threat leading to inconsistent results and could possibly raise multiple challenges in domains like statistics, data mining, signal processing, finance, and networking. The major reasons behind the existence of anomalies on dataset are measurement errors, data entry errors, data processing errors, experimental errors, and sampling errors. Alternatively, anomalies, also known as outliers, are detected by a method popularly known as outlier detection. Outlier detection is the method of discovering fluctuation in a given set of data. The method has a large field of application in various domains; some examples are fault detection, fraud detection and event system, health monitoring detection systems used in computer networks, sensor networks, and so on. The importance of anomaly detection is to enable us to find out outliers without knowing characteristics of possible abnormalities. In machine learning, it includes many classical methods of anomaly detection; some examples are isolation forest, support vector machine, and so on.

A very popular unsupervised algorithm, K-means is often used for detecting outliers from noisy datasets. Internally, it generate group of data points in such a way that points that are similar fall in the same group. Alternatively, points that are different are assigned to different group or cluster. In order to compute similarity between data points, different ways are suggested by different scientists over the period of time. The most popular choice of computing cluster similarity is by measuring Euclidean distance. Other methods suggested and used by researchers to generate clusters using K-means are cosine similarity, Manhattan distance, Mahalanobis distance, etc.

In this manuscript, we aim to build a mechanism that brings out anomalous data points in an intractable stock market data. Volatility of stock markets is always a cause of concern for many investors and traders who have invested a huge amount of capital in stock market and a single mistake may incur heavy monetary loss to them. Our focus is to identify and work on parameters which could not only indicate disparity in current stock market data but also predict information about divergences that may happen in future. TO achieve this, we took a synthetic dataset on stock market, performed K-means to recognize the center mass of the cluster (centroids) and computed dev-score (enhanced z-score) on every data point to identify how far the point is lying from the center mass. If the dev-score of a point is above a threshold, the point is declared as an outlier.

The manuscript is organized as follows. We begin by exploring the previous research conducted in analyzing the effectivity of K-means in anomaly detection. The section (section II) also examines the functional ability of z-score in computing the divergence associated with any data distribution in hand. Section III explains the methodology followed by us in determining anomalies for stock market data. It explores the working of K-means and set down a platform where enhanced z-score formula (termed as dev-score) helps in determining deviating values in stock market fluctuations during trading hours. Next section (Section IV) unveils the complete experimentation done on synthetic stock market data and demonstrates the effectivity of the methodology used by us. We end our manuscript with conclusion section.

2 Literature Review

A lot of research has been conducted for detecting anomalies in variety of dataset. Most of these methods are found to be complex, robust, and capable of providing results on smaller datasets. Mohammad Saeid Mahdavinejad [1], presented in their research how machine learning has helped in providing solutions for the challenges thrown by smart IoT applications. Thudumu et al. [2] have worked on anomaly detection which, according to the authors, is a process of identification of rare items, events, and observation seen in the dataset. Kotu and Deshpande [3] submitted paper work on outlier detection method of finding abnormal behavior in a given record. According to the authors, anomalies are the fluctuations among the normal behavior which may not align with the desired output from the task associated with a given dataset. These anomalies could create undesirable results and may lead to difficulties in obtaining the expected output. Authors further explored that anomaly detection inherently deals with machine learning domain, statistical modeling, k-means clustering, automation, cloud, big data, regression, and clustering. Lei [4] wrote an article on anomaly detection method that identifies outliers by virtue of the neighborhood information to understand the techniques of unsupervised learning. A recent research conducted by Duan et al. [5] indicates that work done by using K-means by incorporating cluster-based, distance-based, and statistical-based mechanism not only generates efficient clusters but also eliminates anomalies from the noisy dataset.

Furthermore, exploring on the same direction to evaluate the efficacy of clustering based methods for detecting outliers, Knapp and Langil [6] suggested a policy-based detection while providing rule-less methods to find abnormalities. At industry level, strong defense acquired by accurate behavior can be calculated, and abnormal behavior can be detected. The authors laid stress on the working principle of the industrial network system and suggested that network behavior is predictable by applying effective detection techniques, and outliers can be determined.

Cardenas and Safavi-Naini [7] showed in their research that anomaly detection, based on network security community is being used since decades. Although they have ability to detect zero-day attack, novel attack, or previously unknown attack, there is always a probability of generating from false alarm (ability to detect false abnormalities). Similarly, Jyothsna and Prasad [8] submitted a survey on anomaly detection and conveyed that, with the invention of anomaly based detection, various techniques were developed in past to track zero-day attack as well as previously unknown attacks (novel attacks). They claimed that maximum detection rate and minimum alarm rate up to 1% can be obtained by applying these methods. Instead of using anomaly-based detection methods, they followed signature-based detection for mainstream implementation of outliers' detection. It was difficult to compare the variety of detection methods purely on the basis of their strengths and weaknesses. Additionally, industries do not approve of using anomaly-based detection by validating the efficiency of all these techniques. To find an optimal solution, the authors survey and reviewed the present state of the trial in the field of outlier-based detection. The research paper presented a detailed analysis of drawbacks, limitation, and future scope of all the recent anomaly detection methods.

Denning et al. [9] submitted an article which is based on a real-time detection expert system model that has the ability to detect penetrations, break-ins, and many forms of computer misuse. The algorithm is based on supposition which breaks the network security can be found out by supervising the dataset for abnormal behavior that system usages. The model acquires knowledge about the abnormal behavior of the dataset with respect to features value in terms of mathematical calculation such as standard deviation, z-score, Euclidean distance, statistical model, or mathematical model to audit records and detect fluctuations on normal behavior. The algorithm is independent for different frameworks, environment, application, and system. In this paper, the general purpose of finding abnormal behavior in given audit records detect fluctuation on normal behavior.

Siddharth Misra et al. [10], wrote an article on detecting outliers for geophysical survey and well logs. Geophysical survey is a collection of geophysical data that has many applications in geology, archaeology, mineral, energy exploration, and oceanography. Well logs is a collection of formation of particle records while drilling wells, and basically it generates an alarm of any irregular activities observed during the drilling of wells. Alongside, it is difficult process, consumes lot of time, requires lot of human labor, and dedicated human expertise is required to get important dataset for well logs. So, to get the expected dataset, authors used ab unsupervised machine learning outlier detection. Oghenekaro Osogbo et al. [10] proposed an outlier detection which is used to find and collect the unusual dataset like geology, archaeology, mineral, and energy exploration. Outlier detection can be found using simple methods such as box plot and z-score. A box plot represents the distribution of samples corresponding to various features using boxes and whiskers, Mark Powers et al. [10] suggested usage of z-score for outlier detection that computes distance between data samples and its mean (μ). It can be interpreted as: if z-score value is greater than or less than $+ 3$ or $- 3$, respectively, the data point can be marked as outlier. z-score is expressed as,

$$z - \text{score} = \frac{x - \mu}{\sigma} \tag{1}$$

In the present paper, we perform anomaly detection for network industries and where he has discussed major problems in detecting irregular activities or issues mainly in network industries. Usman Habib, G. Zucker, M. Blochle, F. Judex, and J. Haase et al. [11] explored on the anomalous data and found that a huge amount of data is recorded by checking various building structures using sensors. These sensors will note down some physical properties of real-life phenomena, for example, temperature, weight, and stream rate. Sensors sometimes give wrong dataset due to weather fault or fault measurement due to defect sensors and hence that leads to various mistakes/errors in the dataset. Hence, before considering the data to obtain some important information, it is important to approve it. So the detection of outlines helps in removing the data that is normal or expected behavior of the machine. S. Katipamula et al. [12] proposed paper is to create a method that can automatically find anomalies in data but checking the data is labor intensive and not achievable. This paper talks about the means required for detecting outliers in the information got from retention chillers utilizing these on/off state information. It explains methods for automatic detection of on/off and/or missing data status of chillers. The initial step of anomaly detection on/off and missing information state is finished by utilizing a two-layer K-mean clustering algorithm calculation. The missing information state will speak to when there is no information recorded for any sensor of the chiller. After discovery of the on/off express, the cycle based z-score standardization is utilized with a clustering calculation to discover the anomalies in the information. In the present paper, we find anomaly detection for a huge database. K-mean clustering algorithm and z-score is used to find anomaly detection.

Shashank Singh Yadav et al. [13] wrote an article on one of the major problems of today's life which is heavy traffic that is caused by an expansion in the quantity of vehicles and substantial traffic needs consistent observation for guaranteeing security of individuals. The very normal reason for traffic is people. Some of the traffic violations are use of mobile phones while driving, signal jumping, drunken driving, not wearing seat belts and helmets, etc. So, video-based traffic observation is the solution considered in this paper. Hence, the aim of this paper is to find anomalies using traffic video surveillance. Anomalies include use of mobile phones while driving, signal jumping, drunken driving, not wearing seat belts and helmets, wrong U-turn, etc. So that traffic can be controlled, and if any accidents take place, ambulances are sent to the accident place quickly. Unsupervised machine learning is used for detecting traffic anomalies. Video recordings are taken as input data, and these recordings are changed into frames from each frame into highlights and from these extracted highlights, a single centroid point is taken with reference to the centroid point from this category of traffic validation taken. Ashok Kumar et al. [14], proposed K-mean clustering which is used to find irregular activities and define them as anomalies.

3 Proposed Methodology for Anomaly Detection

Inspired from the work done by different researchers across various domains for detecting anomalous behavior, we intent to utilize the core idea behind K-means and z-score and introduce a more precise and expressive formula with the name, dev-score. Dev-score, or more formally, deviation score, aims to compute divergence or deflection of data point from its mean in a given distribution of data. The basic idea used in the manuscript involves execution of K-means for identification of mean (also called centroids). Once the centroids are identified, we utilize dev-score to figure out all such points that deviate from centroid beyond a certain threshold, λ. These points are referred as anomalies. Interestingly, computation of dev-score normalizes the distance of every data points from its mean. Dev-score is inspired from the popular method, z-score (extreme value analysis) which is essentially a parametric method for computing outliers. Formally, z-score is represented as:

$$z - score = (X - \mu)/(std) \tag{2}$$

Here, μ (mean) and std (standard deviation) are computed on a single feature value using standard formula. The essence of z-score is it normalizes value of entire dataset into one specific range, eliminating the effect of outliers in the data, if any. Based on the same formula, the extended version of z-score is termed as dev-score. Dev-score works well for two or more that two-dimensional data. It is given as:

$$dev - score = \frac{D}{std_C} \tag{3}$$

where

$$D = \sqrt{(x1 - c1)^2 + (x2 - c2)^2}$$

Here, D is the Euclidean formula for distance computation, that computes distance between X and C in two-dimensional space. Dev-score further computes deviation score of every X, by dividing distance (D) with standard deviation (std_C) for cluster, C. It is important to mention here that dev-score is effectively doing the same thing as z-score, with the difference that it generates a normalized score for data points in higher-dimensional space comprising of two or more than two features.

K-means clustering algorithm—is an unsupervised machine learning method that utilizes the inherent information observed in the dataset to generate meaningful clusters. The algorithm formally divides 'n' data points into k-cluster that follows three essential steps of initialization, assignment, and update. In initialization, k initial centroids (means) are picked-up randomly from the dataset. Next step assigns every data point to the nearest cluster by computing the Euclidean distance of the point and centroids. Once all the points are assigned to nearest cluster, the last step updates centroids of the every cluster. This process continues till algorithm reaches convergence and centroids do not change any further.

For computing distances between sample point and the centroid of the cluster, Euclidean distance is used that is represented as:

$$\text{Euclidean distance } d = \sqrt{(x_1 - y_1)^2 + (x_2 - y_2)^2} \qquad (4)$$

where x_1, x_1.....are the two coordinates of data point X, in two-dimensional space.

3.1 K-mean Clustering Algorithm

Figure 1 depicts formation of four clusters from the available data points in two-dimensional space. The algorithm runs iteratively over each point and assigns it to the nearest cluster. The three basic steps are as follows:

Step 1: Initialization

K initial "means" (centroids) are randomly generated as first step.

Step 2: Assignment

Each sample is assigned to one cluster depending upon its closeness to the centroid. This closeness is evaluated by computing Euclidean distance.

Step 3: Update

The mean of the centroid is calculated by taking mean of all the points in that cluster.

Step 2 and Step 3 are repeated iteratively until convergence is observed. By convergence, we mean that the centroid obtained from previous iteration does not change. After convergence of the algorithm, the points are assigned to each cluster and the sum of squared error is minimized between points and their respective centroids. The

Fig. 1 Illustration of clusters seen on the scatter plot of data points

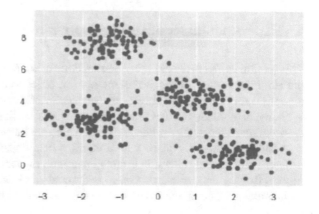

best clustering solution for the given data points is the one picked after repeating the above three steps for ten times.

More detailed explanation of K-mean clustering method: Here are some finer details and aspects of the algorithm. These aspects are discussed on the basis of three basic steps:

1. While choosing the initial centroid values—Once, the decision is made on number of groups/clusters we want to have, there are few options for picking initial centroid values. Select k random points from dataset and run K-means on it. Repeat the process n times and choose the iteration that has minimum least square error. This process is far more reasonable and accurate. We have used this model during executions. Specifically, the number of random initializations to perform K-means clustering model is achieved using sci-kit learn package from Python library.

2. While assigning data points to a cluster—Once the selection of initial centroids is done, the next step is to assign each data point to a cluster. Mathematically, this can be represented as (for ith iteration):

$$S_i^{(t+1)} = \left\{ x_p : \| x_p - m_i^{(t+1)} \ \square^2 \leq \| x_p - m_j^{(t+1)} \ \square^2 \ \forall j, 1 \leq j \leq k \right\} \quad (5)$$

where $S^{(t)}$ indicates tth cluster, $m^{(t)}$ is the mean, and x_p is the data point.

Here, if the distance from the data point (x_p) to the mean (m_i) is smaller than distance to the mean of all other centroids (m_j), then x_p is assigned to $S^{(t)}$.

3. While reassigning data points to new clusters—As soon as all points are allocated to a cluster, there is a need to calculate new centroid values and to reassign data points until the process converges on a cluster segmentation that stops changing. The mean is computed using the formula below

$$m_i^{(t+1)} = \frac{1}{\left| S_i^{(t)} \right|} \sum_{x_j \in S_i^{(t)}} x_j \quad (6)$$

4 Experimental Result and Discussion

This section explores the experimentation done using K-means and also brings forth the admirable results through this method. In order to evaluate the efficacy of the algorithm for bringing out anomalies, we have used synthetic stock market dataset. Working of K-means is explored in the previous section, now we present a brief explanation of the stock exchange data and also put across the working process.

A stock market is a place where equities of company are bought or sold, by different buyers and sellers, in order to maintain ownership on companies or businesses. Buying and selling of shares is done at stock exchanges, which internally maintains list of companies who intent to participate in trading of equities. Buying and selling are formally called as trading which involves transfer of stocks (also known as shares and equities) from seller to buyer after both parties officially agrees upon a price. This is reflected as stock price of a company. The stock exchange not only facilitates trading of securities across companies (or individual investors) but also maintains the entire trading data that gets generated as a result of everyday transactions. It involves the names of companies that are listed for trading, their symbols, stock price, trading volume, opening price, closing price, and highest and the lowest price in a day (or any other time frame). Exchanges also create and maintain charts that indicate movement of stock price for various companies. Generation of charts and the after-market analysis of trading data helps prospective investors in making decisions related to buying (or selling) of equities and thus play a creditable role in making future profits. Let us take a closer look at the stock market dataset used in the model.

Figure 2 shows the input data with distinct features and values with respect to date of transaction of equities. Open, close, high, low, volume, adj_close, and fluctuation are the feature values associated with each listed company stocks. There are 5990 stock values used as dataset for experimental reasons.

- Symbol—The symbol column represents the company name and the symbol of stock assigned.
- Date—The date on which the company's stock prices are observed.
- High—The high column represents the highest price on that date.

1	symbol	date	open	high	low	close	volume	adj_close	Fluctuatioi	f2	f3
2	A	12/30/201	45.76	45.82	45.38	45.56	1215100	45.56	0.438975	0.439999	0.19999;
3	AA	12/30/201	28.97	29.378	28.01	28.08	2651500	28.08	3.169512	1.368	0.88999;
4	AAAP	12/30/201	26.73	26.92	26.6	26.76	42700	26.76	0.112108	0.32	-0.0:
5	AAC	12/30/201	6.88	7.508	6.83	7.24	199100	7.24	4.972376	0.678	-0.3€
6	AAL	12/30/201	47.42	47.66	46.47	46.69	4494100	46.69	1.563502	1.189999	0.72999;
7	AAMC	12/30/201	49.1	53.5	45.62	53.5	17900	53.5	8.224303	7.880001	-4.4
8	AAME	12/30/201	3.95	4.2	3.95	4.1	50400	4.1	3.658537	0.25	-0.1!
9	AAN	12/30/201	32.07	32.29	31.52	31.99	431100	31.99	0.250078	0.770001	0.0€
10	AAOI	12/30/201	24.25	24.25	23.21	23.44	197100	23.44	3.455627	1.040001	0.80999;
11	AAON	12/30/201	33.75	33.75	32.9	33.05	99800	33.05	2.118006	0.849998	0.700001
12	AAP	12/30/201	171.32	172	168.6	169.12	489300	169.12	1.300859	3.399994	2.20001;
13	AAPC	12/30/201	9.84	9.84	9.84	9.84	0	9.84	0	0	(
14	AAPL	12/30/201	116.65	117.2	115.43	115.82	30253100	115.82	0.716631	1.769997	0.83000;
15	AAT	12/30/201	42.41	43.34	42.41	43.08	231300	43.08	1.555251	0.93	-0.6;
16	AAU	12/30/201	1.08	1.08	0.97	0.97	452400	0.97	11.34021	0.11	0.11
17	AAV	12/30/201	6.7	6.85	6.7	6.75	66900	6.75	0.740741	0.15	-0.0!
18	AAWW	12/30/201	52.3	52.85	51.3	52.15	140500	52.15	0.287626	1.549999	0.14999;
19	AAXJ	12/30/201	55.28	55.33	54.84	54.93	860300	54.93	0.637173	0.490002	0.34999;
20	AB	12/30/201	23.8	23.85	23.42	23.45	218300	23.45	1.492529	0.43	0.34999€

Fig. 2 Stock market dataset

- Low—The low column is lowest price on that day.
- Volume—The volume column shows the number of shares traded on that particular date.
- Open—The open column represents the price at which the stock was opened for trading.
- Close—This column indicates the close price at which the stock stopped trading.
- Fluctuations—This column tells about the prices fluctuation due to the difference between profit and loss.
- Dev-score: dev-score column represents the mathematical calculation of deviation from its centroids with respect to standard deviation.
- Distance: Euclidean distance, distance from centroids to a point.
- Cluster0 and cluster1: The column cluster0 and cluster1 represent the two different clusters formed due to K-means for given dataset.
- Volume: This column represents the volume of shares stock trading on particular date.

Working: With all the feature values listed and shown above, we took two critical parameters to highlight anomalous deviations in stock trading, volume, and fluctuations. For a particular company's stock, the deviation seen in the trading volume decides stability or volatility of market and the company. It is an indicator of how large the stock price may vary around its mean thus gauging the risk associated with the equity. This implies that if large fluctuations are observed, the stock price is considered more volatile and risk associated with it is high. In order to conduct the proposed work to detect anomalous variation in the trading volume and fluctuation on a specific day, following steps are followed-

Step 1: Clusters are generated using K-means for the entire dataset, using two parameters, volume, and fluctuations. Figure 3 shows the formation of 2 clusters with their respective centroids marked with star.

Step 2: Store the centroids of the two clusters for computing distance from every point to its centroid.

Step 3: Compute the distance of data point (X) from the two centroids. Let $C1$ and $C2$ be the two centroids obtained after the convergence of K-means on the stock market data. To show calculations, assume $\times 1$ and $\times 2$ to be the two-dimensional representation of X, and c1 and c2 are the two-dimensional representation of C1 (with assumption that points X belongs to cluster C1), then

$$D = \sqrt{(x1 - c1)^2 + (x2 - c2)^2}$$

Step 4: Divide distance D with standard deviation of the cluster. This process generates dev-score of every data point present in that cluster. This is a reflection of the extent to which a point is deviated from the center of the cluster.

$$dev - score = \frac{D}{std_{C1}}$$

Fig. 3 Flowchart of the
methodology used

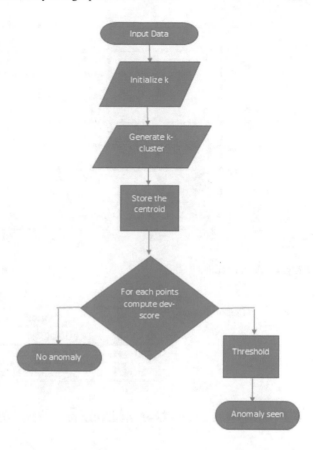

where

dev-score   is the deviation score for point X.
D   is the Euclidean distance of X from C1.
std_{C1} is the standard deviation computed for C1.

Step 5: Compare the dev-score with the pre-computed threshold λ, if the dev-score is greater than λ, the data point is declared as anomaly.

The essence of the method used by authors is to find out all points that are lying farthest from the center of the two clusters. A threshold λ, is computed by identifying all those points which are lying 2-standard deviation away from centroids. It is statistically shown that the data points that are seen with 2 or more standard deviation away from mean are considered isolated from the rest of the data. Once these points are identified, mean of their z-scores is computed and used as threshold. The λ was observed to be 2.0 for the given dataset. Implementation of K-means and dev-score is done using Python 3.6 installed on Intel Core-i3 8145U CPU with 8 GB RAM. Figure 4 depicts the clusters and centroids emerged from the stock market dataset (Table 1).

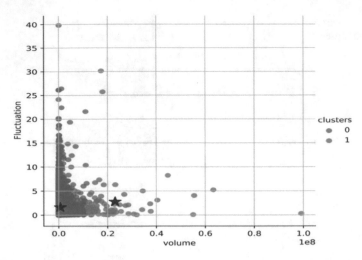

Fig. 4 Cluster outliers

Table 1 Output of K-means

	Centroid position	
	Volume	fluctuation
Cluster 0	455,940.01	1.36
Cluster1	10,746,971.05	2.35

4.1 Analysis of the Output Dataset

As indicated, we have used 5990 points for stock market dataset. The following figure (Fig. 4) shows formation of two clusters as a result of running K-means. In this figure, volume and fluctuation are features used in the algorithm for generating two clusters indicated as 0 and 1. The blue and yellow dots indicate data points that belong to two different clusters. The star sign indicates their centroids. Clusters which are farthest from the centroids are detected as outliers. Once clusters appear, dev-score is computed for all the points present in two clusters. If this score is larger than the threshold set for each cluster, the point is labeled as anomaly. After running these steps on 5990 data points, we observed 5866 points in cluster 0 (shown in blue dots) and 114 points in cluster 1 (yellow points). After the computation of dev-score on each of these points, we could see 97 outliers for cluster 0 and 2 outliers for cluster 1.

In Fig. 5, we presented a screenshot of the output and observe two clusters identified as '0' and '1' for various stocks mentioned in 'SYMBOL', We also observe the value of dev-score for every company, for the companies named 'ABX' and 'FIT', the dev-score value of larger than 2.0. Hence these are declared as outliers for cluster 0 and for cluster 1.

SYMBOL	volume	Fluctuation	cluster	dev-score	distance
AAPL	2293100	0.164068909	0	1.1432529	1609223.9
ABT	128300	1.859678783	0	0.394702	555576.05
ABX	3579700	0.541147602	0	2.0573017	2895823.9
AKX	679600	0.227268182	0	0.0030379	4276.0535
AUY	9800	0.81300813	0	0.4788888	674076.05
BBRY	409800	8.690818457	0	0.1947139	274076.05
C	446200	0.71942446	0	0.168854	237676.05
CEMP	28509500	0.301333615	1	0.4117203	5764081.7
COG	16284700	0.197238659	1	0.4614801	6460718.3
CSCO	18578900	2.379493739	1	0.2976086	4166518.3
F	37662200	0.778210117	1	1.0654849	14916782
FB	15295900	1.138519924	1	0.5321087	7449518.3
FCX	13185800	1.314965519	1	0.6828302	9559618.3
FIT	55097900	0.087489064	1	2.3108926	32352482

Fig. 5 Intermediate results after computing dev-score

5 Conclusion

In this manuscript, we reveal a refined method to detect anomaly by performing K-means on the stock market dataset and later using the centroids generated from K-means for computing a score to detect outliers. We termed the score as dev-score. Additionally, multiple research papers from various domains of machine learning, big data, cloud, automation, data analytics, and data analytics with python were explored thoroughly to evaluate the essence of using standardization techniques like z-score for computing outliers. It is observed that z-score calculates the deviation from its respective centroids, and there is a need to use Euclidean distance and standard deviation in order to appropriately use z-score. Hence, dev-score becomes convenient. A K-mean clustering algorithm is presented in the manuscript in detail. Results obtained from this method are found to be appealing and easily extensible on any kind of time-series data.

References

1. M.S. Mahdavinejad, M. Rezvan, M. Barekatain, *Machine Learning for Internet of Things Data Analysis: A Survey*, vol. 4 (University of Isfahan, Iran Kno.e.sis—Wright State University, USA)
2. S. Thudumu, P. Branch, J. Jin et al., A comprehensive survey of anomaly detection techniques for high dimensional big data. J Big Data **7**, 42 (2020). https://doi.org/10.1186/s40537-020-00320-x
3. V. Kotu, B. Deshpande, *Data Science: Concepts and Practice Published in Cambridge* (Morgan Kaufmann Publishers, MA, 2019)
4. P. Lei, A framework for anomaly detection in maritime trajectory behavior. Knowl Inf Syst **47**, 189–214 (2016). https://doi.org/10.1007/s10115-015-0845-4
5. L. Duan, L. Xu, Y. Liu, J. Lee, Cluster-based outlier detection. Annals Oper. Res. **168**, 151–168 (2009). https://doi.org/10.1007/s10479-008-0371-9
6. E.D. Knapp, J.T. Langill, *Industrial Network Security: Securing Critical Infrastructure Networks for Smart Grid, Scada, and Other Industrial Control Systems* (Waltham, MA: Elsevier, 2015)
7. A.A. Cárdenas, R. Safavi-Naini, *Handbook on Securing Cyber-Physical Critical Infrastructure* (2012). https://doi.org/10.1016/B978-0-12-415815-3.00025-X
8. V. Jyothsna, R.V.V. Prasad, A review of anomaly based intrusion detection systems. Int. J. Comput. Appl. **28**(7), 26–35 (August 2011)
9. D.E. Denning, An intrusion-detection model, in *IEEE Transactions on Software Engineering* (vol. SE-13, issue no. 2, pp. 222–232) (1987). https://doi.org/10.1109/TSE.1987.232894
10. S. Misra, O. Osogba, M. Powers, Unsupervised outlier detection techniques for well logs and geophysical data, in *Machine learning for surface characterization* (Texas A&M University, College Station, TX, United States, 2020). https://doi.org/10.1016/B978-0-12-817736-5.00001-6
11. U. Habib, G. Zucker, M. Blochle, F. Judex, J. Haase, Outliers detection method using clustering in buildings data, in *IECON 2015—41st Annual Conference of the IEEE Industrial Electronics Society*, Yokohama (2015, pp. 000694-000700). https://doi.org/10.1109/IECON.2015.7392181
12. S. Katipamula, M.R. Brambley, Review article: methods for fault detection diagnostics and prognostics for building systems—a review Part I. HVACR Res. **11**(1), 3–25 (2005)
13. S. S. Yadav, V. Vijayakumar, J. Athanesious, Detection of anomalies in traffic scene surveillance, in *2018 Tenth International Conference on Advanced Computing (ICoAC)* (Chennai, India, 2018, pp. 286–291). https://doi.org/10.1109/ICoAC44903.2018.8939111
14. P.M. Ashok Kumar, V. Sathya, V. Vaidehi, Traffic rule violation detection in traffic video surveillance. Int. J. Comput. Sci. Electron. Eng. (IJCSEE) **3**(4) (2015)

Data-Driven Strategies Recommendation for Creating MOOCs for Effective Online Learning Experience

Tanay Pratap and Sanjay Singh

Abstract This paper presents the strategies for creating an online course from the learner's perspective. The paper examines the learners' requirements for signing up and completing an online course. Two hundred eleven online learners provided their responses to a survey request. It had an open-ended verbatim question. The responses were categorized and then re-categorized to get the most common expectation area. The survey found out that the learners care most about having real-life projects, the instructor's delivery style, beginner friendliness of the course, and understanding its purpose in the overall context. In each of these areas, there are suggestions and expectations with high-frequency data to work. This survey was done to an open group of people from across the world with an open-ended question. There was no control on the demographics of the learners. Hence, the next stage of the research should be on controlled demographics as the online courses generally are for a niche area. The findings are useful for online instructors and MOOC platforms alike. With data and computing resources becoming ubiquitous, many more teachers will create courses for online consumption. This paper can thus serve as a checklist for instructors creating such new courses.

Keywords Online learning · Lifelong learning · Pedagogical issues · Social media · Teaching/learning strategies

1 Introduction

While traditional classroom education has been ongoing for quite a few centuries, massive open online course (MOOCs) has just arrived on that timeline. It is far-reaching in the aspect of the opportunities it presents, and everyone from teachers,

T. Pratap
Microsoft India, Bengaluru, India
e-mail: tapratap@microsoft.com

S. Singh (✉)
Department of Information and Communication Technology, Manipal Institute of Technology, MAHE, Manipal 576104, India
e-mail: sanjay.singh@manipal.edu

© The Author(s), under exclusive license to Springer Nature Singapore Pte Ltd. 2022   415
P. Shetty D. and S. Shetty (eds.), *Recent Advances in Artificial Intelligence and Data Engineering*, Advances in Intelligent Systems and Computing 1386,
https://doi.org/10.1007/978-981-16-3342-3_33

students to big-time universities recognize it. As per recent numbers from classcentral.com [19], which lists all the available MOOCs across the internet, 900 universities have a MOOC offering 13.5 K courses enrolling close to 110 M students.

Even though the scale of MOOCs is massive, the completion rate is still alarmingly low. There are multiple reasons for it, but one thing is sure that the key to delivering the perfect online course has not been cracked yet. MOOCs are trying to put traditional classroom on the internet, which could be annoying to the students as even though they get the teachers, they are not getting the peers and support system of a classroom.

This paper aims to relook at MOOCs' best practices from the one that matters the most: the students. We conducted an online survey with a relatively emotional and open-ended question about online courses. Over a decade of MOOCs, most online learners have figured out what works for them and what not. This paper tries to learn from the students and put those best practices for future MOOC creators.

2   Related Work

Work on Online learning, pre-MOOC era: Before MOOCs came into the picture, training and learning were done online. There is some good work in the area of online learning and teaching from the pre-MOOCs era.

Gilly Salmon's [18] work talks about creating an online tool first, then creating an online cohort, and moving on to information exchange. Only in the fourth phase does actual knowledge construction happen to lead to final development. If appropriately applied, these ideas would help in student retention in MOOCs; even though the tools are as old as email, ideas are quite robust.

Terry Anderson [3] focuses on the role of a teacher. This study stresses that the teacher should add his personality in online learning material and "personal reflections, anecdotes, and discussions of teacher's struggles and success." Students have found this inspirational and motivating. It then delves deep into discourse facilitation, assessment, and ideas to build direct instruction provision. This study is beneficial for a teacher looking to expand online and moderate a course. The study concludes with the qualities of e-teacher, pointing that "Internet efficacy" must be effective.

Work related to MOOCs: Oakley and Sejnowski [16] discuss the learnings from the world's most popular MOOC. It stresses heavily on using green screen and video editing techniques for creating content. It goes on to describe the teleprompter and studio lights and everything in the production, of course. Moreover, it is also covering what plays essential roles. It says that instructor presence and putting content on the green screen are a multitude better than using Powerpoint bulleted slides with the instructor on the side frame. It respects the signaling principle. It provides details on how to prepare a script for the course and how to add humor. It makes a bold statement that MOOCs' future is integrating video-making and video-gaming features. Thus, the next generation of "videos" on a MOOC may involve interactive video game-like components.

Based on the interview during the study, Zheng et al. [24] suggest that the completion or success of a MOOC depends on students' motivation. They concluded that few learners are there for certificates and others for learning. Thus, they advise two modules and present the course, which suits both categories' needs. A learning-driven module can remove problems such as lack of time confidence, falling behind the schedule. On the other hand, structured content in a timely fashion can be presented to the students preparing for some job as they need the certificate.

Recent work on MOOCs by Hew et al. [13, 14] study the learning from 3 and 10 most highly rated courses. Emphasis on problem-solving by instructors, their passion for the subject, and availability via various means whether email, forums, or live interactions, is most appreciated by learners. According to these studies, peer interaction and forums can be the most significant asset if appropriately done or the biggest deterrent in experience if left unchecked. Thus, both studies discuss ways to involve course staff in the moderation of content and encouragement of content by posting new threads in forums. Another problem which the studies discuss is superficial reviews in the case of peer to peer review-based assignments. Rating calibration and mandatory minimum (say 200) words comment are a few strategies that the studies suggest. Short and inflexible deadlines are highlighted as a reason for student dissatisfaction. Quality of assignments, unclear wording, or instructors in course work is a few more reasons for dissatisfaction in these studies. Similar findings around flexibility, feedback and peer interaction, and ease of access-to-knowledge being the motivators are presented in [21].

Research around student retention and better learning outcomes point toward one more important aspect: the social or forum presence. Chen et al. [6] show that while instructor to student interaction (on forums) helps in student persistence to learn further in MOOCs, and student interactions with peers are not a significant contributor. In a similar light, another study measures social presence and scores in MOOCs. It argues that perceived social presence is different depending on the size of the posting cohort and the duration of the course. It further stresses that even smaller sized (by the number of participants) courses evaluated the establishment of familiarity, emotional connection, and sense of trust as low. These studies collectively question the need for forums and their effectiveness.

In this paper, we worked with an open survey question presented to thousands of learners. While each of these studies has taken one or more areas/categories and surveyed or researched the data in the same area, our study tried a different approach to invite ideas from all directions possible without any preconceived classification of what makes an incredible MOOC.

Our study thus has holistic ideas for MOOC creation listed entirely by the consumers of these courses. It covers ideas from content design to content delivery and assessment. It provides varied and uncovered ideas around better navigation to getting help in reading the documentation. It has ideas even beyond course completion to encourage continued learning in the area of the MOOC by following thought leaders on social media. These are the problems that learners face while learning or in the real world. After completing the course, they would not have got a forum or checkbox in their work environment to express those. Going through the action items list presented in the study would help any MOOC instructor, designer, or platform creators get a glimpse into learners' psyche as to what they require for active learning.

3 Methodology

This section explains the various steps of our methodology.

3.1 Data Collection

For our study, we collected data through a survey. A set of open-ended questions was asked to all the participants asking for feedback on their expectations if someone created an online course. We asked the following open-ended questions in our survey request:

- What is the one thing you love in an online course?
- What is one thing which you hate or think can be improved?

We asked the same set of questions across many Facebook groups. The groups selected were mostly around web development technologies as these groups mostly learn things online. Courses in universities do not cover web development despite massive demand in the industry, and thus, there are many online courses to fill that gap. These groups have members ranging from 10 K–190 K. Therefore, much traffic was seen on these posts cumulatively. Since this was open to all, people from all geographies and ethnicity participated. Also, volunteer participation meant that only interested learners submitted their responses.

Participants responded to the survey with comments. These comments ranged from a single line to paragraphs of 10 lines describing things in detail. The verbose input was incredibly helpful in understanding the psyche of an online learner.

3.2 Response Analysis

We collated all the comments received through the survey response. We have considered the suggestion given in a blog post by Cho [7] from SurveyMonkey.com to analyze the open-ended survey responses. The below mentioned four-step process is used to gather useful results out of all the collected responses.

1. **Most Frequent Terms**
   Terms which occurred more than once were selected and put in bins. Even though the question was open-ended, the responses had an overarching theme.

Terms like real-life projects, delivery style, video length, assignments appeared in a lot more responses.

2. **Broad Category Creation**

Categories were then created around these terms, and similar responses were put in the same category. Created categories are: Pre-requisites

- Real-Life Projects
- Instructor and Video
- Content
- Testing and Assignments
- Post course Engagement
- Individual Ideas

3. **Individual Response Analysis and Scoring**

Every response was analyzed and categorized accordingly. It ensured creating new broad categories, which were added to the above list. Similar ideas were assigned a higher score inside the category to pay more attention during course creation.

To compute the score, we followed a simple process. There were 211 comments on the forum; we went through each comment and created categories based on groups of similar comments. Some comments were then found out to be echoing the same ideas repeatedly, so we added them together like 1 +1 +1 and then presented the final score on these ideas (which were repeating) in Fig. 1.

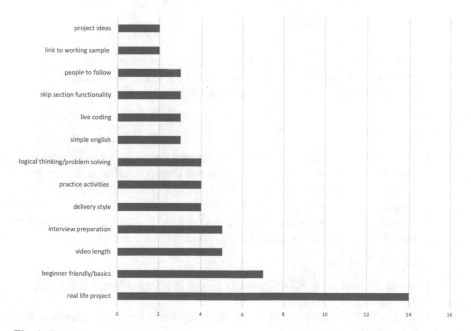

Fig. 1 Term frequency for course ideas

4. **Finalize Ideas**

The ideas with a high score under each category were then finalized as useful. Future online courses can be designed considering these ideas. Other ideas can be taken as suggestions and can be used to create a wholesome learning experience.

4 Results

4.1 Overall Analysis

According to learners, the most important thing to learn is how to *do real-life projects*. There is an overgrowing sentiment that the course content does not relate to the industry's work. There are many ideas around this theme, which we discuss in the next section. The second thing is that the content should cover basic needs. The ideal *video length* and *delivery style* are something that came out quite clearly from this survey, which many would not consider an important factor while creating courses. Points such as using *simple English* in courses came from respondents of non-native speakers. Though the term is not the same, there is much stress on assignments, quizzes, after-course projects, which we discuss in the following sections.

4.2 Category Wise Ideas

As discussed in the methodology part, broad categories were created out of the survey results to get concrete ideas. These ideas would help in making upcoming online courses effective. The ideas in these categories are ranked as per their score in the survey.

1. **Pre-requisites**

   - Considering that the course will end up doing a project, a link to that project's working sample would help students decide their commitment to the course.
   - Explain the reasoning behind choosing a set of tools, i.e., the programming language, the framework, and the hosting.
   - course should have established pre-requisites. If possible, videos of the courses required before doing the course should be provided.

2. **Real-Life Projects**

   - It should work on a project example related to actual software being used by people around the world.
   - It should focus on relating to job interviews and interview questions that relate to the course work.

- Code written should be near to production code, i.e., the code written in industry. Formatting, file naming, and folder structure are what learners expect to get from the course.
- Setting up the development environment, keyboard shortcuts that programmers use, and better documentation navigation are other ideas. These could be an excellent addition to make a course near to real-life scenarios.

3. **Instructor and Video**
Video length should be ideally between 5 and 10 minutes. It means content should be broken into smaller capsules. Most respondents commented that it gets sleepy and boring in one long video.

- The instructor should explain with simple words and keeping beginners in mind. Many MOOCs are information and abbreviation heavy. However, it could be a good practice to have an introductory video explaining the terminologies used.
- Respondents stressed the delivery style of the instructors as well with "should not be monotonous," "at least try to be funny," "fun, energetic, and engaging." High energy instructors with a dash of humor are appreciated everywhere, even outside the MOOC world. However, in MOOCs, the learner interacts only with one personality, i.e., the instructor. Thus, it is essential to have a session high on energy and engaging.
- As a learning medium, students prefer an "interactive live coding" platform, which would present the students a chance to try out what is being taught themselves.
- While most MOOCs have a Professor on screen, there are few online course sites like egghead [10], which has the only computer screen recording. Respondents to the survey expressed a strong opinion against slide or code editor with just voice over. Preference is to have an instructor on screen.

4. **Content**

- Course should start from the very basics. If the basics are not covered, then links to learning should be given. The assumption that a learner would know something before-hand is not advisable.
- Adequate time should be spent on the usefulness of the content. Answer to, "Why are we learning this language/framework/software library/concept?" It helps the learner relate to the content better.
- When learning a new language or a framework, it is better to build small projects along. This practice builds muscle memory as well as gives a sense of accomplishment after doing it.
- Lessons should be in small videos and sessions. It would enable learners to save time by skipping sections that they already know. Also, it makes learning more systematic and more comfortable to resume.
- Many online courses are screencasts with the "DO-AS-I-DO" approach. In this case, learners see and type the code and pick up concepts from it. However, survey respondents voted against this approach. According to

the responses, an ideal course will have these three ingredients apart from screencasts:

- Conceptualization process: thinking a project from the concept and covering the entire thought process
- Logic building
- Problem-solving approaches

• Interview Readiness: Learners do most of these online courses to get better jobs or employment purposes. Therefore, respondents have a big ask that the course covers some interview preparation tips related to course material. Some programming languages/frameworks have age-old interview questions/ patterns. It is then advisable to cover those as part of the course.

5. **Testing and Assignments**

• Actual exercises and challenges after each section/ lesson to test the concepts covered.
• Respondents were vocal about the need for lots of exercises, mini-projects, and major projects. It is because once a learner grasps a new language or framework, and he/she wants to try it out. However, it is unsure as to what all things can be created using the concepts. A given problem/project would scope the learning to it as the instructor has a much broader idea.
• The question inquired things which learners dislike; one thing that came out quite strong is being tested on something not covered in the course. It should be taken into consideration while creating assignments and quizzes.
• Capstone projects are the idea in online courses where everything taught over many lessons are brought together to solve one big problem and develop a practical application. Survey respondents were quite optimistic about such assignments at the end of a vast course.

6. **Post course Engagement**

• Course completion should be done with project ideas to practice upon and build expertise.
• One thing that respondents dislike about online courses is that some do not provide a book or some material to follow through. A supplement reading material would help in cementing the concepts further.
• In a technology spaced world, it is easier to follow thought leaders in any industry to know more about an area and stay on top of the trends. Respondents demand that courses should list out people or blogs to follow for a wholesome learning experience.

7. **Individual Ideas**

• Some of the respondents said that teaching via gaming is an effective technique. This technique is used in small tutorials and is very famous for learning Cascading Style Sheet (CSS) [4, 11, 20]. It may be that this could be explored more for learning other disciplines as well.

- Few of the respondents pointed out that the online course should be accessible to the physically challenged crowd. It is an area that needs much thought. How can one make videos accessible more? Moreover, that too videos with coding content/text content?
- Some respondents pointed out that they do not mind paying for suitable courses, but the course should be at the price point of Udemy [23]. Udemy is an online marketplace for courses that are incredibly professional, short, and to the point. It provides a completion certificate, and since the platform is mostly on "sale," the courses' prices are way less than competing Coursera and Udacity courses. It needs further research to ascertain whether the universities offering online courses through the MOOC platforms are being viewed as value for money by the course takers.
- The survey question asked for things people love in online courses. Survey responses are freedom of time, place, and what to learn is much appreciated, and the reason why MOOCs are being consumed at this rate.

5 Discussion

The responses from the survey were pinned into sections depending on the number of occurrences. There is an entire section about instructors, and learners had given feedback on their presence, energy, tone, and sense of humor.

Adamopoulos [1] has analyzed student retention in MOOCs. He has suggested that professors/instructors are the most important in online course retention and have the largest positive effect on a course. This trend is more supported on sites like Udemy, where multiple courses are offering the same subject, but instructors with higher ratings keep getting more students. New courses from these instructors see multiple signups, and the students have tuned themselves to the teaching style.

The idea of democratized learning via online resources seems exciting because it opens many doors for the learner. One of these doors is a job, a recurrent theme in the survey is about learning to build logic and getting sections dedicated to interviews. In another study, by Christensen et al. [8] have concluded that 21.3% of the responders participate in MOOCs to "gain specific skills to get a new job" while the other 49% said that they do it to "gain specific skills to do any job better." The results from these surveys are a clear indicator that the course should be job-oriented. It makes sense seeing in computer science learning people are paying $200 for egghead.io [10] or $400 for Frontend Masters [12] subscription every year. These sites produce courses for job-specific requirements, and most of the time, the employers end up buying bulk licenses for employees. To summarize, adding job-related projects would help students to commit more time and money to the course, knowing that knowledge gained will be compensated later.

Respondents have asked for exercises, assignments, and small projects between courses and even after the course. Quizzes have proved an effective way of increasing retention of the topic covered in memory. Quizzes require retrieval and thus makes a student's brain to retain the content better. According to Brown et al. [5], while quizzes serve this purpose, coding or work assignments help cement the learning even further. Koedinger et al. [15] support this as well, going as far as to claim that the students benefit as much as six times when doing things rather than watching only video content. While multiple-choice questions are easy to evaluate, coding, or work assignments pose a challenge in evaluation when the number of students is massive. To solve this problem, let us go back to one of the survey questions: teaching via games. That suggestion, if implemented properly, has two benefits:

(a)   it would encourage a student to "do more" rather than just "watching" the video, thus increasing participation and retention, and

(b)   it makes grading assignments automated, just in time, and thus easier to scale the course as well [22].

While the survey covers most of the points, there is one crucial point that respondents have not included. This miss could be attributed to the fact that the tool they did not talk about was the tool they were using, i.e., the online learning forums. Almost all MOOCs today provide a forum for discussions and Q&A. These groups are quite helpful and should not be missed if creating a new course today. Students need these forums to ask for help when facing difficulty in the absence of which they are most likely to cease participation in the course [9]. However, in today's age choosing between a forum, a Facebook group, a Twitter handle, and many more choices can be exhausting. The work of Alario-Hoyos et al. [2] is quite helpful as it points out that students' most preferred tool is a forum. It goes forward to state that too many choices can, in fact, "hinder the real learning process" since participants can get let lost in too many posts without the proper filtering of relevant material.

The human brain responds to motion better. It is the reason all user interfaces today use animation to draw user's attention. The same is true while teaching. Therefore, survey respondents pointed out that they prefer to live code compared to slides. Learners also prefer professor/instructor on screen rather than just a voice over with screencast. It follows the feedback from the edX MOOC platform, which pointed out that participants would prefer a hand, writing an equation or sentence on paper than stare at with writing already on it [17].

In this study, we came out with learners' points to make the learning experience better following some previous research. However, a few of the ideas around making courses more accessible are unexplored. Accessible in the sense of getting it to people with disabilities. Furthermore, a price point, which is sensitive to the learner's spending capabilities are both excellent points to work.

Having blogs and people to follow at the end of a course is something we have not seen in any online course or MOOC. However, it points to the apparent need for people to stay updated using social media tools.

5.1 Limitations of Study

This survey was conducted on an open platform, and volunteers responded out of enthusiasm. Therefore, it suffers from response bias and does not have many dislike points around MOOCs.

Out of 211 responses, most responses were from men, and almost all (barring one) are from English speakers. While this group is the primary consumer of MOOCs, it would be good to include people from marginalized representation and work toward their inclusion. The survey was shared in groups that had computer science or STEM learners only.

Since the survey questions were open-ended, there are areas not covered in responses but useful when creating a MOOC. There were not responses around learning groups. Although all viewed the answers, there is a possibility some did not participate because their ideas were already voiced. Running an open for all, visible to all survey has its advantages and disadvantages as well. The absence of anonymity on open social media groups might be detrimental to many prospective participants.

6 Conclusion

The survey was conducted to understand the learner's psyche and need to create better courses. It is enough to say that the outcomes help any new course creator and make common sense.

When creating a course, two things are universal: (1) either the instructor is obsessed with getting the material so right that they forget about the generic points, about presentation and course structure (2) or, the general idea is to take an already taught classroom course, record videos and put it online trying to create the same experience. Both overlook the fact that MOOC is different and needs to be handled differently. With the reach presented by current MOOCs (the massive word is there for a reason), even special care should be taken while creating one, especially when you have some university or other credentials backing it.

This paper's findings will serve as a checklist of things to do before and after preparing an online course. The discussion has points to go through before even start preparing the course to one after the content is entirely consumed, i.e., the capstone project and providing whom to follow to read the next list.

References

1. P. Adamopoulos, What makes a great mooc? An interdisciplinary analysis of student retention in online courses, in *ICIS* (Milano, Italy, 2013), pp. 1–21
2. C. Alario-Hoyos, M. P´erez-Sanagust´ın, C. Delgado-Kloos, M. Mun˜oz-Organero, A. Rodr´ıguez-de-las Heras, et al.: Analysing the impact of built-in and external social tools in a

mooc on educational technologies, in *European Conference on Technology Enhanced Learning* (Springer, Berlin, 2013), pp. 5–18

3. T. Anderson, Teaching in an online learning context. Theory Pract. Online Learning **273** (2004)
4. C. Battle, Css code-golfing. https://cssbattle.dev/ (2020) [Online]. Accessed 29 June 2020
5. P.C. Brown, H.L., Roediger III, M.A. McDaniel, *Make it Stick* (Harvard University Press, 2014)
6. H. Chen, C.W. Phang, C. Zhang, C., S. Cai, What kinds of forum activities are important for promoting learning continuance in moocs? in *PACIS* (2016), p. 51
7. S. Cho, How to analyze open-ended survey responses. Available: https://www.surveymonkey.com/curiosity/open-response-question-types/. Online; Accessed 10 Apr 2020
8. G. Christensen, A. Steinmetz, B. Alcorn, A. Bennett, D. Woods, E. Emanuel, The mooc phenomenon: who takes massive open online courses and why? Available at SSRN 2350964 (2013)
9. D. Coetzee, A. Fox, M.A., Hearst, B. Hartmann, Should your mooc forum use a reputation system? in Proceedings of the 17th ACM conference on Computer supported cooperative work & social computing (ACM, 2014), pp. 1176–1187
10. O. Veblen.: Learn to code—egghead.io, video tutorials for badass web developers. Available https://egghead.io/ Online. Accessed 30 Apr 2020
11. Froggy: Flexbox Froggy. https://flexboxfroggy.com/ (2020), [Online]. Accessed 29 June 2020
12. FrontendMasters: Master the important JavaScript and front-end development skills—FrontendMasters, web development course providers. Available https://frontendmasters.com/ [Online]. Accessed 15 May 2020
13. K.F. Hew, Promoting engagement in online courses: What strategies can we learn from three highly rated moocs. Br. J. Edu. Technol. **47**(2), 320–341 (2016)
14. K.F. Hew, Unpacking the strategies of ten highly rated moocs: implications for engaging students in large online courses. Teachers Coll. Rec. **120**(1), n1 (2018)
15. K.R. Koedinger, J. Kim, J.Z. Jia, E.A. McLaughlin, N.L. Bier, Learning is not a spectator sport: doing is better than watching for learning from a mooc, in *Proceedings of the second (2015) ACM conference on learning scale* (ACM, 2015), pp. 111–120
16. B.A. Oakley, T.J. Sejnowski, What we learned from creating one of the world's most popular moocs. NPJ Sci. Learn. **4**, 1–7 (2019). https://doi.org/10.1038/s41539-019-0046-0
17. L. Pappano, *The New York Times*, 2 November 2012. Available https://www.nytimes.com/2012/11/04/education/edlife/massive-open-online-courses-are-multiplying-at-a-rapid-pace.html [Online]. Accessed 26 Apr 2020
18. G. Salmon, *E-tivities: The Key to Active Online Learning* (Routledge, 2013)
19. D. Shah, By the numbers: MOOCs in 2019. *Class Central*, 11 December 2018. Available https://www.classcentral.com/report/mooc-stats-2019/ [Online]. Accessed 26 Mar 2020
20. Sid: Learn react hooks by building a game. https://gumroad.com/l/hooks-team (2020) [Online]. Accessed 29 June 2020
21. R. Sujatha, D. Kavitha, Learner retention in mooc environment: analyzing the role of motivation, self-efficacy and perceived effectiveness. Int. J. Educ. Dev. Using ICT **14**(2) (2018)
22. N. Tillmann, J. De Halleux, T. Xie, S. Gulwani, J. Bishop, Teaching and learning programming and software engineering via interactive gaming, in *Proceedings of the 2013 International Conference on Software Engineering* (IEEE Press, 2013), pp. 1117–1126
23. Wikipedia contributors: Udemy—Wikipedia, the free encyclopedia. https://en.wikipedia.org/w/index.php?title=Udemy&oldid=906447764 (2020) [Online]. Accessed 29 June 2020
24. S. Zheng, M.B. Rosson, P.C. Shih, J.M. Carroll, Understanding student motivation, behaviors and perceptions in moocs, in *Proceedings of the 18th ACM Conference on Computer Supported Cooperative Work & Social Computing* (ACM, 2015), pp. 1882–1895

A Neural Attention Model for Automatic Question Generation Using Dual Encoders

Archana Praveen Kumar (ID), Gautam Sridhar, Ashlatha Nayak (ID), and Manjula K Shenoy (ID)

Abstract In the field of education, framing right questions is an important way of measuring the knowledge of an individual, and automation of the generation of questions would reduce the strain on educators and thus increase efficiency in learning. Literature survey reveals that while there are several promising methods for testing comprehension generating questions, but none of them can be used reliably for generating meaningful questions all the time, and thus, further research is required. A novel method of generating questions is presented in this paper from input text, by encoding the answers into the model and using attention that learns the dependencies between both input text and answers and between input text and questions, thus generating questions which are relevant to the answer given. The dataset used is the SQuAD dataset, and the model produces fluent questions which have been evaluated with the required metrics.

Keywords Comprehension question generation · Long short term memory · Word embeddings

A. P. Kumar (✉) · A. Nayak
Department of Computer Science and Engineering, Manipal Institute of Technology,
Manipal Academy of Higher Education, Manipal 576104, India
e-mail: archana.kumar@manipal.edu

A. Nayak
e-mail: asha.nayak@manipal.edu

G. Sridhar
Department of Electronics and Communication Engineering, Manipal Institute of Technology,
Manipal Academy of Higher Education, Manipal 576104, India
e-mail: gautam.s2@learner.manipal.edu

M. K. Shenoy
Department of Information Communication and Technology, Manipal Institute of Technology,
Manipal Academy of Higher Education, Manipal 576104, India
e-mail: manju.shenoy@manipal.edu

© The Author(s), under exclusive license to Springer Nature Singapore Pte Ltd. 2022   427
P. Shetty D. and S. Shetty (eds.), *Recent Advances in Artificial Intelligence and Data Engineering*, Advances in Intelligent Systems and Computing 1386,
https://doi.org/10.1007/978-981-16-3342-3_34

1   Introduction

Education is continuously evolving, along with this, the way we go about educating is also tremendously changing. Questions are a fundamental tool in gauging the skills of a student, and the ways of generating questions need not only a degree of skill, but also good amount of time. One most important application of generating questions in the field of education is to generate comprehension questions for passages. Question generation deals with the problem of generating or "asking" question, given a sentence or a paragraph. Figure 1 as shown has three manually generated questions from a given corpus. Such reading comprehension question has variety of uses, as the ability to ask questions from a given statement, which is a defining characteristic of intelligence. The initial methods of question generation used specific heuristics, which while not completely flawed, could not be generalized to different domains. These approaches have also created fairly simple questions, which were almost completely from the input text. Most recent research in this direction has been done using deep learning. One of the biggest advantages of this study contributed to the creation of labelled datasets. One such labelled standardized dataset is Stanford Question Answering Dataset (SQuAD) which was created by Rajpurkar et al. [1] .

The literature review shows all the models which have been implemented in this direction wherein the initial works were done by Suskever et al. [2]. The research had proposed a sequence-to-sequence model which would work on any sentence making minimal assumptions on the structure of the sequence. This required mapping the input to a fixed dimension vector using a multilayered long short-term memory (LSTM) and using another LSTM which could decode the target sequence sentence from the input vector. Though the experimental system was used from translation of English language to French, it paved the right path for making advancements in

---

Sentence:

Delhi is also widely believed to have been the site of Indraprastha, the legendary capital of Pandavas during the time of Mahabharata.

Questions:

Which was the legendary capital of Pandavas during the time of Mahabharata?

Indraprastha

Whose legendary capital was Indraprastha during the time of Mahabharata?

Pandavas

Which city is believed to have been the site of Indraprastha?

Delhi

Fig. 1 Example of a sentence picked from the first paragraph of the article Delhi, along with the manual generated questions and answers

the current question generation which is what has been used in this paper as well. Several papers since the work of Sutskever have reused the researcher's work in various ways; however, the base has remained the same. The inputs to the encoder and decoder architecture, however, have varied from approach to approach.

Xingwu et al. [3] encoded the position of the answer and used a copy mechanism; however, this caused the words to move far away from the answer. Xiyao Ma et al. [4] encoded lexical features along with the answer position as the input to the encoder. Yanghoon Kim et al. [5] replaced the answers with a special token and added a second encoder in order to obtain the information from the answers. Vrindavan and Marilyn [6] also encoded lexical features, but used more than just the basic named entity recognizer (NER) and parts of speech (POS) features, e.g. case feature, answer position, etc. Qingyu Zhou et al. [7] built upon the base model, added a copy mechanism to it and used lexical features to generate questions, while Du et al. [8] used a simple sequence-to-sequence model with attention for the same task.

The previous works do an excellent job of generating questions; however, the fluency of the questions is not the best in all types of questions, and they struggle with paragraphs. Our model is simpler and produces almost equal or better results, generating fluent questions even from the paragraph input. Our model uses two encoders, one to encode the answer, and one to encode the input text. The model uses the attention score between these two encoders and the decoder, to model the probability distribution of the vocabulary, in order to generate the questions. The model thus outperforms other previous works and therefore generates fluent questions which pertain to the answer and produces diverse questions.

The rest of the article is subdivided as follows. The related works are presented in Sect. 2. Section 3 gives brief explanation of background knowledge required to solve the problem. Methodology is explained in Sect. 4. Section 5 provides the analysis and discussions of the results. Finally, we conclude our work in Sect. 6.

2 Related Work

Reading comprehension, a challenging task for machines requires not only the natural language understanding but also the real-world knowledge of the context used in the paragraphs or sentences as stated by [1]. Moreover, in this direction, the researchers have been contributing in making many datasets available so that the training of this task to the machine could be easily done. Weston et al. [9] have created bAbI, a synthetic dataset which has features for twenty different tasks. Hermann et al. [10] have created a corpus for cloze questions by replacing the required entities with their placeholders in abstractive summaries generated from CNN/Daily Mail news articles. Chen et al. [11] have claimed that the former dataset is easier and the system performs very well using the CNN/Daily Mail dataset. Richardson et al. citeRich has created MCTest consisting of natural questions paired by four options. But this dataset is too small to be used for training answering models. Of recent advancements, Rajpurkar

[1] has relased SQuAD dataset which overcomes all the aforementioned problems in terms of size and quality.

Question generation is the most sought research of this generation and has attracted many of the researchers to actively promote experiments in this direction. Most works tackle the methods by imposing rules and patterns which transform the input into a representation which could be then transformed into an interrogative approach. The related works target at reviewing the research which have made use of models to train the machine and then automatically generate the questions. Xingwu Sun et al. [3] for question generation used a sequence-to-sequence approach, along with a pointer generator model. It combined a position aware and answer focused model. In position aware model, they directly incorporated relative word position embedding. This gave the model information about where the model is and what context words are close to it. In answer focused model, they added a restricted set of question word vocabularies in order to explicitly model the question words. The model produces a question word based on the answer embedding. The two models are combined using their context vectors, and an output is generated. Due to the encoding of answers, the model is better at generating a question word relevant to the answer. However, as the position of the answer is paid more attention to, the other words which are far from the answer in the context, but still relevant, are not taken into consideration all the time.

Xiyao Ma et al. [4] use a sequence-to-sequence model, with the inputs as the named entity recognition (NER) features, the parts of speech (POS) features and answer position features along with the context. The output of the encoder is passed through an answer aware gated function and then passed through the decoder. An attention mechanism along with a copy mechanism is implemented which performs sentence semantic matching using the last hidden state of the decoder output and the output of the answer gated function. The researcher then uses an attention flow network, in order to improve answer position inferring. The model improves upon previous work by improving the question words and copying only the parts of the context that are of consequence. Yanghoon Kim et al. [5] use a sequence-to-sequence model with answer separation, which instead of adding features to the input passage replaces the corresponding answer with a special token. It then uses another encoder to encode answer information. An answer separated decoder is used along with a keyword net, which extracts the keyword feature from the given passage. Finally, it uses a retrieval style word generator, which then replaces the output layer of the decoder.

A sequence-to-sequence approach is again made use by Vrindavan and Marilyn [6]. The inputs to the encoder are varied, wherein not only are the tokens passed to the encoder but also the features regarding the data are given along with details about the answer. The answer signals are fed by the use of a binary signal, along with the tokens. Case features (whether or not the token has a capital letter), NER features and coreferences feature are also added to the input of the encoder. After the encoder completes the calculations, the output is concatenated with the token level encoding, which allows the answer encoding to affect the final output more. For training initially, an encoder–decoder model is given the question as the input. After the training, the model learns to "copy" the input as the target. The encoder is then decoupled from the model. A new encoder is trained with the sentences, and the previous encoder's embeddings are used as the ground truth, training the model to maximize the similarity between the embeddings of the new and the old encoders. The model generates factual questions only and also requires answers from the context. Therefore, future work on generating answer tokens from context is required.

One more research contribution in this direction was from Zhou et al. [7] who build upon the neural question generation (NQG) framework for their study, then went on making NQG++ which was similar to NQG system with a difference that it encorporated a feature rich encoder into the sequence-to-sequence architecture. It also used a maxout layer as stated by Goodfellow et a. [12], along with a copy mechanism which was implemented by Gulcehre et al. [13]. This research in terms of question generation improves BLEU scores over the previously used NQG framework. Likewise Xinya Du et al. [8] used a simple sequence-to-sequence mechanism by incorporating an attention mechanism to the model. This caused the model to pay more attention to the relevant words in the context given and hence improves performance. The researcher placed an attention layer between the encoder and decoder, to generate a context vector and used it for the final prediction. It, however, does not produce the best kind of question for all types of questions and thus can be improved more.

In the model developed by Wang et al. [14], linguistic features have been used, along with answers encoded with the context as the input to the network. It uses an attention mechanism and a copy mechanism in order to generate readable questions. The model generates human readable questions, which are better in quality than the work before it. However, it only generates factual questions. More work is required for more advanced questions. To the best of our knowledge our research given potential results and has outperformed some of the baseline models in this direction. We use SQuAD dataset in our work and similarly focus on the generation of natural questions for reading comprehension materials through automatic approach.

3   Task Definition

Given an input text, we have to generate a meaningful question which pertains to the input text and is coherent in its grammatical structure. The question length can be arbitrary, and the lengths chosen for the same are elaborated further in the implementation details section. In the next section, we will elaborate on the architecture used for generating questions.

3.1   Model

Our model which is partially inspired by the sequence-to-sequence model done by Sutskever [2] is just similar to a human who could solve the task of reading comprehension. In order to generate an interrogative sentence from the declarative, there is

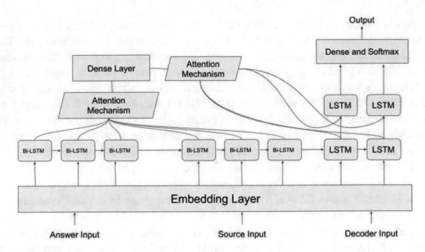

Fig. 2 Proposed model

a need to pay attention to certain parts of the input sentence as well as to the context in which the sentence has been written. Hence, along with the attention mechanism with the encoder and decoder architecture, the model is able to focus on certain parts of the input sentence while generating the word during decoding.

Model as shown in Fig. 2 is inspired from the seq2seq model Sutskever et al. [2]. The model has two encoders instead of the usual one, where one encoder is used to encode the paragraph, while the other is used to encode answers. The decoder takes hidden state from the second encoder (paragraph encoder) and is trained using a shifted version of the sequence.

3.2 Sequence-to-Sequence Network

The RNN encoder–decoder model is used for generating task-specific sequential output given sequential input. The architecture is widely used for problems such as machine translation, text summarization and dialogue models. In this case, we are using an input of paragraphs and generating questions from the paragraph which are relevant to it. There are two parts to the model, namely the encoder and the decoder. The encoder, an RNN network, takes in a variable length input and gives an output which encodes the data in the input text into a single vector. The decoder uses this output as the input along with a shifted version of the target to generate the questions.

3.3 LSTM

LSTM networks are RNNs which can learn long-term dependencies. They have three gates, namely input, forget and output. Forget gates are used in order for the LSTM network to decide whether or not the information has to be passed on to the next cell, or not. Information from previous hidden state and the current input is passed through a sigmoid function, and depending on the value received, the forget gate is set to 0 or 1. Input gate is used to update the cell state. First the input and the hidden state are passed through a sigmoid. Both the two are also passed through a tanh activation function. Lastly, the tanh output and the sigmoid output are multiplied together to give the input gate output. Then, the cell state is updated. This is done by first multiplying the cell state with the forget vector. Then, the input gate output is added to the cell state, thus updating its values. Finally, the output gate is used to decide the hidden state to be passed along to the next LSTM cell. We pass the previous hidden state and the current input through a sigmoid function. The cell state is passed through a tanh. The output of the previous two calculations is then multiplied together to give us the hidden state. A bidirectional LSTM uses the output of two LSTMs to form a final vector, with both the LSTMs going in opposite directions of each other and then are combined (concatenated).

3.4 Embeddings

Word embeddings are used to model language data. Each word is represented by a list of numbers, which correspond to some information about the word. They are made such that words which are similar have similar "embeddings" that is the lists of numbers are closely related. The closer the embeddings are, the similar the words. A word embedding was used to represent each of the words in the vocabulary. The dimension used was 512, and it was trained from scratch on the data.

3.5 Answer Encoder

The answer encoder is a bidirectional LSTM with hidden size and two layers. At any timestep, the LSTM cell of the encoder takes in the input from the embedding, w_t, at that timestep, and the hidden states of the previous timestep. The final output of the encoder as given in Eq. 3 is the concatenated output of both the forward as shown in Eq. 1 and the backward LSTMs as given in Eq. 2, giving us an output with a hidden size of 2*hidden size of input. The output vector is denoted as h_t.

$$\overrightarrow{h_t} = \text{LSTM}(w_t, \overrightarrow{h_{(t-1)}}) \tag{1}$$

$$\overleftarrow{h_t} = \text{LSTM}(w_t, \overleftarrow{h}_{(t+1)}) \tag{2}$$

$$h_t = [\overrightarrow{h_t}; \overleftarrow{h_t}] \tag{3}$$

3.6 Paragraph Encoder

The context encoder is also a bidirectional LSTM with the same hidden size and two layers. The input to this encoder is the paragraph. The hidden state is initialized by the output hidden state of the previous encoder. The final output of this encoder is denoted by p_t. The output of this encoder is passed through the attention layer.

3.7 Attention

Attention is used to allow the model to learn the dependencies between the input texts used and the questions generated. It is usually used to give the decoder information about each of the encoder hidden states and learn which of those are important; however, we have used it to also find the area of importance between the answer and the input text. It learns the alignments, that is, the relation between the two vectors. This is especially useful for long input data. A scoring function is first used on each of the hidden vectors as shown in Eq. 4, which gives the importance of each of the vectors. This is then passed through a softmax layer, forming the alignment vectors given in Eq. 5. Then, alignment vector is multiplied with the hidden vector output of the encoder to form the context vector given in Eq. 6. The final context vector is then fed into the decoder, by various ways depending on architecture to architecture.

$$\text{score} = (h_t, h_s) = h_t^T \cdot h_s \tag{4}$$

$$\alpha_t = \text{Softmax}(\text{score}(h_t, h_s)) \tag{5}$$

$$c_t = h_t^T \cdot \alpha_t \tag{6}$$

3.8 Answer Attention

The hidden states from the answer encoder, ht, and the paragraph encoder, pt are passed through the attention layer, to give the context vector $c^1{}_t$. The context vector is concatenated with the hidden state output of the paragraph encoder. This is then passed through a dense layer, to align the dimensions with the decoder. This gives us the context vector $c^1{}_t$ as shown in Eq. 7.

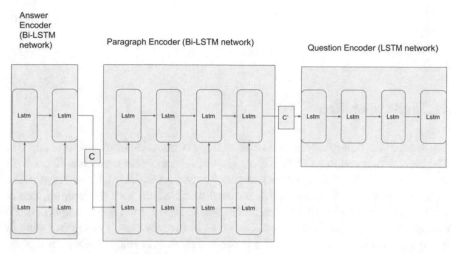

Fig. 3 Vanilla encoder–decoder architecture without attention

$$c_t^1 = \text{Attention}(h_t, p_t) \tag{7}$$

3.9 Decoder

The decoder is an LSTM with hidden size twice the hidden size of the encoders (for correct dimensions). It is initialized with the hidden state of the paragraph encoder. The questions, shifted by one, are fed into the decoder as inputs, giving us the output vector d_t as shown in Eq. 8. The output of the first LSTM layer of the decoder and the context vector c_t^1 is passed through the attention layer, giving us context vector c_t as given in Eq. 9. Then, c_t and the output of the decoder d_t are concatenated together and passed through a second LSTM layer as given in Eq. 10. The output of which is passed through dense layer with a softmax activation shown in Eq. 11 to finally get the output in the shape of (batch_size, sequence_length, vocab_size). P_v is the probability distribution of the predicted word over the entire vocabulary, and W_v and b_v are parameters. The architecture of encoder and decoder has been adapted from the vanilla encoder–decoder architecture without attention shown in Fig. 3. As shown in Fig. 3, C and C' are answer context vector and paragraph context vector, respectively.

$$d_t = \text{LSTM}(q_t, d_{(t} - 1) \tag{8}$$

$$c_t = \text{Attention}(c_t^1, d_t) \tag{9}$$

$$s_t = \text{LSTM}||([c_t; d_t], s_{(t} - 1)) \tag{10}$$

$$P_v = \text{Softmax}(W_v \cdot s_t + b_v) \tag{11}$$

4 Implementation

The source vocabulary size is set to 50k, and the target vocabulary size is set to 35k. The rest are mapped to an OOV token. The model was implemented on Keras with tensorflow as backend. Dropout between layers was set to 0.4 implemented as in the work of Zaremba et al. [15]. Optimizer used was Nadam, with the learning rate at 0.001, with a batch size of 32. The model was trained for ten epochs. Sparse cross entropy was chosen as the loss function so as to save memory. During decoding, a beam search with a beam width of 5 is used. Decoding is stopped when an <eos> token is generated or the maximum permissible length of the question is reached.

4.1 Parameters

Number of paragraph tokens : 130
Number of question tokens : 13
Number of answer tokens : 5
Training triplets : 79488.

5 Evaluation

We have used automatic evaluation wherein the metrics used are BLEU Papineni et al.[16], METEOR Denkowski and Alon [17] and ROUGE-L Lin et al. [18]. BLEU 1–BLEU-4 gives a score to the hypothesis text, by counting the matching 4-gram in the hypothesis to the 4-gram in the reference text. METEOR—This metric uses synonym matching and stemming and is the harmonic mean between the 1-gram precision and recall, with more weightage being given to recall. It was used to find correlation at the sentence level. ROUGE-L measures the longest matching subsequence between the reference and the hypothesis text.

6 Results and Analysis

We have compared our model perfomance with two models of Quinyu et al. [7] and of Yanghoon et al. [5] using METEOR, BLEU-4 and ROUGE-L metrics. Table 1 shows the comparison of the models on the above metrics.

As shown in Table 1, the proposed model works better with sentence input; however, paragraph input also gives results which are better than the SOTA algorithms. Our proposed model outperforms all the previous NQG models in all metrics except ROUGE-L. There is a significant increase in the BLEU-4 and the METEOR metric. Figure 4 shows the examples of questions generated by our model. Here, source is the input text; reference text is the gold standard question, and the generated question is the question generated by our proposed model. The text in bold is the answer for the question. The model predicts questions which are grammatically coherent and fluent in the language, even if the gold standard is somewhat incorrect as shown in the first example of Fig. 4.

The questions generated by the model are coherent and take the answer into account. However, the model predicts a different question than the one in the test set, thus bringing down the BLEU and METEOR score even when the question in and of itself is acceptable. In some cases, however, the model recognizes a date or a quantity

Table 1 Comparison of the models on the METEOR, BLEU-4 and ROUGE-L metrics

Models	BLEU-4	METEOR	ROUGE-L
NQG++ [7]	13.29	17.59	40.75
ASs2s [5]	16.20	19.92	43.96
Our model	18.27	23.08	42.97
Our model with paragraph	17.26	20.90	42.22

Source	The Aare also contains the waters from the 4,274 m (14,022 ft) summit of **Finsteraarhorn**,the highest point of the Rhine basin.
Reference	Where is the highest point of the rhine basin?
Generated	What is the highest point of the rhine basin called?
Source	In 1960 **Maurice Hilleman** of Merck Sharp Dohme identified the sv40 virus, shown to cause tumors in many mammalian species .
Reference	Who identified the sv40 virus?
Generated	Who identified the sv40 virus?
Source	The word insect comes from the latin word **insectum** meaning with a notched or divided body or literally cut into from the neuter singular perfect passive participle of insectare (to cut into to cut up from in – into) and secare (to cut) ; because insects appear cut into three sections. .
Reference	What is the latin term for insect?
Generated	What is the latin term for insect?

Fig. 4 Examples of questions generated by the model

Source	Apollo and his sister **Artemis** can bring death with their arrows.
Reference	It was believed that this woman could bring death with her
Generated	Who is Apollo's sister?
Source	Indic scripts such as tamil and devanagari are each allocated only **128 code points** matching the ISCII standard .
Reference	What is the ISCII standard?
Generated	How many code points are Tamil and Devanagari allocated?

Fig. 5 Examples where the generated questions gets a lower score

and provides a question which has a separate date or quantity as an answer, but is not the same as the one provided. The model works well for sentence input; however, its efficacy reduces if the input length becomes larger. Fig. 5 shows examples where the model predicts fluent questions; however, it gets a very poor score as the generated question is not the one given as the gold standard. This indicates a need for a better way to score questions, as there exists no metric solely for scoring questions. The model also generates questions which are not grammatically correct, but the meaning does come through sometimes. However, it does not produce the best question all the time and fails to copy the correct words from context on several occasions.

7   Conclusion and Future Work

In this paper, we presented a novel method of generating questions from input text, which is one of the key indications of intelligence. We use a dual attention-based model to generate the questions. The model works with input sentences to automatically generate questions for reading comprehension. Using an attention-based neural network model for this task has proved to be useful to generate fluent and coherent questions which has been compared with automatic evaluation metrics. The model produces questions which are grammatically coherent and relevant to the text input. It outperforms state-of-the-art models and produces good questions even with paragraph input. As there is no effective metric that can classify questions, it is hard to train the model.

Attention implemented was global; however, as the paragraph size is large, the model would benefit from local attention. The model will also benefit with a copy network, as it tends to not get the correct word from the context. Giving the model which type of questions to predict would also allow it to generate specific questions and not just random ones the model tries to. Future work is to extend the model to predict specific questions rather than the random question which the model generates. We plan to extend our research by generating options in an automatic mode in future. Benefit with respect to local attention and copy network as the size of paragraph increases needs to be checked as a future extension of the research.

References

1. P. Rajpurkar, J. Zhang, K. Lopyrev, P. Liang, SQuAD: 100,000+ Questions for machine comprehension of text, in *Proceedings of the 2016 Conference on Empirical Methods in Natural Language Processing (EMNLP)Association for Computational Linguistics* (2016), pp. 2383–2392
2. I. Sutskever, O. Vinyals, Q.V. Le, Sequence to sequence learning with neural networks. Adv. Neural Inf. 3104–3112 (2014)
3. S. Xingwu, L. Jing, L. Yajuan, H. Wei, M. Yanjun, W. Shi, Answer-focused and position-aware neural question generation, in *Proceedings of the 2018 Conference on Empirical Methods in Natural Language Processing* (2018), pp. 3930–3939
4. X. Ma, Q. Zhu, Y. Zhou, X. Li, Improving question generation with sentence-level semantic matching and answer position inferring, in *Proceedings of the AAAI Conference on Artificial Intelligence* (2020), 8464–8471
5. K. Yanghoon, L. Hwanhee, S. Joongbo, J. Kyomin, Improving Neural Question Generation using Answer Separation (2018). arXiv preprint arXiv:1809.02393
6. H. Vrindavan, W., Marilyn, Neural generation of diverse questions using answer focus, contextual and linguistic features, in *International Conference on Natural Language Generation (INLG)* (2018), pp. 296–306
7. Z. Qingyu, Y. Nan, W. Furu, T. Chuanqi, B. Hangbo, Z. Ming, Neural question generation from text: a preliminary study. CoRR, abs/1704.01792, 2017. http://arxiv.org/abs/1704.01792
8. D. Xinya, S., Junru, C. Claire, Learning to ask: neural question generation for reading comprehension, in *Proceedings of the 55th Annual Meeting of the Association for Computational Linguistics, ACL 2017* (1: Long Papers), pp. 1342–1352
9. J. Weston, B. Antoine, C. Sumit, R. Alexander, M. Bart, M. Armand, M. Tomas, Towards AI-complete question answering: a set of prerequisite toy tasks (2016)
10. M. Hermann, K. Tomas, G. Edward, E. Lasse, K. Will, S. Mustafa, B. Phil, Teaching machines to read and comprehend, in *Advances in Neural Information Processing Systems (NIPS)* (2015), pp. 1693–1701
11. D. Chen, D., B. Jason, M. Christopher, A thorough examination of the cnn/daily mail reading comprehension task, in *Proceedings of the 54th Annual Meeting of the Association for Computational Linguistics* (vol 1: Long Papers). Association for Computational Linguistics (2016), pp. 2358–2367. http://www.aclweb.org/anthology/P16-1223
12. I. Goodfellow, D. Warde-Farley, M. Mirza, A. Courville, Y. Bengio, Maxout networks, in *Proceedings of The 30th International Conference on Machine Learning* (2013), pp. 1319–1327
13. Ç.S. Gülçehre, S. Ahn, R. Nallapati, B. Zhou, Y. Bengio, Pointing the unknown words. CoRR abs/1603.08148 (2016)
14. W. Zich, A. Lan, W. Nie, A. Waters, P. Grimaldi, R. Baraniuk, QG-Net: a data-driven question generation model for educational content, in *Proceedings of the Fifth Annual ACM Conference on Learning at Scale* (2018), pp 1–10. https://doi.org/10.1145/3231644.3231654
15. W. Zaremba, I. Sutskever, O. Vinyals, Recurrent neural network regularization (September 2014). arXiv preprint arXiv:1409.2329
16. K. Papineni, R. Salim, W. Todd, W. Wei-Jing and Z.: Bleu: a method for automatic evaluation of machine translation, in *Proceedings of 40th Annual Meeting of the Association for Computational Linguistics*. Association for Computational Linguistics (2002), pp. 311–318. https://doi.org/10.3115/1073083.1073135
17. M. Denkowski, L. Alon, Meteor universal: language specific translation evaluation for any target language, in *Proceedings of the Ninth Workshop on Statistical Machine Translation*. Association for Computational Linguistics (2014), pp. 376–380. http://www.aclweb.org/anthology/W14-3348
18. C. Lin, Rouge: a package for automatic evaluation of summaries, in ed. by S. Szpakowicz, M.-F. Moens, *Text Summarization Branches Out: Proceedings of the ACL-04 Workshop*. Association

for Computational Linguistics (2004), pp. 74–81. http://aclweb.org/anthology/W/W04/W04-1013.pdf

A Comparative Study of Efficient Classification Models

Roopashri Shetty⬤, M. Geetha⬤, Dinesh U. Acharya⬤, and G. Shyamala

Abstract The lung cancer is a principle cause of deaths in both women and men. Diagnosis and treatment depend on the cancer type, its stage, and the performance status of status. Depending on the cancer stage, treatments such as chemotherapy, radiotherapy or surgery will be decided. Patient's survival can be determined based on his overall health, stage and other factors. Only 14 out of 100 people survive around five years after the diagnosis. This paper gives a comparative study of various existing efficient classification models carried out for diagnosing lung cancer and K-Nearest Neighbor classifier outperforms other classifers.

Keywords Accuracy · Classification · Lung cancer · Precision · Prediction

1 Introduction

The classification is a main subfield of data mining [1] which builds a model for predicting the classes. The model is first trained, and it is tested for new data. Decision tree, probability based, k-nearest neighbor, logistic regression, etc., are some of the classification models. The main application of the classification algorithms is in the

R. Shetty (✉) · M. Geetha · D. U. Acharya
Department of Computer Science and Engineering, Manipal Institute of Technology,
Manipal Academy of Higher Education, Manipal, Karnataka 576104, India
e-mail: roopashri.shetty@manipal.edu

M. Geetha
e-mail: geetha.maiya@manipal.edu

D. U. Acharya
e-mail: dinesh.acharya@manipal.edu

G. Shyamala
Department of Obstetrics and Gynaecology, Kasturba Medical College,
Manipal Academy of Higher Education, Manipal, Karnataka 576104, India
e-mail: shyamala.g@manipal.edu

© The Author(s), under exclusive license to Springer Nature Singapore Pte Ltd. 2022   441
P. Shetty D. and S. Shetty (eds.), *Recent Advances in Artificial Intelligence and Data Engineering*, Advances in Intelligent Systems and Computing 1386,
https://doi.org/10.1007/978-981-16-3342-3_35

area of medical science for predicting different diseases [2]. Lung cancer is one such disease which can be predicted using classification.

Some of the symptoms of lung cancer include [3]:

1. Bronchitis or pneumonia
2. Cachexia (fatigue, loss of appetite, weight loss)
3. Change in regular coughing pattern or Chronic coughing
4. Dysphasia(difficulty swallowing)
5. Shortness of breath with activity or Dyspnea
6. coughing up blood or Hemoptysis
7. Pain in shoulder ,chest , arm, abdomen
8. Wheezing.

With the increase in the use of tobacco, mortality and morbidity rate also increase. Lung cancer usually develops in the epithelium of the bronchial tree, and it may affect any part of the respiratory system. Lung cancer often takes many years to develop and people between the ages of 55 and 65 are most likely affected.

This work undertakes the experimental study to investigate the influence of symptoms and associated risk factors in the lung cancer detection in the absence of experts so that the rate of mortality and prevalence could be reduced. A comparative study of various existing efficient classification models such as K-Nearest Neighbor (KNN), Logistic Regression (LR), Naive Bayes (NB) and Support Vector Machine (SVM) is carried out for the diagnosis of lung cancer and KNN classifier outperforms other classifers. Lung cancer dataset from UCI repository is taken as input data for all the classifiers. The data is preprocessed which drops the tuples with missing values and selects the attributes which contribute to the lung cancer. Then, different classification algorithms are used to classify the lung cancer dataset into 2 classes: Yes and No.

2 Literature Review

A model was proposed by Krishnaiah et al. [4] for correct and early diagnosis of lung cancer. Naive Creedal Classifier 2 (NCC2) and One Dependency Augmented Naïve Bayes classifier (ODANB) are used in data preprocessing and its classification. The data is preprocessed by replacing the missing values with the modes and mean. Using the symptoms such as age, pain in arm and chest, sex, shortness of breath, wheezing, etc., it can predict the probability of patients getting the disease.

Hsu et al. [5] compared different methods for multiple class support vector machines which are basically designed for classifying into two classes. This model extended to obtain multi class binary classifiers by combining several SVM models. However in large dataset, generating multiple classes using SVM is difficult to achieve hence is restricted to be used only to classify smaller data set.

Yadav, Tomar [6] developed a foggy k-means algorithm for accurate clustering of real-time datasets. This algorithm considers the lung cancer data set provided by

SGPGI, Lucknow which contains nine attributes like age, sex, BMI, family history, tumor size, tuberculosis, smoking, lymph node involvement, and radiation and forms two clusters, one cluster containing the records of the patients suffering from lung cancer and the other cluster containing the record of those who are not suffering from cancer. This method also handles outliers. Using lung cancer data set, foggy k-means algorithms selects some prominent attributes and depending on the value of these attributes number of clusters are decided. The attributes are plotted and clusters are obtained.

Bala and Dr. Kumar [7] explained a classification method ANN (Artificial Neural Networks) and its variants. ANN is a collection of input-output connections having an input layer, one or more number of hidden layers associating weights and an output layer. Weight of the connection could be iteratively modified to get the proper output from ANN. And it is found that ANN is not time-efficient for large datasets.

The performance of the KNN classifier [8] uses Euclidean distance as a standard distance metric to determine the k nearest neighbors of the query data points. To make the diagnosis based on historical data,large storage of information is used in this study. The probability of occurrence of a particular ailment is computed. The accuracy of such diagnosis is increased using KNN.

Christopher and Banu [9] carried out a study of classification algorithms such as the Bayesian network,48 and Naïve Bayes, algorithms for lung cancer prediction. The dataset used has 100 records with 25 attributes. This paper analyzes the performance of classifiers to provide an early warning to the users. The Naïve Bayes algorithm outperformed other classification algorithms.

It is understood from the literature review that

1. There is a lack of efficient diagnostic method for diagnosing lung cancer.
2. The prediction of the disease becomes complex as the symptoms and impact of risk factors associated with lung cancer are not implicit and vary.

2.1 Objectives

The objective of this work is to make a comparative study of various efficient classification models with respect to their classification accuracy for diagnosis of lung cancer.

3 Methodology

3.1 Database Description

Dataset used in this study is taken from UCCI repository. 568 records with 32 initial attributes are chosen in the dataset. The attribute such as, Age, Air Pollution, Balanced Diet, Chest pain, Chronic Lung Disease, Coughing of blood, Dry Cough, Dust Allergy, Fatigue, Frequent Cold, Gender, Genetic Risk, Occupational Hazards, Obesity, Passive smoker, Shortness of breath, Smoking, Snoring, Swallowing difficulty, Use of alcohol, Weight loss, Wheezing, etc., are considered for predicting the lung cancer.

3.2 Preprocessing

The dataset is preprocessed using iPython Jupyter Notebook to drop the attributes that do not contribute to lung cancer and to drop the duplicate records and records with missing values. Finally 22 prominent attributes are considered for the prediction.

3.3 Experimental Setup

The preprocessed data is then given as input to different classifiers such as NB, KNN, SVM and LR. For the experimental analysis, and the classifiers are tested with 80:20, 50:50 and 70:30 data distributions for training and testing and performances of the algorithms are analyzed.

4   Analysis and Results

All the models have been built using Python. The performance of each model has been compared with respect to the accuracy. Accuracy is calculated as

$$\text{Accuracy} = \frac{(\text{True}_P + \text{True}_N)}{(\text{True}_P + \text{True}_N + \text{False}_P + \text{False}_N)} \tag{1}$$

where True_P = True Positive, True_N =True Negative, False_P=False Positive and False_N=False Negative

Accuracy of the models with different data distribution is shown in Table 1.

Table 1 Accuracy of the algorithms

Data distribution	SVM (%)	NB (%)	KNN (%)	LR (%)
80:20	84	82.5	91	87
50:50	78	80	87	85
70:30	81	86	89	85

Fig. 1 Accuracy of the models in 80:20 distribution

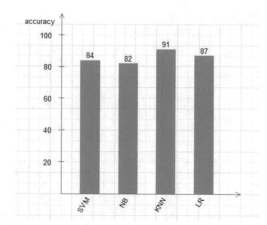

Fig. 2 Accuracy of the models in 50:50 distribution

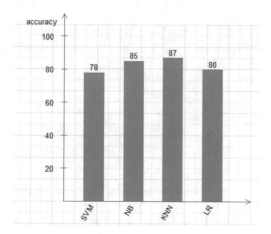

From Table 1, it is clear that KNN algorithm outperformed when compared with other classifiers. Out of three different data distributions, 80:20 gave the better result.

The graphical representation of the accuracy values with different data distribution are shown in Figs. 1, 2 and 3.

Table 2 shows the Specificity ($True_N$ Rate), sensitivity ($True_P$ Rate), Precision, Recall and F - Measure of different classifiers with respect to 80:20 distribution.

$$Sensitivity(Recall) = True_P/(True_P + False_N) \qquad (2)$$

$$Specificity = True_N/(True_N + False_P) \qquad (3)$$

$$Precision = True_P/(True_P + False_P) \qquad (4)$$

Fig. 3 Accuracy of the
models in 70:30 distribution

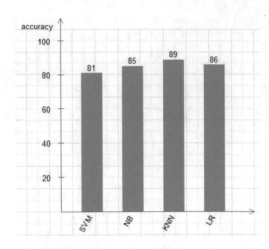

Table 2 The performance evaluation of various classifiers

Classification algorithm	Specificity	Sensitivity/recall	Precision	F-measure
SVM	0.609	0.955	0.866	0.908
KNN	0.655	0.972	0.855	0.91
LR	0.564	0.938	0.877	0.907
NB	0.645	0.869	0.866	0.867

$$\text{F-Measure} = 2 * (\text{Recall} * \text{Precision})/(\text{Precision} + \text{Recall}) \qquad (5)$$

From Table 2 it is clear that KNN classifier performed better than other three classifiers for the Lung cancer dataset considered.

5 Conclusion and Future Scope

A comparative study of various classification techniques for predicting the lung cancer is carried out. The classifiers take data from lung cancer disease database and data is preprocessed and classified using various classifiers. Finally, K-Nearest Neighbor classification model came out to be the effective model in predicting the lung cancer in patients.

This comparative study uses a lung cancer dataset with 568 records. This can be extended for large datasets. Other data mining techniques, like Association Rules, Clustering and Time Series and Ensembled classification algorithms can be used to improve the accuracy.

References

1. N. Rikhi, Data mining and knowledge discovery in database. Int. J. Eng. Trends Technol. (IJETT) **2**(23) (2015)
2. J. Han, J. Pei, M. Kamber, *Data mining: concepts and techniques* (Elsevier, Amsterdam, 2011)
3. N. Singh, S. K. S. Bhadauria, Early detection of cancer using data mining. Int. J. Educ. Manag. Eng. **1**(9), 47–52 (2016)
4. V. Krishnaiah, G .Narsimha, N. Subhash Chandra, Diagnosis of lung cancer prediction system using data mining classification techniques. Int. J. Comput. Sci. Inf. Technol. **1**(4), 39–45 (2013)
5. C.-W. Hsu, C.-J. Lin, Comparison of methods for multiclass support vector machines. IEEE Trans. Neural Netw. **2**(13)(2002)
6. A.K. Yadav, D. Tomar, Clustering of Lung cancer using Foggy K-means, in *International Conference on Recent Trends in Information Technology (ICRTIT)* (2013), pp. 13–18
7. R. Bala, D. Kumar, Classification using KNN. Int. J. Comput. Intell. **7**(13) (2017)
8. H.S. Khamis, K.W. Cheruiyot, S. Kimani, Application of k-nearest neighbour classification in medical data mining. Int. J. Inf. Commun. Technol. Res. **4**(4) (2014)
9. T. Christopher, J.J. Banu, Study of classification algorithm for lung cancer prediction. Int. J. Innov. Sci. Eng. Technol. (IIJISET) **2**(3) (2016)

Sentiment Classification on Twitter Media: A Novel Approach Using the Desired Information from User

B. Shravani, Chandana R. Yadav, S. Kavana, Dikshitha Rao, and Sharmila Shanthi Sequeira

Abstract All around the globe brimming with the IT companies, each company has different data that needs to be analyzed. It is impractical to build models for different data available. Hence, the work carried out serves for this need to build a model that works for different data efficiently. An application will be created, wherein a search box is included that takes an input (a word or any phrase) and gives the related tweets. This application solves the problem in two sections. Primarily, the training section is where we train the given dataset with Naive Bayes, KNN, and random forest algorithms, and in the second section, this trained model is used to test the dynamically extracted tweets from the Twitter media. The results are visualized with the help of a confusion matrix and the pie chart.

Keywords Naive Bayes · KNN · Random forest · Confusion matrix · Pie chart

1 Introduction

The Web 2.0 communities have given rise to plenty of publicly accessible Web sites that highlighted the user-generated content. User-generated content on the Web matters a lot to know how the information is being circulated in the real-time. These endless applications laid down a path for many studies and researches in the area of text mining, machine learning, and natural language processing. Sentiment analysis (SA) is one of the recent developments in the research area used in mining and natural language processing. Sentiment analysis targets identifying, extracting, and categorizing the sentiments from user-generated data [1]. In the task carried out, sentiment analysis is performed on Twitter data. The desirable advantage of Twitter is its accessibility that is the ease to use nature. This made Twitter a principle source for data extraction [2]. Also, the Twitter Web site restricts the user-tweets to 140 characters, thereby forcing the user to stay focused on the point and provides us with the qualitative insights about a topic.

B. Shravani (✉) · C. R. Yadav · S. Kavana · D. Rao · S. S. Sequeira
Department of Computer Science and Engineering, Nitte Meenakshi Institute of Technology, Bengaluru, India

© The Author(s), under exclusive license to Springer Nature Singapore Pte Ltd. 2022   449
P. Shetty D. and S. Shetty (eds.), *Recent Advances in Artificial Intelligence and Data Engineering*, Advances in Intelligent Systems and Computing 1386,
https://doi.org/10.1007/978-981-16-3342-3_36

Twitter media is an open-source platform where one can find ample amount of data that are different from each other (movie review, brand endorsement, etc.). Every data communicated is dominant concerning the associated organization or company. But the regretful reality is that a single trained model is not sufficient to analyze the different types of data available. Is it possible to create separate models for all the data domains? It is highly impossible to design an enormous number of classifier models. It is impractical to achieve accurate models with high-performance rates. The fundamental issue here is the domain of the text. The work developed takes a progression to solve this issue. A simple, interactive sentiment-based approach that opens up the possibility to operate on different data domains is established. The problem discussed above can be solved by developing a simple interactive environment which collects the user responses and performs the training according to the specified algorithm on the selected data domain. In addition to this, many sentiment analyzers developed out in the market works with static data. In practical, one has to work with the real-time data. Hence, there is a need for a dynamic sentiment analyzer. The project developed also serves for this purpose, which works with the dynamic data by extracting the tweets from Twitter API. These dynamically extracted tweets are considered as the test data and are analyzed with the trained classifiers, and the results are visualized with pie chart.

2 Architecture

The block diagram of project work carried out is displayed in Fig. 1. The brief overview of work flow to solve the task is demonstrated sequentially.

3 Methods

- *Data Collection for analysis*
  The labeled datasets from different domains (airlines, movie,…) are collected, which are used to train the classifiers on different data domains. The US Airline Tweets [3] is one such labeled dataset which is collected to train the model for airlines domain. This dataset contains 14,640 rows and 15 columns.
  The columns which are not useful for analysis are removed and only the required columns are considered. The airline dataset which is reduced to two columns (airline sentiment and text) is displayed in Fig. 2.

- *Preprocessing Steps*

  The purpose of the work incorporates raw textual data from Twitter sites. This is often considered as unstructured form of data. Thus, there is a need to convert the unstructured format of text into a structured fashion, which is formally known

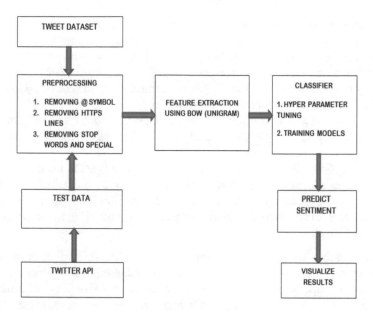

Fig. 1 Representing the flow of classification process

Fig. 2 Dataset used in training phase

as cleaning the data [4]. To clean the data and make it more machine-readable, one should follow preprocessing steps. These steps help us with transformation of textual data to extract meaningful information from the textual tweets. The preprocessing steps adopted for cleaning the data are:

- Removing @ Symbols: The '@' symbols are one of the commonly used signs while tweeting online. However, these symbols will not give any useful information in analyzing the tweet. Thus, it is required to eliminate all these unwanted symbols. One can accomplish this task with the help of regular expressions.
- Negation Handling: Negation handling is the replacement of the contracted words with its full form. It is a common practice to use apostrophes for connecting words. For example, do not is written as don't using an apostrophe. It is suggested to split these negated words to maintain the uniform structure throughout the process.
- Removing Http Links: The raw form of textual data contains huge number of http links which does not add any weight in analysis part [5]. Hence, http links should be removed.

- Replacing Emojis: The electronic messages on the Web pages encourage the use of emoticons which are shortly called as emojis. These add meaning to the text [6], for example, smiling face, thumbs up, and red heart represents positive emotions; angry face and cross mark represents negative emotions. Notably, emojis can clearly describe what a person is trying to communicate. These emojis should be replaced with their respective English phrases.
- Removing URLs: The raw form of textual data contains a lot of URL mentions. This does not add any weight in analysis part [5]. Hence, the URLs should be removed.
- Removing Stop Words: The analysis of textual tweet is at the word level. Hence, there is a need to remove frequently occurring unwanted words which are generally known as stop words [4]. The, is, and, for, of, etc., are few stop words. One can remove stop words using the inbuilt libraries or by collecting the stop words list.
- Feature Extraction

Feature extraction is a dimensionality reduction technique that decreases the number of elements to be processed without eliminating the originality from it. In addition, to this, feature extraction also solves the problem of over-fitting, helps in increasing the speed of training model, and increases the accuracy rates. Bag-of-words (BOW) is one of the frequently used techniques for feature extraction and is easy to understand. In simple words, BOW is the storehouse of collection of words irrespective of their orders [2]. This method serves our purpose by extracting the words from different tweets. The number of occurrences which is mathematically called as frequency of each word is considered as the feature for training the classifier. In BOW technique, there are three tasks: 1. Tokenization, 2. Vocabulary of Words, and 3.Generating the Vectors. Tokenization aims at dividing the given text into pieces of words called as tokens. A vocabulary of words is formed by gathering all the unique words from the dataset and recording their respective occurrences. Eventually, the numerical vectors are generated which can serve as input for machine learning algorithms.

- *Training The Model*

Three different ML algorithms—Naive Bayes, K-nearest neighbors (KNN), and random forest are used to train the classifier.

Naive Bayes
Naive Bayes is the most frequently used ML algorithm. It is simple and easy to code. It involves two main mathematical concepts from probability [5]. They are conditional probability and Bayes theorem.

Conditional probability of an event, Y is the probability that an event will occur given the knowledge that an event X has already occurred. The conditional probability of y given X is easily computed by:

$$P(Y|X) = P(X \cap Y)/P(X) \tag{1}$$

The Bayes theorem is frequently used probability theorem. This calculates the unknown probability with the other know probabilities. This theorem is based on the knowledge of prior probabilities to calculate the posterior probabilities.

The prior probability is the estimation of the probability of event with current knowledge that is, before the new data is collected, while posterior probability is the reformed probability of an event considered after the occurrence of new data and is calculated by updating the prior probabilities using Bayes theorems [10]. The Bayes theorem is given by:

$$P(Y|X) = P(X|Y) * P(Y)/P(X) \qquad (2)$$

Description of terms used:

$P(Y|X) \rightarrow$ The posterior probability of class Y given the evidence X is seen.

$P(X|Y) \rightarrow$ The probability of likelihood for evidence X given class Y.

$P(Y) \rightarrow$ The prior probability of class Y (probability before the evidence X is seen).

$P(X) \rightarrow$ The probability of evidence X.

Naive Bayes is the probabilistic ML algorithm that can be used in prediction problems, spam filtering, and sentiment analysis. The major assumption adopted by the Naive Bayes is that features are independent that is one feature does not affect the other feature [5]. The Bayes rule is simplified with naive assumption that Xs (features) are independent to each other. The simplified calculation is shown below.

$$P(Y|X) = P(X1|Y) * \cdots * P(Xn|Y) * P(Y)/P(X) \qquad (3)$$

As the probability of evidence remains same for all the target classes, this denominator can be ignored during the computation. Hence, the naive equation is given by:

$$P(Y|X) = \pi_i P(X_i|Y) P(Y) \qquad (4)$$

Terms used in the algorithm are described below:

- $X = (X_1, X_2, ..., X_n)$ represents the set of features or attributes
- Y represents the target class (either positive or negative sentiment)
- P represents the probability function

Naive Bayes Algorithm

Step1: Compute prior probabilities

The prior probability of each target class (positive and negative) is calculated. It can be examined as the probabilistic measure of portion of each sentiment out of all the sentiments from the populated dataset. This can be calculated by:

Prior probability of class Y = total number of examples belonging to class Y/total number of examples in the set.

Step 2:   Compute the probability of likelihood

The probability of likelihood is calculated for each target class. This can be mathematically performed by applying the dot product for the conditional probabilities of all features. This can be formulated as:

$$P(X|Y = k)\ P(X1|Y = k) * \cdots * \ P(Xn|Y = k) \tag{5}$$

Step 3:   Compute the posterior probability

The posterior probability of each target class is calculated by using the naive Bayes formula. The task is accomplished by substituting the above two computations in naive equation. Mathematically, it is calculated as:

$$P(Y = k|X) = \pi_i P(X_i|Y)\ P(Y = k) \tag{6}$$

Step 4:   Predict the sentiment

Once the posterior probability of each target class is computed, a max function is applied over these posterior probabilities to predict the sentiment. The one with maximum posterior probability is considered as the predicted class.

K-Nearest Neighbors (KNN)

KNN is the most simple and easy-to-use ML algorithms. It is used to solve the classification problems. KNN stands for K-nearest neighbors. The whole algorithm depends on the parameter 'K'. The target class is predicted based on the K-nearest neighbors of the test instance. It is termed as Lazy learner since it works on dataset at the time of classification. Before that it just stores the entire dataset [5]. The distances between train and test data are calculated. The k-most similar instances are considered to predict the sentiment of the class. The most frequent class label among the K instances is considered to be the predicted class. A step-by-step procedure involved in performing the KNN algorithm is discussed.

K-Nearest Neighbors Algorithm

Step 1:   Handling the data

KNN is lazy learning algorithm, thus it starts the process from the scratch. Initially, the dataset is loaded, and the split operation is performed on the entire dataset which results in partitioning the complete dataset into train and test data.

Step 2:   Initialize k

The value of K should be selected in such way that it improves the effectiveness of the classifier. The value of K represents the number of nearest neighbors which are used in voting process to predict the class label of new test instance.

Step 3:　Compute the distance

For each example in the training data.

3.1　The distance between the test instance and each training instance is calculated. This calculation is performed with the help of Euclidean distance formula. The formula used to find the distance between two data points $a(x_1,y_1)$ and $b(x_2,y_2)$ is:

$$d(a, b) = \sqrt{\{(x_2 - x_1)^2 + (y_2 - y_1)^2\}} \qquad (7)$$

3.2　The calculated distance is appended to the collection, which stores all the calculated Euclidean distances between the points.

Step 4　Sort the collection

The collection of calculated distances from all the training instances is ordered in an ascending fashion based on the values of computed distance.

Step 5:　Selecting the points

The first K entries from the sorted collection are considered for the predicting class label of test instance.

Step 6:　Predict the sentiment

The K class labels are pulled out from the K entries, and the test instance belongs to the most frequent class label.

Random Forest

The random forest algorithm is introduced to overcome the 'over-fitting' drawback of decision tree algorithm. The concept is based on the collection of multiple decision tree classifiers randomly, hence the name random forest. As the number of trees increases, the accuracy of the classifier also increases [5]. The prediction is done by considering the majority voting from all the decision tree classifiers. It is an ensemble learning method, which learns from multiple decision trees and thereby solves the problem of over-fitting by averaging the result. A step-by-step procedure involved in performing the random forest algorithm is discussed.

Random Forest Algorithm

Step 1:　Select samples

In random forest algorithm, the very first step is to select random number of samples from the entire dataset.

Step 2:　Construct decision tree

A decision tree is constructed for each randomly selected sample which results into multiple decision trees.

Step 3:   Collect the predictions

A prediction result is collected from each decision tree, and a voting process is performed for each predicted result.

Step 4:   Predict the sentiment.

A prediction result with more number of votes is considered as the final prediction.

- *Performance Tuning*

Hyperparameter Using GridSearchCV
The machine learning models are mathematical methods where one can learn about parameters by working with data. But there are some specified parameters which are tuned to obtain better predictions. Hyperparameter is one such parameter whose value is decided before the training process starts and is given to the model [6]. These parameters should be tuned to improve the performance of our model. This can be achieved by performing trails with different parameter values. The value that increases the performance of model and reduces the complexity of operation is selected as the hyperparameter value. GridSearch is one of the simplest methods used for tuning the hyperparameters. To simplify, the task a grid of hyperparameters is created, and this will allow us to try all the combinations of parameter values.

Cross Validation
Cross validation is the test performed on the model to check if the classifier predicts the sentiment correctly for unknown data (i.e., other than the training data). K-fold cross validation is widely used cross-validation techniques that works on splitting the overall data (train/test split) [7]. As the name indicates, this method contains a single parameter 'K' which is selected randomly or by previous experiences. Generally, the value of 'K' ranges from 5 to 10. Here, the data is shuffled in a random manner and is split into 'K' groups which are technically called as 'K' folds. Next, each unique group is taken as a test data (i.e., the holdout data) and rest (K-1) groups are considered as trained data. The process is repeated k times, and the evaluation score is noted for each group by fitting 'K' different models. These evaluation scores can describe the skill of the trained model.

- *Twitter Streaming API*

Twitter API (Application Programming Interface) serves for extracting the tweets dynamically from the Twitter sites. Tweepy is the Python tool used for extraction process [8]. This process generates the credential keys. By using these keys, one can search the required tweets using a hash-tag (Ex: #airlines). The output tweets of this extraction process will serve as the test data for our project. Once the dynamic data is available for testing, the data undergoes the same steps of preprocessing, feature extraction, and so on. One of the three trained classifiers is used to classify the sentiment. The results of test data are visualized with a pie chart, which is discussed in the results section.

Fig. 3 ROC curve of random forest classifier

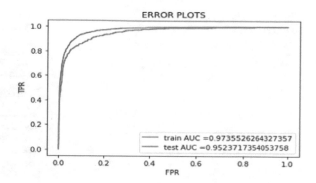

- *Performance Evaluation*

In machine learning, use of an AUC (area under the curve) ROC (receiver operating characteristics) curve to visualize the performance of classification model. AUC-ROC curve aims at calculating the performance of classification model at various thresholds [9]. Area under the graph measures how better the model classifies the sentiments. The graph is plotted with false positive rate (FPR) on X-axis and true positive rate (TPR) on Y-axis. A model is said to have a good measure of separability, when AOC of plot nears to 1 and poor if it nears to 0. The ROC curve of random forest classifier is shown in Fig. 3.

4 Flowchart

The sequence of steps required to perform sentiment analysis is illustrated in this chapter. The control flow and inter connection between the different modules involved in entire process is visualized (Fig. 4).

5 Results

- *Confusion Matrix*

A confusion matrix which is also called as an error matrix represents the performance of classifiers on the test data and allows visualizing the confusion made by the classifier while predicting the sentiments (i.e., the miss-classification of positive to negative and negative to positive). In this matrix, the principle diagonal entries represent the true predictions, while the secondary diagonal entries represent the false predictions made by the model [10]. As shown in Fig. 5, error matrix is depicted as a table with four entries of actual and predicted sentiment combinations, where

Fig. 4 Flowchart of
sentiment analysis

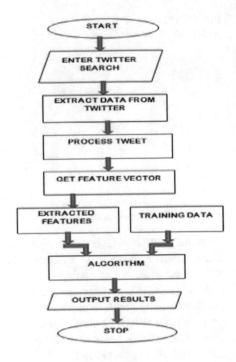

Fig. 5 Confusion matrix
representation

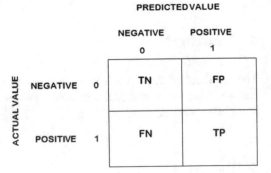

the row elements belong to the actual class while the column elements belong to the
predicted class.

Describing the terms from the matrix:

- TN: It stands for true negative, the case in which the sentiment predicted by the
  classifier is negative and the actual sentiment is also negative.
- FP: It stands for false positive, the case in which the sentiment predicted by the
  classifier is positive, but the actual sentiment is negative. The classifier made an
  error by miss-classifying the negative sentiment as positive.

- FN: It stands for false negative, the case in which the sentiment predicted by the classifier is negative, but the actual sentiment is positive. The classifier made an error by miss-classifying the positive sentiment as negative.
- TP: It stands for true negative, the case in which the sentiment predicted by the classifier is negative and the actual sentiment is also negative.

Confusion matrix allows us to perform calculations for finding different measuring rates that evaluates effectiveness of the model. In the work carried out, confusion matrix is used to calculate the accuracy of classifiers. It is the measure of how often classifier classifies the sentiment correct (i.e., the positive sentiment is predicted as positive and the negative sentiment is predicted as negative).

- *Accuracy of the Classifiers*

It is the measure of how often classifier classifies the sentiment correct (i.e., the positive sentiment is predicted as positive, and the negative sentiment is predicted as negative). Accuracy is calculated by performing the division operation on number of true predictions (TN & TP) and overall predictions (TN, FP, FN& TP). Accuracy is calculated as:

$$\text{Accuracy} = (TN + TP) - (TN + FP + FN + TP) \tag{8}$$

As soon as the classifiers are trained, there is a need to find accuracy of classifiers. Here, confusion matrix is used to compute the accuracies.

The confusion matrix of Naive Bayes classifier is shown in Fig. 6
Accuracy of Naive Bayes classifier is calculated as:

$$\text{Accuracy} = (1808 + 211) - (1808 + 40 + 250 + 211) = 87\%$$

Accuracy of naive bayes classifier $=$ 87%

Fig. 6 Confusion matrix of Naive Bayes classifier

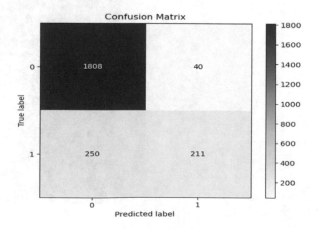

The confusion matrix of KNN classifier is shown in Fig. 7
Accuracy of KNN classifier is calculated as:

$$\text{Accuracy} = (1506 + 383) \div (1506 + 361 + 59 + 383) = 82\%$$

Accuracy of KNN classifier $= 82\%$

The confusion matrix of random forest classifier is shown in Fig. 8
Accuracy of random forest classifier is calculated as:

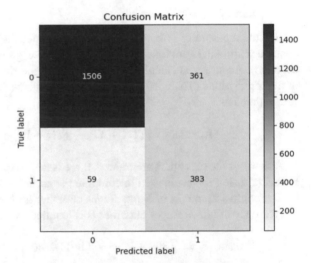

Fig. 7 Confusion matrix of KNN classifier

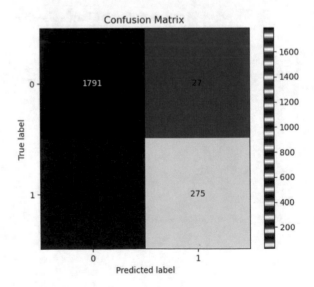

Fig. 8 Confusion matrix of random forest classifier

Fig. 9 Extracted tweets
from the Twitter

21-04-2020 12:30	b'United Airlines sees $2.1 billion loss\n\nSave your time by reading/listenin
21-04-2020 11:15	b'#UnitedAirlines expects to report a pretax loss of $2.1 billion in Q1.\n#covic
21-04-2020 11:10	b'United Gives First Glimpse Of How Ugly First Quarter Was For U.S. Airlines V
21-04-2020 10:40	b'#unitedAIRLINES Airlines: Airlines - Refund all ticket costs due to #COVID-1!
21-04-2020 08:58	b'United Airlines reports huge losses for first quarter #United #UnitedAirlines
21-04-2020 08:55	b'In #Israel, #lawsuits have been filed against 6 #airlines (#UnitedAirlines, #UI

$$\text{Accuracy} = (1791 + 275) \div (1791 + 27 + 216 + 275) = 89\%$$

Accuracy of Random Forest classifier = 89%

- *Pie chart of test data*

As discussed earlier, dynamic data is used in testing phase. The tweets are extracted from Twitter based on certain hash-tags as per the requirements. The tweets gathered from extraction process are shown in Fig. 9.

The dynamically extracted tweets of united airlines have more of negative feedback because the flights have been canceled due to corona pandemic situation. These extracted tweets are analyzed with the three different classifiers developed in the training phase. The results are observed in the form of pie charts.

The prediction obtained from Naive Bayes classifier is visualized with a pie chart in Fig. 10

The prediction obtained from random forest classifier is visualized with a pie chart in Fig. 11

The prediction obtained from random forest classifier is visualized with a pie chart in Fig. 12.

- *Comparison of predicted results*

Fig. 10 Pie chart of Naive
Bayes classifier

Fig. 11 Pie chart of KNN
classifier

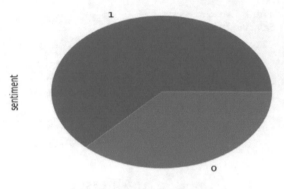

Fig. 12 Pie chart of random
forest classifier

Table 1 Accuracy rates of
classifiers

Proposed classifiers	Dataset	Accuracy (%)
Naive Bayes	US Airline Tweets	87
KNN	US Airline Tweets	82
Random forest	US Airline Tweets	89

The accuracy measurements of three different machine learning algorithms are calculated and listed in the Table 1.

6 Conclusion

In the work carried out, a classifier is developed that successfully classified the sentiments from different data domains. The model is trained with three different machine learning algorithms—Naive Bayes, KNN, and random forest. The results show that random forest gives better performance than the other two ML algorithms. The reason is KNN always depends on the parameter 'K', and it behaves abruptly with irrelevant attributes; and on the other hand, precision of Naive Bayes algorithm is not makeable with smaller dataset, and the naive assumption makes it fall behind as the features may not be independent in real-time data. According to the calculations made, random forest is the best among three machine learning algorithms with 89% accuracy.

7 Challenges and Future Enhancements

The classifier was not able to classify the tweets which had both negative and positive sentiments. The processing speed becomes slower, due to the hidden noise in the text, with the increase in dimension of data. In recent times, sentiment analysis has gained significance, but it is restricted to English texts. The work can be extended to perform sentiment analysis on other languages. There is a huge requirement in improving the analysis on sarcastic sentences. There is an on-going research with deep learning as they produce good accurate models.

References

1. N. Jivane, Twitter Sentiment Analysis of Movie Reviews Using Machine Learning Techniques, 20 Nov 2018 [Online]. Available https://medium.com/datadriveninvestor/twitter-sentiment-analysis-of-movie-reviews-using-machine-learning-techniques-23d4724e7b05
2. R. Wahome, This Is How Twitter Sees The World: Sentiment Analysis Part One, 8 September 2018. [Online]. Available https://towardsdatascience.com/the-real-world-as-seen-on-twitter-sentiment-analysis-part-one-5ac2d06b63fb
3. Figure Eight, "Twitter US Airline Sentiment" [Online]. Available https://www.kaggle.com/crowdflower/twitter-airline-sentiment
4. S. Bhansal, Steps for effective text data cleaning (with case study using Python, November 16, 2014 [Online]. Available https://www.analyticsvidhya.com/blog/2014/11/text-data-cleaning-steps-python/
5. S. Ray, Commonly used Machine Learning Algorithms (with Python and R Codes), September 9, 2017 [Online]. Available https://www.analyticsvidhya.com/blog/2017/09/common-machine-learning-algorithms/
6. E. Lee, An Intro to Hyper-parameter Optimization using Grid Search and Random Search, June 6, 2019 [Online]. Available https://medium.com/@cjl2fv/an-intro-to-hyper-parameter-optimization-using-grid-search-and-random-search-d73b9834ca0a
7. Srinivasan, The Importance Of Cross Validation In Machine Learning November 23rd, 2018, November 23rd, 2018 [Online]. Available https://www.digitalvidya.com/blog/cross-validation-in-machine-learning/
8. A. Moujahid, An Introduction to Text Mining using Twitter Streaming API and Python, July 21, 2014 [Online]. Available http://adilmoujahid.com/posts/2014/07/twitter-analytics/
9. S. Narkhede, Understanding AUC-ROC Curve, June 26, 2018 [Online]. Available https://towardsdatascience.com/understanding-auc-roc-curve-68b2303cc9c5
10. J. Brownlee, What is a Confusion Matrix in Machine Learning, November 18, 2016. [Online]. Available https://machinelearningmastery.com/confusion-matrix-machine-learning/

Printed in the United States
by Baker & Taylor Publisher Services